Automatic Milking

a better understanding

Automatic Milking

Edited by:
A. Meijering
H. Hogeveen
C.J.A.M. de Koning

Wageningen Academic
P u b l i s h e r s

CIP-data Koninklijke Bibliotheek
Den Haag

ISBN 9076998388

hardcover

Subject headings:
Automatic milking
Dairy farming

First published, 2004

Wageningen Academic Publishers
The Netherlands, 2004

Printed in The Netherlands

PREFACE

In 2000 the book **Robotic Milking**, reflecting the proceedings of an International Symposium, which was held in The Netherlands, came out. At that time, commercial introduction of automatic milking systems was no longer obstructed by technological inadequacies. Particularly in a few
West-European countries, systems were being installed at an increasing rate. However, it was recognised that the changeover from "traditional" to automatic milking affected the farming operation, herd management and control of milk quality profoundly and that many of the implications were still unknown. So, new challenges in various fields of dairy farming and new research areas emerged.

Since the International Symposium in 2000, much has happened. In general, automatic milking has been adopted as a realistic alternative for milking in the "traditional" milking parlour. Systems have gradually been improved and, maybe even more importantly, farmers have become more familiar with their potential and limitations, both technically and in herd management. The number of farms milking with an automatic milking system has worldwide increased from approximately 500 in 2000 to more than 2.200 by the end of 2003. The majority is still concentrated in north and west Europe, but interest from the south of Europe, Japan, Canada and the USA is increasing. In Sweden and Denmark up to half of the farmers planning the replacement of their milking system, decide for automatic milking. In The Netherlands about 25% do so.

From 2000 to now, the level of scientific knowledge on various aspects and consequences of automatic milking has increased largely as well because of research efforts all over the world. A significant share of these efforts has been made within the framework of a EU-granted project on the implications of the introduction of automatic milking on dairy farms. Some seven research institutes and six industrial companies from six countries joined their expertise and experience in order to facilitate a widespread adoption of automatic milking without undesirable side effects.

In the latter days of the EU-project a follow-up symposium on **Automatic Milking** was organised to present the results from this project and to merge these with knowledge generated in studies performed all over the world. The motto of the symposium, **for a better understanding,** reflects the aim of all efforts made in this field in recent years, i.e. to acquire knowledge on all consequences and issues related to the introduction of automatic milking on dairy farms.

This book reflects the proceedings of this symposium, again held in Lelystad, The Netherlands, in March 2004. Its contents may be considered as the present state of knowledge in the field of automatic milking, in particular with respect to the following topics:
- Economic aspects and social implications
- Perception of the technology by society at large
- Impact on milk quality
- (Re)definition, detection and separation of abnormal milk

- Impact on the excretion of antibiotic residues
- Effects of farm hygiene and teat cleaning on milk quality
- Demands with respect to cleaning of automatic milking systems
- Effects of converting to automated milking on animal health
- Automatic milking and animal welfare
- Integrating automatic milking and grazing
- Demands and opportunities for operational management support

These topics are addressed by 47 full papers and some 80 poster abstracts, originating from 16 countries from Europe, Asia, Oceania and North America, giving an excellent coverage of the topics mentioned. A special paper is dedicated to the "state of the art" in automatic milking in North America and in Europe.

The efforts by the members of the Scientific Committee to select the right papers for oral and poster presentation are gratefully acknowledged. The work by the members of the Organising Committee, the "draught horses", to make the symposium a successful event deserves particular acknowledgement. Last but not least, the crucial financial support by the many sponsors is highly appreciated.

Albert Meijering
Henk Hogeveen
Kees de Koning
Editors

SYMPOSIUM ORGANIZATION

Organizing Committee

Dr Albert Meijering   chairman
Ir André van der Kamp   administrator
Ing. Kees de Koning   treasurer and business liaison
Dr Henk Hogeveen   scientific programme
Ing. Anita Wolsing   organisation and publicity

Scientific committee

Dr Henk Hogeveen (chairman) Wageningen University, The Netherlands
Dr Christel Benfalk   Swedish Institute of Agricultural and Environmental
                             Engineering, Sweden
Ing. Kees de Koning   Animal Sciences Group, Wageningen UR, The Netherlands
Dr J. Eric Hillerton   Institute for Animal Health, United Kingdom
Dr Karin Knappstein   Federal Dairy Research Centre, Germany
Dr Erik Mathijs   Katholieke Universiteit Leuven, Belgium
Dr Albert Meijering   Animal Sciences Group, Wageningen UR, The Netherlands
Dr Morten D. Rasmussen   Danish Institute of Agricultural Sciences, Denmark
Dr Keith Roe   Katholieke Universiteit Leuven, Belgium
Dr Hans Wiktorsson   Swedish Agricultural Univesity, Sweden

CONTENTS

FARM AND SYSTEM HYGIENE

ANIMAL HEALTH

ABNORMAL MILK

GRAZING

MILK QUALITY

WELFARE

FARM AND HERD MANAGEMENT

OVERVIEW

AUTOMATIC MILKING: STATE OF THE ART IN EUROPE AND NORTH AMERICA

Kees de Koning[1] & Jack Rodenburg[2]
[1]Applied Research, Animal Sciences Group, Wageningen UR, Lelystad, The Netherlands
[2]Ontario Ministry of Agriculture and Food, Woodstock, Ontario, Canada

Abstract

An overview of historical development of automatic milking, as well as the current situation and perceived challenges and opportunities for future development are discussed. Since the first commercial systems appeared in 1992, Automatic milking systems (AM-systems) have been installed at an increasing rate. No other new technology since the introduction of the milking machine, has aroused so much interest and expectations among dairy farmers and the periphery. Reduced labour, a better social life for dairy farm families and increased milk yields due to more frequent milking are recognised as important benefits of automatic milking.

Automatic milking changes many aspects of farm management since both the nature and organisation of labour is altered. Manual labour is partly replaced by management and control, and the presence of the operator at regular milking times is no longer required. Visual control on cow and udder health at milking is, at least partly, taken over by automatic systems. Facilities for teat cleaning and separation of abnormal milk are incorporated into the automatic system and adaptation of conventional cleaning schemes and cooling systems is needed to accommodate continuous milking. Cow management including routing within the barn, the opportunity for grazing and the use of total mixed rations is altered. A high level of management and realistic expectations are essential to successful adoption of automatic milking. Results from commercial farms indicate, that milk quality is somewhat negatively effected, although bacterial counts and somatic cell counts remain well below penalty levels. Automatic milking systems require a higher investment than conventional milking systems. However increased milk yields and reduced labour requirements may lead to a decrease in the fixed costs per kg milk. Automatic milking is gaining widespread acceptance and is estimated to be in use on more than 2200 farms in over 20 countries worldwide.

Introduction

Interest in fully automated milking began in the mid-seventies, and was initially driven by the growing costs of labour in Europe. Since machine milking, and automatic detaching, and teat spraying were already in common usage, automatic cluster attachment became the focus of European work. Although various prototypes demonstrated this capability, it took a decade before fully integrated and reliable automatic milking became a reality.

The term "Automatic Milking System" refers to a system that automates all the functions of the milking process and cow management undertaken in conventional milking, by a mix of manual and machine systems. In contrast to conventional milking, where humans bring the cows to be milked at regular times (usually twice a day), automatic milking places emphasis on the cows motivation to be milked in a self-service manner several times a day by a robotic system without direct human supervision.

In modern society consumer concern about methods of food production include food safety, as well as ethical questions related with animal welfare, animal health, housing conditions and access to grazing. Because unsupervised, automatic milking, raised a number of questions, an extensive EU research project was started at the end of 2000 (www.automaticmilking.nl). This project focussed on farm-level adoption determinants of automatic milking, on-farm social-economic and environmental implications, societal acceptance, impact on milk quality, impacts on animal health and welfare, including the combination of automatic milking with grazing and requirements for management information systems. Other research groups around the world have also contributed substantially to progress in our understanding of automatic milking and related management considerations.

Automatic milking systems

AM-systems include single stall systems with integrated robotic and milking functions and multi-stall systems with a transportable robot device, combined with milking and detachment devices at each stall. Single stall systems milk 55-60 cows, while multi-stall systems with 2 to 4 stalls milk 80 to 150 cows up to three times per day. Automatic milking strongly relies on the cow's motivation to visit the AM-system voluntarily. The main motive for this is the supply of concentrates dispensed in a feed manger in the milking box during milking. An automatic milking system has to take over the 'eyes, ears and hands" of the milker. Such a system includes electronic cow identification, cleaning and milking devices and computer controlled sensors to detect abnormalities in milk, in order to meet international legislation and hygiene rules from the dairy industry.

Teat cleaning systems include brushes or rollers, inside teat-cup cleaning or a separate 'teat cup like' cleaning device. Several trials showed that, while cleaning with a device is better than no cleaning (Schuiling, 1992, Knappstein et al., 2004), it is not as good as manual cleaning by the herdsman.

AM-systems are also equipped with sensors to observe and to control the milking process. Data are automatically stored in a database and the farmer has a management program to control the settings and conditions for cows to be milked. Attention lists and reports are presented to the farmer by screen or printer messages. The AM-system also provides remote notification to the farmer if intervention is required.

Farms with Automatic Milking Systems

The first AM-systems on commercial farms were implemented in The Netherlands in 1992 primarily in response to expensive labour and the farm structure of family farms. Increasing costs of inputs while milk prices decreased, forced farmers to increase their output per man-hour. After the introduction of the first AM-systems, adoption went slowly, until 1998 (Figure 1). From that year on automatic milking became an accepted technology in The Netherlands and other European countries, and also Japan and North America. At the end of 2003, worldwide some 2200 commercial farms used one or more AM-systems to milk their cows. While the majority of farms are family run with 1 to 3 milking boxes, the largest operation

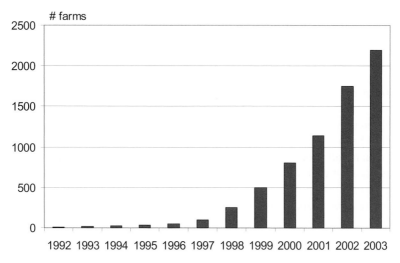

farms

Figure 1. Number of farms using an automatic milking system at the end of 2003.

at the end of 2003, located in California, USA, includes 20 milking boxes. After completion, this facility will include 32 milking boxes, in 4 interconnected barns each with a central cluster of 8 milking units. More than 80% of the world's automatic milking farms are located in north-western Europe.

Automatic milking and management aspects

Switching from a milking parlour to automatic milking results in big changes for both the herdsman and the cows and can cause stress to both. Although with AM-systems immediate supervision of milking is eliminated, new labour tasks include control and cleaning of the AM-system, twice or three times a day checking of attention lists including visual control of the cows and fetching cows that exceeded maximum milking intervals. Field data on labour savings is limited, but model studies (Sonck & Donkers, 1996) showed physical labour savings of 30 to 40% compared with conventional milking. Labour demands for AM-systems from 32 minutes up to 3 hours per day (Ipema *et al.,* 1998, Van't Land *et al.,* 2000) have been reported On average a 10% reduction in total labour compared with the conventional twice daily milking is recorded, but big variations between farms can be found.

However, the biggest change is the nature of labour. The physical work of machine milking, is replaced with management tasks such as frequent checking of attention lists from the computer and appropriate follow up. Since this work is less time bound than parlour milking the input of labour is more flexible. This is attractive on family farms. But because milking is continuous, and system failures can occur anytime there must be a person "on call" at all times. System failures and associated alarms typically occur about once in two weeks although this varies with the level of maintenance and management.

In terms of the impact on cows, the AM-system is not suitable for all cows. Poor udder shape and teat position may make attachment difficult and some cows may not be trainable to attend for milking voluntarily. In new installations, the number of cows found to be unsuitable is generally reported to be less than 5-10%. In the transition from conventional to automatic milking, cows must learn to visit the AM-system at other than traditional milking times. Training and assistance in the first weeks should involve quiet and consistent handling, so they adapt to the new surroundings and milking system.

Milking frequency

In practice, the average number of milkings per cow day varies from 2.5 till 3.0, but rather big differences in milking intervals are reported by commercial farms. A typical figure is presented in figure 2 (De Koning and Ouweltjes, 2000). Almost 10% of the cows realised a milking frequency of 2 or lower over a two year period milking with an single stall AM-system. This occurred even though cows with a too long interval were fetched three times per day.

Such cows will not show an increase in yield and may even show a production loss. By changing the milking parameters of the AM-system, it is quite easy to prevent cows from being milked at low yields or short intervals. But it is much more difficult to prevent cows from being milked with long intervals. This means it will be necessary to manage the intervals by fetching cows that have exceeded a maximum interval. Usually this is done several times per day at fixed times around the cleaning procedures of the AM-system. In a large study on 124 farms in The Netherlands, Van der Vorst & Ouweltjes (2003) found that in farms where cow numbers were 25% or more under the capacity of the system, cows were rarely fetched. On almost 50% of farms cows were fetched twice a day when the interval exceeded 12 hours and on 35% cows were fetched 3 or 4 times per day. These studies also showed that too long intervals cannot be prevented completely. Fetching cows three times per day that have exceeded an interval of 12 hours, means that the maximum interval will be 20 hours. Fetching cows with intervals shorter than 12 hours is time-consuming and moreover may lead to some habituation of the cows.

Increase in milk yield

One of the benefits of automatic milking is increased milk yield from more frequent milking. An increase from 6 to 25% in complete lactations has been shown when milking frequency increases from two times to three times per day (Erdman & Varner, 1995). French data show an average 3% increase in milk yield and up to 9% for farms that utilized the AM-system for more than 2 years (Veysset et al., 2001). In the study of Van der Vorst & Ouweltjes (2003) an average increase of 5% with a range of - 16% to + 35% was reported.

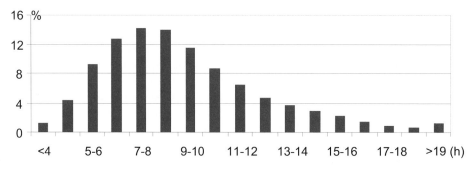

Figure 2. Frequency distribution of milking intervals in hours over a 2-year period (De Koning & Ouweltjes, 2000).

Automatic milking – A better understanding

In many larger US herds with highly automated conventional parlours, 3 times daily milking is commonplace. For 3x herds adopting automatic milking, a production decrease of 5 to 10% would be expected.

Attitude and expectations

One important factor in successful implementation of an AM-system is the attitude and expectation of the dairy farmer (Hogeveen *et al.*, 2001, De Koning *et al.*, 2002, Ouweltjes, 2004). While there is considerable variation in level of satisfaction with different types of systems, an estimated 5-10% of owners have switched back to conventional technology. In some cases expectations were not realistic, in others farmers were unable to adapt to the different management style, and in some cases a high rate of failures on the AMS resulted in ongoing high labour input for manual intervention. During the start up period, automatic milking requires a high input of labour and management. Key factors of a successful implementation of AM-systems are:

- Realistic expectations
- Good support by skilled consultants before, during and after implementation
- Flexibility and discipline to control the system and the cows
- Ability to work with computers
- Much attention to the barn layout and a good functioning cow traffic
- Good technical functioning of the AM-system and regular maintenance
- Healthy cows with good feet and 'aggressive' eating behaviour

Barn layout, capacity of the AM-System and milk cooling

Since these systems depend on voluntary attendance, a well laid-out freestall barn is essential to success. The main motive for a cow to visit the AMS is the concentrate provided in the manger of the milking box. The routing in the barn should be according the Eating - Lying - Milking principle. Cows should have easy access to the milking stall and selection gates, long alleys, steps and other obstructions, should be minimised. A central location for the AM-system minimises walking distances of the cows. In many countries regulations require that the AM-system be located close to the milking room and that it be accessible to the operator via a clean route.

After visiting the milking system, the cow should have access to the feeding area. In "forced traffic" systems one way gates, often with cow identification and selection capabilities, restrict cows so they must go through the milking box en route from the resting to the eating area, whenever the interval since the last milking exceeds the pre-set minimum. In "free cow traffic" systems access to the feeding area is unrestricted and only the concentrate fed in the AM system is used to attract cows. In practice, differences in average milking frequency between these systems are relatively small. Forced cow traffic does minimise the number of cows that must be fetched (Hogeveen *et al.*, 1998) and is an effective training aid, however, there is a consensus that for animal welfare, free cow traffic is preferred (Wiktorsson *et al.*, 2003). Cows spend more time in the waiting area in forced cow traffic systems. Especially for AMS with a high occupancy rate, this might affect the number of visits to the roughage station and result in reduced intake of roughage.

On large farms the benefits of locating milking boxes in clusters, and the reduced need for manger space have led to barn designs with as many as 8 to 12 rows of free stalls and

1 to 3 drive through feed alleys. While these barns will function well in terms of cow and operator traffic and milk handling, the impact of barn widths of 40 to 65 meters on in-barn climate and ventilation will present new challenges.

In most European countries, grazing during summer time is routine (Van Dooren *et al.*, 2002) or in some Scandinavian countries even compulsory. Moreover, from an ethological point of view, many consumers in North Western Europe believe grazing is essential for cows and one Dutch dairy pays a premium for milk from grazed herds. In The Netherlands grazing is common practice (>80%). However, about 52% of the farms with an AM-system apply grazing, showing that grazing in combination with AM is less common, but still possible (Van der Vorst & Ouweltjes, 2003). Grazing is critical to low cost milk production in New Zealand, and while there is no commercial use of AMS in that country at this time, the "Greenfield" project uses automatic milking in a 100% grazing system under very different circumstances than those found in Europe.

In North America, the more continental climate favours diets of whole plant corn and alfalfa silage, which by their growth characteristics, are more easily fed from storage. To accommodate these feeds, cows are housed year round in freestall barns. Grain, often grown on the farm, is inexpensive and investment in housing is high, therefore diets with a high concentration of grain, which support high milk production per cow are favoured. Mixing all grain, forage and supplement ingredients into a total mixed ration, fed at the manger provides the accuracy of formulation and control of fibre level needed to minimise the risk of digestive disturbance with these diets.

The important role of concentrate in the manger of the milking box, to attract cows for voluntary milking, conflicts with traditional feeding practices in the US and Canada. Field data (Rodenburg and Wheeler 2002) suggests that in AMS where a high grain TMR is fed at the manger, cow traffic is decreased, frequency of voluntary milking is lower and more cows must be fetched. Since fetching cows represents the main labour expense for milking in AMS, low voluntary attendance is recognised as an impediment to adoption of automatic milking in North America. Research is needed to optimise these diets and methods of feed delivery so that both high production and high voluntary milking are achieved.

Capacity of the AM-system

The capacity of an AMS is often expressed as the number of milkings per day. This number will depend on a number of factors including the number of stalls, the use of selection gates, milking frequency, machine on time, herd size and cow traffic system. Increasing the number of milkings per cow per day does not necessarily contribute to a higher output of milk per day for the AMS. This is due to the more or less fixed handling time per milking and the decreasing yield of milk per milking when cows are milked more frequently. Milk flow rate and yield have a large impact on capacity in kg per day (De Koning & Ouweltjes, 2000). By changing the milking criteria for individual cows, the AM-system can be optimised to achieve maximum capacity in kg per day.

Milk cooling systems

Milk should be cooled to below 4 °C within 3 hours of milking. Four different systems are being used (De Koning *et al.*, 2002b) to adjust the cooling system to automatic milking; a) indirect cooling with an ice-bank tank, b) combination of bulk and buffer tank, c) storage tank with modified cooling system and d) instant cooling. For an ice bank tank and modified cooling system, it may be useful to have an additional buffer tank which is able to store the milk when the bulk tank is emptied and cleaned. This enables the AM-system to continue milking, thus increasing the capacity of the system.

Milk quality

Milk quality is a critical concern on modern dairy farms because milk payment systems are based on milk quality and consumers expect a high level of quality and safety from the milk products they buy. Although automatic milking uses the same milking principles as conventional milking, there are major differences. Results from commercial farms in Europe (Klungel *et al.*, 2000, Van der Vorst & Hogeveen, 2000. Pomies et Bony, 2001, Van der Vorst *et al.*, 2002) and North America (Rodenburg and Kelton 2001) indicate, that milk quality is somewhat negatively effected after introduction of automatic milking. In general data show an increase in bacteria counts, although the levels are still relatively low and well within the penalty limits. A recent study (Helgren and Reinemann, 2003) determined that SCC and bacteria counts in the US were similar to conventional milked herds. Both the cleaning of the milking equipment and milk cooling are critical factors in controlling bacteria counts. Also cell counts are not reduced after the change to automatic milking, despite the increased milking frequency.

With increasing milking frequency a small decrease in fat and protein percentage and an increase in the free fatty acids levels has been reported (Ipema and Schuiling, 1992; Jellema, 1986; Klei *et al.*, 1997). Van der Vorst *et al.* (2003) found both technical and management factors influencing FFA levels. Wiking and Nielsen (2003) found relations with FFA levels and fat globule size and showed that feeding and cooling strategies affect FFA levels. In studies from Van der Vorst *et al.* (2002) and Svennersten & Wiktorsson (2003) increased FFA levels were also found with increased milking frequencies using conventional milking methods.

The general conditions of hygiene in milk production in the EU are currently defined by the Commission Directive 89/362/EEC (1989) but not all elements apply to automatic milking (Rasmussen, 2003). The following text is proposed to be included in the coming EU Hygiene Directive: "Milking must be carried out hygienically ensuring in particular, that milk from an animal is checked for abnormalities by the milker or by a method achieving similar results and that only normal milk is used for human consumption and that abnormal, contaminated, and undesirable milk is excluded".

Regulatory status of automatic milking in other parts of the world varies widely. In Canada, the province of Ontario has specific guidelines that permit automatic milking, while in other jurisdictions each AM-system has temporary approval permitted for experimental technology. A recent decision by the US regulatory authority clears the way for state regulators to formally permit automatic milking.

AM-systems have accurate cow identification and this also means less chance of human errors than in conventional milking, which might have a positive effect on lowering the presence of inhibitors in milk, as reported from North America. In this way automatic milking also potentially enhances food safety and quality.

Economical aspects

Investment required for AM-systems are much higher than for conventional milking systems and thus the fixed costs of milking are higher. However more milk with less labour means that the costs of milking per kg of milk will decrease. Theoretically, with an AM-system more cows can be kept with the same labour force than with conventional milking, but this may involve additional investments in buildings, land or feed and perhaps milk quota. On a farm with more than one full time worker the possibility exists to reduce labour input and thus costs. Quite often that does not happen and the time saved as a result of lower labour requirement is used for personal activities. Mathijs *et al.* (2002) found that two third of AM-farmers state social reasons for investing in automatic milking, such as increased labour flexibility, improved social life and health concerns. Little economical information is available from commercial herds using an AM-system. The high-tech farm at Waiboerhoeve experimental station realised a cost price, which was approximately € 1,50 per 100 kg higher compared with a the cost price of a reference group of farms using conventional milking. The small plus on the cost price is mainly due to increased machinery costs per kg of milk, despite the decreased costs of labour (Van der Kamp *et al.*, 2002). It was calculated that an extra 10% more milk production per year would lead to a reduction in cost price of approximately 3 € per 100 kg milk.

Several simulation models have been developed to calculate the economic effect.

The "Room for Investment" model computes the amount of money that can be invested in an AMS, without a decrease in net return compared with conventional milking (Arendzen & van Scheppingen, 2000). The RFI-value calculates the annual accumulated return from increased milk yield, savings in labour, and savings in not investing in a milking parlour and divides this by the annual costs of the AM-system. The model can use farm specific factors and circumstances to calculate the RFI-value. Figure 3 shows the results of a combined sensitivity analysis illustrating that increased milk yield and labour savings are

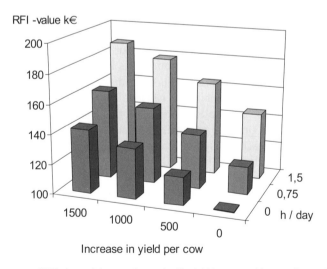

Figure 3. Room for Investment (RFI) due to labour saving and milk yield increase with annual costs for AM-system of 25% of investment. Comparison made with an highly automated milking parlour.

essential factors regarding the economy of automatic milking. The RFI-value for the basic farm with 500 kg per cow yield increase, 0,75 hour net labour saving per day (~10% labour saving), compared with a highly automated milking parlour and 25% annual costs of the AM-system amounts € 136,942. Both labour saving and yield increase have a large effect on the RFI value.

Two North American models (Rodenburg 2002, Rotz *et al.*, 2003) suggest that on large farms, as the hours of use of large automated milking parlours increase, their relative economic advantage over AM-systems increases. These studies found that on farms with 60 to 180 cows, capital investment in automatic milking is only slightly higher than conventional milking systems, in part because of smaller space requirements than for parlours with large holding areas. Lower labour input for AM-systems makes them competitive in this herd size range. For herds with more than 240 - 270 cows, extending the hours of use of conventional parlours, without additional capital investment made parlour milking much more efficient. It follows that in the central and western United States, where herds of 500 to several thousand cows predominate, widespread adoption of automatic milking will not occur unless capital cost of these systems decrease or labour costs increase substantially. Since both of these parameters are highly likely to trend in this direction, more widespread adoption of automatic milking in nearly all areas of the developed world would appear to be only a matter of time.

Conclusion

The number of farms milking with automatic milking has increased significantly since 1998. In areas where labour is expensive or in short supply, automatic milking is a valid alternative to traditional parlour milking. However if labour is available, and particularly where herd sizes are large conventional milking, often with rotary or rapid exit parlours equipped with features to increase throughput per man hour will remain popular.

The introduction of automatic milking has a large impact on the farm and affects all aspects of dairy farming. Because milking is voluntary there is large variation in milking intervals. Both farm management and the lifestyle of the farmer is altered by automatic milking. AM-systems require a higher investment than conventional milking systems but increased milk yields and reduced labour may lead to lower fixed costs per kg milk. Successful adoption of automatic milking depends on the management skills of the farmer and the barn layout and farming conditions.

A better understanding of the characteristics of automatic milking systems will help farmers to make the right decision. Both conventional and automatic milking will be used on dairy farms in modern dairy countries in the foreseeable future.

References

Arendzen I, A.T.J van Scheppingen, 2000, Economical sensitivity of four main parameters defining the room for investement of AM-systems on dairy farms, in: Robotic Milking, Proceedings of the international symposium held in Lelystad, pp 201-211.

De Koning C., and W. Ouweltjes, 2000, Maximising the milking capacity of an automatic milking system, in: Robotic Milking, Proceedings of the international symposium held in Lelystad, pp 38-46.

De Koning C.J.A.M., Y. van der Vorst, A. Meijering, 2002, Automatic milking experience and development in Europe, in: Proceedings of the first North American Conference on Robotic Milking, Toronto, Canada, pp I1 - I11.

De Koning C.J.A.M., J.A.M. Verstappen-Boerekamp, H. Schuiling, 2002b, Milk cooling systems for automatic milking, in: Proceedings of the first North American Conference on Robotic Milking, Toronto, Canada, pp V25 - V35.

Erdman, R.A. and M. Varner (1995) Fixed yield responses to increased milking frequency. Journal of Dairy Science 78: 1199-1203.

Helgren, J.M. and D.J. Reinemann (2003) Survey of Milk Quality on US Dairy Farms Utilizing Automatic Milking Systems, ASAE paper # 033016, St. Joseph's, Michigan USA.

Hogeveen, H., A.J.H. van Lent and C.J. Jagtenberg (1998). Free and one-way cow traffic in combination with automatic milking, in J.P. Chastain (editor) Proceedings of the Fourth International Dairy Housing Conference, ASAE 01-98, pp 80-87, Michigan, U.S.A.

Hogeveen H., Y. van der Vorst, C. de Koning, B. Slaghuis (2001) Concepts et implications de la traite automatisée, in: Symposium sure les bovines laitiers, pp 104-120, CRAAQ, Canada

Ipema, A.H., Schuiling,E., 1992. Free fatty acids; influence of milking frequency. in Proceedings of the Sumposium Prospects for Automatic Milking November 23-25, 1992, EAAP Publ. 65, Wageningen, The Netherlands, pp 491-496.

Ipema A, A. Smits and C. Jagtenberg, 1998, Praktijkervaringen met melkrobots, in: Praktijkonderzoek, 98-6, pp 37-39 (in Dutch).

Jellema, A., 1986. Some factors affecting the susceptibility of raw cow milk to lipolysis. Milchwissenschaft. 41:553-558.

Kamp, A. van der, A.G. Evers, B. Hutschemakers, 2003, Three years high-tech farm, cost price, labour and mineral balance score, Scientific report 30, Animal Sciences Group WUR, Applied Research.

Klei, L. R., Lynch, J. M., Barbano, D.M., Oltenacu, P.A., Lednor, A.J., Bandler, D.K., 1997. Influence of milking three times a day on milk quality, J. Dairy Sci. 80: 427-436.

Klungel, G.H., B.A.Slaghuis, H. Hogeveen, 2000, The effect of the introduction of automatic milking on milk quality, Journal of Dairy Science, 83:1998-2003.

Knappstein, K., N. Roth, H.G. Walte, J. Reichmuth, B.A. Slaghuis, R.T. Ferwerda-van Zonneveld, A. Mooiweer, 2004, Report on effectiveness of cleaning procedures applied in different automatic milking systems, Deliverable D14 from EU project Implications of the introduction of automatic milking on dairy farms (QLK5 2000-31006), www.automaticmilking.nl

Land, A. van 't et al., 2000, Effects of husbandry systems on the efficiency and optimisation of robotic milking performance and management, in: Robotic Milking, Proceedings of the international symposium held in Lelystad, pp 167-176.

Meskens, L., E. Mathijs, 2002, Socio economic aspects of automatic milking, Motivation and characteristics of farmers investing in automatic milking systems, Deliverable D2 from EU project Implications of the introduction of automatic milking on dairy farms (QLK5 2000-31006), www.automaticmilking.nl

Ouweltjes W., 2004. Demands and opportunities for operational management support. Deliverable D28 from EU project Implications of the introduction of automatic milking on dairy farms (QLK5 2000-31006), www.automaticmilking.nl

Pomies D., J. Bony (2001) Comparison of hygienic quality of milk collected with a milking robot vs. with a conventional milking parlour, in: H. Hogeveen and A. Meijering (editors) Robotic Milking, pp. 122-123, Wageningen Pers, Wageningen, The Netherlands.

Rasmussen, M.D., 2003. Redefinition of acceptable milk quality, Consequences of definitions of acceptable milk quality for the practical use of automatic milking systems, Deliverable D6 from EU project Implications of the introduction of automatic milking on dairy farms (QLK5 2000-31006), www.automaticmilking.nl

Rodenburg, J. and D.F. Kelton (2001) Automatic milking systems in North America: Issues and challenges unique to Ontario; in: National Mastitis Council Annual Meeting Proceedings, pp 163-169, NMC, Madison, WI, USA.

Rodenburg, J, (2002) The Economics of Robotic Milking - Good Enough to be Interesting, Hoard's Dairyman, March 10, pg 181.

Rodenburg, J., and B. Wheeler (2002), Strategies for Incorporating Robotic Milking into North American Herd Management, in Proceedings of the first North American Conference on Robotic Milking, Toronto, Canada, pp III-18 - III-32.

Rotz, C.A., C.U. Coiner and K.J. Soder (2003) Automatic Milking Systems, Farm Size and Milk Production, J. Dairy Sc. 86:4167-4177.

Schuiling H.J., 1992, Teat cleaning and stimulation, in: Proceedings of the international symposium on prospects for automatic milking, EAAP publication 65, pp 164-168.

Sonck, B.R. and J.H.W. Donkers 1995. The milking capacity of a milking robot. Journal of Agricultural Engineering Research 62: 25-38.

Svennersten-Sjaunja K., H. Wiktorsson, 2003, Effect of milking interval on FFA in milk from cows in an AMS. Italian Journal of Animal Science, vol. 2, p. 314.

Veysset P., P. Wallet, E.Prognard (2001) Automatic milking systems: Characterising the farms equipped with AMS, impact and economic simulations, in: A.Rosati, S. Mihina, C. Mosconi (editors), Physiological and Technical Aspects of Machine Milking, pp 141-150, ICAR TS 7, Rome, Italy.

Van Dooren, H.J., E. Spordnly, H. Wiktorsson, 2002, Automatic milking and grazing, Applied grazing strategies in Europe, Deliverable D25 from EU project Implications of the introduction of automatic milking on dairy farms (QLK5 2000-31006), www.automaticmilking.nl

Van der Vorst, Y., and H. Hogeveen, 2000, Automatic milking systems and milk quality in The Netherlands, in: Robotic Milking, Proceedings of the international symposium held in Lelystad, pp 73-82.

Van der Vorst, Y., K. Knappstein, M.D. Rasmussen, 2002, Effects of Automatic Milking on the Quality of Produced Milk, Deliverable D8 from EU project Implications of the introduction of automatic milking on dairy farms (QLK5 2000-31006), www.automaticmilking.nl

Van der Vorst, Y. and W. Ouweltjes, 2003. Milk quality and automatic milking, a risk inventory. Report 28 - Research Institute for Animal Husbandry, Lelystad, The Netherlands.

Van der Vorst, Y., K. Bos, W. Ouweltjes, J. Poelarends, 2003. Milk quality on farms with an automatic milking system; Farm and management factors affecting milk quality. Report D9 of the EU Project Implications of the introduction of automatic milking on dairy farms (QLK5-2000-31006), www.automaticmilking.nl

Wiking, L., J.H. Nielsen, 2003, Can strategies for cow feeding and cooling of milk in automatic milking systems improve milk quality, Italian Journal of Animal Science, vol. 2, p. 316.

Wiktorsson, H., G. Pettersson, J. Olofsson, K. Svennersten-Sjaunja, M. Melin, 2002. Welfare assessment of dairy cows in automatic milking systems, Welfare status of dairy cows in barns with automatic milking, Deliverabel D24 from EU project Implications of the introduction of automatic milking on dairy farms (QLK5 2000-31006), www.automaticmilking.nl

SOCIO-ECONOMIC ASPECTS

MILK IN THE NEWS

Ulrike Maris & Keith Roe
Department of Communication, Catholic University Leuven, Leuven, Belgium

Introduction

One of the main objectives of work package 2. was to investigate the state of public opinion with regard to automatic milking. In order to assess the central role of the mass media in this respect a descriptive content analysis of newspapers from six countries was conducted in order to establish which events become news and to assess the ways in which newspapers cover agricultural and food issues. Content analysis is, "a research technique for the objective, systematic and quantitative description of the manifest content of communication" (Berelson, 1952). The analysis was designed to examine news coverage in each country with regard to food production (especially milk and dairy products) and to identify specific national issues and concerns.

Method

In total 14 newspapers were analyzed - two from each country with the exception of Belgium where four (two from each of the main language communities) were included. The period covered was 1-11-2001 to 31-10-2002 with the exception of the two Danish newspapers where, for administrative reasons, the period covered was 1-12-2001 to 30-11-2002. In total 1,385 news items were recorded. The newspapers were:
Belgium (Flemish Community)
- De Morgen
- De Financieel Economische Tijd

Belgium (French Community)
- La Dernier Heure
- La Libre Belgique

Denmark
- Jyllands Posten
- Politiken

Germany
- die Welt
- Frankfurter Allgemeine Zeitung

The Netherlands
- De Telegraaf
- NRC Handelsblad

Sweden
- Dagens Nyheter
- Svenska Dagbladet

The United Kingdom
- The Independent
- The Times

Results

Belgium (Flemish Community)

In the two Dutch language Belgian newspapers examined 313 articles were recorded (De Morgen 163, De Financieel Economische Tijd 150). In neither newspaper were any news items related to automatic milking found and only four (3 on investment in the dairy industry and 1 on fraudulent sale of fake butter) dealing with dairy production in general. In De Morgen the most frequently covered specific issue (9.8% of all items) was the scare over PCB's (Polychlorinated biphenyls), followed by MPA's (meddroxyprogesterone acetaat - 8.0%) and Nitrofen (6.1%). In De F.E.T reports centering on the Belgian Federal Food Agency (19.3% of all items) dominated the category of specific issues followed by farming economics (16.1%) and MPA's (8.7%). In both newspapers the dominant tone of the articles in each of these categories was negative, especially in De Morgen where all of the 13 articles on MPA's, 13 of the 16 articles on PCB's and 8 of the 10 articles on nitrofen were negative. In terms of general categories, by far the greatest number of items in the two Flemish newspapers fell under the heading 'food scares/safety' (53.7% of all items in De Morgen and 55.5% in De F.E.T.), followed by articles dealing farming in general (12.1% of all items), and the economics of farming (10.5% - not surprisingly with the great majority in De FET). In both newspapers the dominant tone of around 60% of all news items was negative, although De F.E.T. had more positive items (29%) than did De Morgen (17%).

Belgium (French-Speaking Community)

In the two French language Belgian newspapers 105 articles were recorded (52 in Le Soir and 53 in La Libre Belgique). In the period studied neither featured a single news article on milk or dairy production. In both newspapers articles concerning GMO's clearly constituted the most frequently reported specific issue (30.8% in Le Soir and 18.9% in La Libre Belgique), with no other single issue receiving significant coverage. In terms of general categories, news items covering food scares/safety were most frequent (32.2% of all items), followed by GMO's (30.5%) and genes technology (17.1%). In both newspapers the dominant tone of the reports was neutral (58% in Le Soir and 59% in La Libre Belgique), although few were positive (less than 6% in each case).

Denmark

In the Two Danish newspapers no less than 427 items were recorded (174 in Politiken and 253 in Jyllands Posten), easily the highest number in the six countries studied. Two large articles relating to automatic milking were found, both in Jyllands Posten and both positive. In comparison with the other countries dairy products were also much more frequently dealt with: Politiken featuring 17 items and Jyllands Posten 24. The great majority of these articles dealt with the dairy company Arla and in particular it's 'butter war' over Lurpack with a small competitor. Other aspects dealt with the companies' monopoly position and ambitions and its alleged sexist milk advertising campaign. In general the coverage of Arla tended to be negative, placing the company in the role of a monopolistic 'Goliath' being heroically fought by various small 'David's. Another issue in the Danish newspapers was the fight to allow Danish producers to use the cheese label Feta. In other sectors the most frequent specific topic in Politiken was GMO's (6.9%), followed by animal transport (4.6%) while in Jyllands Posten it was newcastle disease (10.7%), followed by salmonella outbreaks (3,2%). In terms of general categories, the most frequently reported in both newspapers was food scares/safety (overall 28.3% of items). In Politiken the second most frequent category was the economics of farming (17.8%), followed by animal rights (10.3%) and GMO's (8.6%). In Jyllands Posten, after food scares/food safety, the list consisted of the economics of farming (16.6%), agricultural (particularly E.U) policy (8.7%), and organic farming/food (7.9%). In both newspapers the dominant tone of the reports was neutral (58% in Politiken and 55% in Jyllands Posten), followed by positive (21% and 24%, respectively).

Germany

In the two German newspapers 95 articles were recorded (43 in die Welt and 52 in Frankfurter Allgemeine Zeitung (F.A.Z), the lowest number in the six countries studied. No articles dealing with dairy production were found. The most frequently reported specific issue in die Welt was Nitrofen (36.2%) and in F.A.Z. it was biotechnology (28.8%), followed by Nitrofen (25.0%). In terms of general categories food scares/safety received the most attention in both the German newspapers, with 49.5% of all items. Thereafter in die Welt came genetic manipulation (11.6%) and in F.A.Z. new agricultural technologies (28.8%). The dominant tone of the articles tended to be negative, though more so in die Welt (63%) than in F.A.Z. (42%). In both newspapers around 21% of the articles were positive.

The Netherlands

In the two newspapers from The Netherlands 131 articles were recorded (De Telegraaf 71, NRC Handelsblad 60). Only one news item dealing with dairy issues was found - dealing with labeling of milk from free grazing cows. Otherwise, in both newspapers reports about MPA constituted the most frequent specific category, accounting for 11.3% of items in De Telegraaf and 15.0% in NRC Handelsblad. In both cases the dominant tone of the articles covering this issue was negative (100% and 69%, respectively). The second most frequent specific topic dealt with in both newspapers was cloning (11.3% in De Telegraaf and 11.7% in NRC Handelsblad), although here the dominant tone was somewhat more balanced, with one-third of the articles positive. In terms of general categories in the two Dutch newspapers, food scares/safety received the most attention (37.4%), followed by gene technology (14.5%) and farm economics (8.4%). Overall, 32% of the coverage in the two Dutch newspapers was positive, 10% neutral, and 58% negative.

Sweden

In the two newspapers from Sweden 186 articles were recorded (103 in Dagens Nyheter and 83 in Svenska Dagbladet). One large article about automatic milking was found in Svenska Dagbladet and it was positive. Moreover, one article dealt (positively) with the fact that new farm technology provides farmers with more free time. In addition 7 other items dealt with dairy production, although the subjects involved were very diverse ranging from, 'ginseng makes cows healthier' and 'dead cows to be used for central heating' to 'radio-active goats milk in Norway'. In both newspapers the most frequently reported single issues were the same: first the akrylamid scare (Dagens Nyheter 18.4%, and Svenska Dagbladet 26.5%), followed by BSE in cows (5.8% and 7.2%, respectively), and Salmonella (Dagens Nyheter 5.8%, and Svenska Dagbladet 6.0%). In both the most reported general category was food scares/safety, accounting for 49.5% of all items, followed by animal welfare/rights with 24.2%. Overall in the two newspapers the dominant tone of the coverage was neutral (60% in Dagens Nyheter and 55% in Svenska Dagbladet), while in each newspaper around 27% of the items were negative.

The United Kingdom

In the two British newspapers 128 articles were recorded (101 in the Independent and 27 in The Times). No articles dealing specifically with dairy production were found in either newspaper In both the most frequently reported single issue was foot and mouth disease, including its social implications (22.8% in The Independent and 29.6%, in The Times), followed by gene therapy (8.9%, and 11.1%, respectively). Overall, food scares/safety formed the most frequent general category (28.9% of all items) and genes technology was also frequently reported (around 15% of items). In both newspapers around 60% of the coverage was negative, although 26% of the items in the Independent and 33% of those in The Times were positive.

News Items: The Wider Picture

The analysis in each country shows varying particular national concerns, reflecting the importance of local events (e.g. foot and mouth disease in the U.K., nitrofen in Germany, akrylamid in Sweden, newcastle disease in Denmark). However, it is also useful to aggregate the results across the six countries in order to obtain an indication of which issues are of more general concern. Table 1. shows the ten most frequently dealt with topics in the 14 newspapers analyzed. Interestingly, the issue of GMO's tops the list with 15.3% of all items, followed by BSE and akrylamid (although this last named is mostly accounted for by the large number of items in the Swedish newspapers). The economics of farming comes in fourth place.

Table 2 contains the distribution of news items across more general categories. It shows that one issue in particular dominated the news coverage of the sector: food scares/safety which accounted for just over a half of all news items and which, as we have seen, was dominated by akrylamid, nitrofen, PCB and FMD. In second place comes animal welfare which was dominated by stories of animal mistreatment, especially in relation to transport of livestock.

Table 1. The most frequent specific contents covered.

Topic		Number of items	Percentage
1.	GMO's	56	15.3
2.	BSE	44	12.1
	Akrylamid	44	12.1
4.	Economics of farming	41	11.2
5.	Nitrofen	39	10.7
6.	Newcastle disease	33	9.0
7.	PCB's	30	8.2
	FMD	30	8.2
9.	Cloning	28	7.7
10.	Organic farming	20	5.5
Total		365	100

Table 2. The most frequent general categories covered.

Topic	Number of items	Percentage
1. Food scares/safety	561	50.8
2. Animal welfare/rights	102	9.2
3. GMO's (general)	97	8.8
4. Genes technology/cloning	96	8.7
5. economics (general)	84	7.6
6. farming (general)	67	6.1
7. policy	34	3.1
8. Crime (e.g. illegal use hormones)	30	2.7
9. New technologies	25	2.3
10. Science	9	0.9
Total	1105	100

Discussion

In the year studied automatic milking was virtually absent from the news contained in the 14 European newspapers studied. Of the 1,385 news items concerning agriculture and the food chain recorded only 3 dealt with this new technology. In terms of informing and sensitizing the public to the advantages offered by automatic milking this result is not favourable. On the other hand, given the fact that much news is 'bad', in the sense of events being negatively evaluated by journalists (41.3% of all items were evaluated as negative, compared to 21.5% positive), it also implies that the development is not an issue and is perhaps best left alone. Moreover, milk and dairy products as a whole appear only infrequently in the news and seem largely to have escaped the bad press received by some other sectors. However, while milk as such escapes, issues surrounding livestock do receive significant (and negative) coverage - particularly BSE, foot and mouth disease, GMO's and pesticides in fodder, and animal welfare. GMO's and animal welfare in particular appear to be widespread concerns in the press being kept visible, at least in part, by powerful pressure groups. What is evident from these results is that the press is extremely sensitive to scares over food safety. Consequently, if for any reason one should ever break loose over milk quality, especially in relation to automatic milking, the damage to the imago of the sector in the short and medium term might be considerable and difficult to redress. .

References

Berelson, B., 1952. Content A!nalysis in Communication Research. New York. Free Press.

SOCIO-ECONOMIC ASPECTS OF AUTOMATIC MILKING

Erik Mathijs

Afdeling landbouw- en milieueconomie, K.U.Leuven, Leuven, België

Abstract

This paper reports on a survey among 107 farmers who have recently invested in an automatic milking (AM) system in Belgium, Denmark, Germany and The Netherlands. A questionnaire was designed to capture the characteristics and motivations of AM farmers, farm characteristics and the implications of the introduction of an AM-system. This paper summarizes the motivation and characteristics of farmers investing in an AM-system in a first part and the implications for labour use and the farmers' quality of life in a second. The average AM farmer is male, married, 44 years old and has about 3 children. On average, AM farms have about 87 dairy cows. Two thirds of the interviewed AM farmers state social reasons for investing in an AM-system, such as increased labour flexibility, improving social life and health concerns. AM farmers reported an average labour saving of 19.8%, which increases to 21.3% when only farms that have kept their herd size more or less constant are considered. Only few farmers disagree that AM has improved their health, while about one third has experienced no changes on their health. Two thirds of all farmers state that their quality of life has increased. Noteworthy is that the improvements in quality of life are relatively unaffected by the chosen growth strategy.

Introduction

Socio-economic research on the adoption of a new technology usually follows two lines. One line investigates who innovates and why they do so, in other words the determinants of innovation. Such studies are either concerned with farmers having proper access to the new technology or are interested in the pattern of adoption. Often, they find that non-economic motives play an important role. The second line is concerned with the implications of the innovation, both at the farm and at the sector level. We refer to Meskens *et al.* (2001) for a literature review of past research on the determinants and implications of automatic milking (AM). This paper reports on a survey among 107 farmers who have recently invested in an AM-system in Belgium, Denmark, Germany and The Netherlands.

A questionnaire was designed to capture the characteristics and motivations of AM farmers, farm characteristics (such as farm size, grazing system, etc.) and the implications of the introduction of an AM-system. Addresses of AM farmers were assembled with the assistance of the AM manufacturers, and made up the frame out of which a sample was taken. A random sample was then drawn from the frame to yield a sample of 13 AM users in Belgium, 57 in The Netherlands, 13 in Denmark and 24 in Germany, totalling to 107 respondents. Farmers were interviewed face-to-face between November 2001 and November 2002.

This paper summarizes the motivation and characteristics of farmers investing in an AM-system in a first part and the implications for labour use and the farmers' quality of life in a second. We refer to Meskens and Mathijs (2003) and Wauters and Mathijs (2004a)

respectively for more details. This paper does not deal with the financial implications of investing in an AM-system, which is covered by Wauters and Mathijs (2004b) in this volume and which can also be found in Wauters and Mathijs (2004a).

Motivation and characteristics of farmers investing in AM-system

Profile of AM farmers

Characteristics of AM adopters
    Almost all (98.1%) farm managers in the sample are men (Table 1). The average age of the interviewed farm managers is 43.9. AM farmers are younger in Belgium, with an average age of 41.6 and only 8 percent are older than 50. Farm managers are relatively old in

Table 1. Characteristics of AM users and their business*.

	B	NL	D	DK	Total
Age (in years)	41.6	44.3	43.5	45.1	43.9
Education					
Primary	0.0	1.7	0.0	28.6	4.7
Secondary	84.6	92.9	95.6	71.4	89.7
Higher	15.4	5.3	4.3	0.0	5.6
Workshops or courses > 4x/year	46.1	22.8	34.7	64.2	33.6
Management clubs > 4x/year	30.8	19.2	4.3	42.8	20.6
Personal contact to AM farmer	38.5	42.1	30.4	57.1	41.1
External consultant	38.5	52.6	43.5	64.3	50.5
One-person operation	30.7	29.8	43.5	57.1	36.4
Successor	38.4	19.2	52.1	35.7	30.8
Other enterprises	61.5	29.8	65.2	28.6	41.1
Self declared styles of farming					
Fine tuner	61.5	52.5	30.4	7.1	43.0
Cost saver	7.7	22.8	4.3	7.1	15.3
Grower	7.7	12.3	8.7	21.4	12.1
Labour saver	23.1	12.3	43.4	64.3	27.1
Diversifier	0.0	0.0	13.0	0.0	2.8
Statements**					
Innovation is important	30.8	45.7	65.8	64.2	50.5
Leisure is important	61.6	52.6	86.9	64.3	67.2
Opinion of others is important	7.7	5.2	13.0	14.1	8.4

Notes:
* Figures are percentages of the sample, either per country (B = Belgium, NL = The Netherlands, D = Germany, DK = Denmark) or in total
** The question was asked on five-item scale ranging (totally disagree, disagree, neutral, agree, totally agree). Reported here is the share of the sample that agreed or totally agreed with the statement.

Denmark, with an average of 45.1 and where almost 36 percent of all AM farmers are older than 50. The majority of farmers (90.6%) is married or living together. Farm households have on average 2.7 children, with somewhat larger families in Belgium and Denmark with an average of 3.1 and 2.9 children respectively. In short, the average AM farmer is male, married, 44 years old and has about 3 children.

Most respondents (89.7%) had secondary education. The educational level in Denmark is somewhat lower, as 29 percent had only primary education, but this can be explained by the relatively higher age of Danish farmers. The majority of farmers (89.7%) acquired their degree in a specialised agricultural school. Further, only few farmers followed a special training at a dairy farm, with the notable exception of Denmark where most farmers (92.8%) have been engaged in an apprenticeship. Quite a large share of the farmers (31.7%) never attends courses, lectures and workshops. Again, the Danish farmers score high: 64.2% of them attend lectures 4 times a year or more, a figure that is twice the average score (33.6%). Management clubs are not regularly attended: 54.2% of the interviewed farmers never go to these clubs. Overall, 41.1 percent already knew someone with an AM-system, a percentage which is somewhat lower in Germany (30.4%) where the penetration of AM-systems is relatively low and higher in Denmark (57.1%). About half of the farmers consulted an external consultant before buying an AM-system. Again, this percentage is higher in Denmark (64.3%) and lower in Germany and also in Belgium.

Farmers were asked to select the style of farming that best fits their situation. They could choose only one of five options depending on which aspect they find most important in their management: a "fine tuner" wants to earn a good revenue with a lot of dedication and craftsmanship, a "cost saver" focuses on the minimisation of production costs, a "grower" wants to expand his farm through new investments, a "labour saver" seeks to increase output and save labour through mechanisation, and a "diversifier" realises a surplus by diversification such as on-farm processing and selling of own products. In The Netherlands, the option of profit maximiser was added, but we merged this category with cost saver. Overall, most AM farmers (43.0%) see themselves as a fine tuner. Most German (43.4%) and Danish (64.3%) AM farmers claim to be labour savers. Pure growers can be found most in Denmark (21.4%) and The Netherlands (12.3%). Diversifiers are virtually absent with the exception of Germany where 13 percent of the AM farmers declare diversification as their main style. This is not surprising since part-time farming is more important in Germany.

Farmers had to give their opinion to several statements to get an idea about how they think about new technologies, free time and what other farmers think of them. Farmers who agree with the statement that "it is important to have new technologies in an early stage on the farm" are more likely to be early innovators. About half of the AM farmers (50.5%) agree with this statement. Particularly the Danish (64.2%) and the German (65.8%) farmers think it is important to be an early innovator, while only 30.8% of the Belgian farmers do. Almost two thirds (62.7%) of the respondents agree or strongly agree with the statement that "it is important to have some free time and/or to go on holiday every year". Particularly German farmers agreed with this statement (86.9%). Dutch farmers agree the least with this statement. The majority of respondents (80.4%) strongly disagree with the statement "it is important what other farmers think of me". The Danish farmers are somewhat more sensitive to what others think of them.

Most AM farmers (47.6% overall) work in a co-operation with their wife (or husband). This is the dominant form in Belgium and The Netherlands. In Denmark and Germany farms usually operate as single-person businesses. Co-operations with a partner other than spouse or child is scarce (7.4% overall). Most farmers (68.1% overall) do not have a successor or do not know yet. It is very likely that this is related with the relatively young age of the AM farmers. A relatively high amount (30.4%) of German farmers is sure to have a successor for the farm. Succession is the least sure in The Netherlands.

Characteristics of the AM farms

First, for almost all farmers, dairy farming is the main profession (98.1% overall) (Table 1). Exceptions include for example a veterinary doctor in Belgium who keeps dairy cows as a second profession. Second, 41.1 percent of all AM farmers have at least one other enterprise on the farm. Particularly in Belgium and Germany other enterprises are well represented (61.5% and 65.2% respectively). In Belgium, this may be explained by the fact that milk quota are not tradable and only very small amounts can be purchased from a quota fund. Hence, expanding the dairy business is very difficult. The nature of the other enterprises is very diverse. We have grouped the answers into six categories: products made of milk (e.g., ice cream, butter); vegetables, grains and fruit; other livestock (e.g., chickens, goats, sheep and pigs); contract work for other farmers; veterinary and advisory services; other enterprises that cannot be placed under the other categories; and all possible combinations of the above named categories. Belgian AM farmers that have diversified tend to process their dairy products on the farm or contract out their services to do machinery work. Dutch diversified AM farmers rather keep other livestock. German diversified AM farmers tend to have other livestock and a combination of various activities. Danish diversified AM farmers tend to have crops as a second enterprise. In general, keeping other animals is the main diversification strategy of AM farmers (almost half of the farms that have other enterprises).

To get an idea about the average farm size in the different countries, we considered the average number of milking cows. Most AM farms (41.1%) have between 70 and 100 cows, with an average number of 87.2 cows. The smallest herds are in Belgium (81.1 cows), the largest in Denmark (107.5 cows). Belgium is the only country where there are AM users with fewer than 40 cows. In general, AM users are considerably larger than average specialised dairy farmers. For example, FADN data shows that the average number of dairy cows per farm in 1999 was 34.3 in Germany, 45.7 in Belgium, 56.5 in The Netherlands and 61.4 in Denmark.

Motivation for AM adoption

We asked AM farmers why they have installed an AM-system in two different ways. First, the farmer was asked to state the most important reason in an open-ended question. Second, the farmer had to indicate the three most important reasons from a closed list of options. Overall, both questions give similar results. We grouped the answers to the open-ended question into two categories:

- Social reasons: to decrease labour intensity, to spend more time on other activities, to have more flexibility, not to have to milk anymore, health problems, challenge, to improve social life, animal welfare

- Economic reasons: to increase milk production, to produce less manure, to have more management information, to improve cow and udder health, to expand the farm, because a labour unit has fallen away, to milk more then twice a day, because the old stable had to be replaced, to optimise labour, because it is difficult to find hired labour.

Overall, more of the AM farmers interviewed declared social reasons (67.3%) than economic reasons (32.7%). However, the differences between countries are relatively large: social reasons are more important in Denmark (85.7%) and Germany (82.6%) than in Belgium (61.5%) and The Netherlands (57.9%).

Table 2 reports the results of the closed question. Overall, most farmers stated labour reduction as the most important reason (28.9%), immediately followed by increasing labour flexibility (27.1%). Labour reduction is more important than in The Netherlands, while in Belgium and Denmark increasing labour flexibility is more important. The third most important reason (14.9% overall) is to get rid of a hired hand, as hired labour is considered to be more expensive than having an AM-system. This reason can be found primarily in The Netherlands and Denmark, as hired labour is much less common in Belgium and Germany. Improving technical parameters is again a reason primarily found in The Netherlands and Denmark. To be able to spend more time on other activities, in other words to be able to diversify more, is the second most important reason in Belgium (30.7%) and the third reason in Denmark (17.4%).

Table 2. Reason to install an AM-system*.

	B	NL	D	DK	Total
Labour reduction	7.7	33.3	34.7	21.4	28.9
Labour flexibility	38.5	19.3	34.7	35.7	27.1
Get rid of hired labour	7.7	17.5	8.7	21.4	14.9
Improving technical parameters	7.7	14.0	4.3	21.4	12.1
Future, challenge	7.7	12.3	0.0	0.0	7.5
Other activities	30.7	3.5	17.4	0.0	9.3

Notes:
* Figures are percentages of the sample, either per country (B = Belgium, NL = The Netherlands, D = Germany, DK = Denmark) or in total

Implications of automatic milking

Implications on herd size and grazing strategies
A large proportion of the farmers changes grazing strategy after adoption: only 10.3% of the AM farmers apply unlimited grazing versus 28% before the adoption (Table 3). Slightly less than half of the AM farmers apply in stable summer feeding, while before the adoption, only 14.9% of the farmers chose this grazing strategy. Although we see some small movements between different size classes, average herd size is hardly affected by adoption of an AM-system. Average herd size increases with only 0.5%.

	B	NL	D	DK	Total
Before introduction					
Unlimited grazing	69.2	15.8	30.4	35.7	28.0
8 to 12 hours	23.1	52.6	34.7	21.4	41.1
Less then 4 hours	0.0	0.0	4.3	7.1	1.8
2 times less then 4 hours	0.0	1.7	0.0	0.0	0.9
In stable zero-grazing	0.0	4.0	4.3	7.1	4.7
In stable summer feeding	7.7	12.3	17.4	28.6	14.9
Walking out zero-grazing	0.0	0.0	0.0	0.0	0.0
Walking out summer feeding	0.0	1.7	8.7	0.0	2.8
2 times 8 hours	0.0	8.8	0.0	0.0	4.7
After Introduction					
Unlimited grazing	38.4	5.3	8.3	7.7	10.3
8 to 12 hours	23.1	35.1	0	15.4	23.4
Less then 4 hours	0.0	0.0	8.3	15.4	3.7
2 times less then 4 hours	15.4	1.8	0.0	0.0	2.8
In stable zero-grazing	0.0	8.8	4.2	7.7	6.5
In stable summer feeding	23.1	42.1	58.3	46.2	43.9
Walking out zero-grazing	0.0	0.0	0.0	0.0	0.0
Walking out summer feeding	0.0	3.5	20.8	7.7	7.5
2 times 8 hours	0.0	3.5	0.0	0.0	1.9

Notes:

* Figures are percentages of the sample, either per country (B = Belgium, NL = The Netherlands, D = Germany, DK = Denmark) or in total.

Implications on labour

AM farmers report, on average, a 19.8% labour saving after investing in an AM-system (Table 4). The labour savings are largest in Belgium (28%) and smallest in Denmark (11.5%). What farmers do with the labour saved is difficult to assess. In addition, some farmers use the extra time to let their farm grow by increasing herd size, which biases the results. Therefore, we recalculated the average labour saving by considering three herd size change categories: group(-) are farmers who decrease in size, group(0) contains all farmers whose size stays constant within a certain range and group(+) includes all farms who grow after adoption of an AM-system. We excluded farms for which we do not have information about changes in herd size. In each country, the group(0) is by far the largest group: 79% of the farmers in the sample do not combine the adoption of an AM-system with a change in size of their farms. Only 19% of the farms have grown in size, while there are only 2 farms - in The Netherlands - who have decreased in size. Average labour savings for the group(0) farmers is 21.26%.

Table 4. Changes in labour allocated to the dairy enterprise.

	Before	After	Change
	hours/week	hours/week	%
All farms			
B	124.55	89.45	-28.18
NL	94.59	77.34	-18.24
D	109.54	84.63	-22.74
DK	96.04	84.99	-11.51
Total	101.47	81.37	-19.81
Group (0) farms			
B	127.90	92.1	-27.99
NL	94.86	75.90	-19.99
D	114.72	88.44	-22.91
DK	93.57	84.07	-10.15

Implications on health

Only 6.5% of the farmers does not agree with the statement that their physical health has improved since the purchase of the AM-system (Table 5). More than one third of the farmers say their physical health did not change, whereas a third totally agrees with the statement. There were no Belgian nor German farmers who disagreed. Although only a minority of the questioned farmers does not agree with this statement, the improvement of the psychic health does not seem as clear as the improvement of the physichal health. In total 23.3% does not agree with the statement. Almost half of the farmers have no change in their sleeping quality since the purchase of the AM-system. In Germany there are remarkably more farmers who agree with the statement (average score of 3.3).

Implications on leisure and quality of life

Farmers were asked about their opinion about certain statements concerning friends, hobbies and family time. The answers to statements concerning leisure may be affected by the number of cows before and after investing in the AM-system. Therefore, we will also report the results for the three herd size change categories. The large majority of the farmers (86%) agree with the statement of being able to spend more time with their family. This percentage is the highest in The Netherlands (89%) and the lowest in Denmark (71%). The agreement seems to be independent from herd size category, as also the large majority of the group(+) farms (89%) agree. Only 7.5% says to have less time for hobbies. In Belgian and Germany no farmer disagrees with the statement. In total 62% of the farmers says to have more time for hobbies. Again, the effect seems independent of herd size change category. Two thirds of the farmers say the quality of life of their family has improved. The Dutch farmers see their quality of life least improved (54% agree), the German farmers most improved (91% agree). This time there are clear differences between the herd size change categories: while two thirds of both groups agree with the statement, agreement is stronger

Table 5. Results of statements related to health (in %).

	B	NL	D	DK	Total
my physical health has improved					
Totally not agree	0	10.7	0	0	5.6
Partly not agree	0	0	0	7.1	0.9
No change	53.8	41.4	37.5	14.3	38.3
Partly agree	30.8	8.9	37.5	21.4	19.6
Totally agree	15.4	39.3	25.0	57.1	35.5
Average score	3.6	3.7	3.9	4.3	3.8
my psychic health has improved					
Totally not agree	15.4	19.6	0	14.3	14.0
Partly not agree	15.4	8.9	8.3	7.1	9.3
No change	23.1	39.3	33.3	35.7	35.5
Partly agree	30.8	17.9	33.3	35.7	25.2
Totally agree	15.4	14.3	25.0	7.1	15.9
Average score	3.2	3.0	3.8	3.1	3.2
my sleeping quality has improved					
Totally not agree	15.4	16.4	8.3	21.4	15.1
Partly not agree	15.4	5.5	16.7	7.1	9.4
No change	38.5	50.9	33.3	50.0	45.3
Partly agree	30.8	5.5	25.0	7.1	13.2
Totally agree	0	21.8	16.7	14.3	17.0
Average score	2.8	3.1	3.3	2.9	3.1

for group(0) than for group(+). To conclude, the quality of life has largely improved, also for farmers who have chosen for a growth strategy and thus less leisure time.

Conclusions

This paper reported the results of a survey of 107 AM users carried out in Belgium, Denmark, Germany and The Netherlands concerning the personal characteristics of AM users, their motivation to install an AM-system and its implications on farm management and the quality of life of the farmer.

The average AM farmer is male, married, 44 years old and has about 3 children. He had secondary education in an agricultural school. Most AM farmers did not follow a practical training (except in Denmark) and only one third has attended additional training activities on a regular basis. Most AM farmers declare to be fine tuners in Belgium and The Netherlands and labour savers in Germany and Denmark. External advice has been consulted only by about half of the AM farmers before installing an AM-system. AM farmers are relatively neutral towards innovation, but rather positive towards more leisure. Most AM farms only have one enterprise, while those that are more diversified tend to have other livestock.

Table 6. Results of statements related to leisure and quality of life (in %)*.

	1	2	3	4	5	S
I have more possibilities to spend time with my family						
All farms						
B	0	7.7	7.7	53.8	30.8	4.1
NL	3.6	1.8	5.5	30.9	58.2	4.4
D	0	0	12.5	50.0	37.5	4.3
DK	7.1	7.1	14.3	21.4	50.0	4.0
All	2.8	2.8	8.5	36.8	49.1	4.2
Herd size change categories						
Group(-)	0	0	0	0	100	5.0
Group(0)	1.4	4.1	10.8	37.8	45.9	4.2
Group(+)	5.6	0	5.6	38.9	50.0	4.3
I have more time for hobbies						
All farms						
B	0	0	30.8	38.5	30.8	4.0
NL	8.9	1.8	30.4	21.4	37.5	3.8
D	0	0	26.1	47.8	26.1	4.0
DK	14.3	0	35.7	35.7	14.3	3.4
All	6.6	0.9	30.2	31.1	31.1	3.8
Herd size change categories						
Group(-)	0	0	0	50	50	4.5
Group(0)	6.8	1.4	29.7	27.0	35.1	3.8
Group(+)	5.6	0	27.8	50.0	16.7	3.7
the quality of life from our family has improved						
All farms						
B	0	7.7	23.1	46.2	23.1	3.8
NL	9.3	0	37.0	22.2	31.5	3.7
D	0	0	8.7	43.5	47.8	4.4
DK	7.1	7.1	7.1	50.0	28.6	3.9
All	5.8	1.9	25.0	33.7	33.7	3.8
Herd size change categories						
Group(-)	0	0	50.0	0	50.0	4.0
Group(0)	6.8	1.4	23.0	29.7	39.2	3.9
Group(+)	5.6	0	27.8	55.6	11.1	3.6

*1 = totally not agree, 2 = partly not agree, 3 = no change, 4 = partly agree, 5 = totally agree, S = average score

Automatic milking – A better understanding

Most AM farms practiced some kind of grazing system before introducing an AM-system. On average, AM farms have about 87 dairy cows, which is considerably larger than the average specialized dairy farms that range from about 34 cows in Germany to 61 cows in Denmark. Two thirds of the interviewed AM farmers state social reasons for investing in an AM-system, such as increased labour flexibility, improving social life and health concerns.

A large proportion of the AM farmers has changed their grazing strategy after adoption, primarily towards in stable summer feeding. Average herd size is hardly affected by adoption of an AM-system. AM farmers reported an average labour saving of 19.8%, which increases to 21.3% when only farms that have kept their herd size more or less constant are considered. Only few farmers disagree that AM has improved their health, while about one third has experienced no changes on their health. For 55% physical health has improved, for 41% psychic health has improved and for 30% sleeping quality has improved. Two thirds of all farmers state that their quality of life has increased. For example, 86% of all farmers spend more time with their family, while 62% also spends more time on hobbies. Noteworthy is that the improvements in quality of life are relatively unaffected by the chosen growth strategy.

Acknowledgements

This paper is produced within the EU project Implications of the introduction of automatic milking on dairy farms (QLK5-2000-31006) as part of the EU-program 'Quality of Life and Management of Living resources'. The content of this publication is the sole responsibility of its publisher, and does not necessarily represent the views of the European Commission. Neither the European Commission nor any person acting on behalf of the Commission is responsible for the use, which might be made of the following information. The author thanks Erwin Wauters for research assistance.

References

Meskens, L. and E. Mathijs, 2002. Motivation and characteristics of farmers investing in automatic milking systems. EU Project Automatic Milking (QLK5-2000-31006), deliverable 2.

Meskens, L., Vandermersch, M. and E. Mathijs, 2001. Literature review on the determinants and implications of technology adoption. EU Project Automatic Milking (QLK5-2000-31006), deliverable 1.

Wauters, E. and E. Mathijs, 2004a. Socio-economic implications of automatic milking on dairy farms, EU project, EU Project Automatic Milking (QLK5-2000-31006), deliverable 3.

Wauters, E. and E. Mathijs, 2004b. The economic implications of automatic milking: a simulation analysis for Belgium, Denmark, Germany and The Netherlands. This volume.

MOTIVATIONS OF DUTCH FARMERS TO INVEST IN AN AUTOMATIC MILKING SYSTEM OR A CONVENTIONAL MILKING PARLOUR

Henk Hogeveen[1,3], Kees Heemskerk[1] & Erik Mathijs[2]
[1]Business Economics Group, Wageningen University, Wageningen, The Netherlands
[2]Department of Agricultural and Environmental Economics, Catholic University Leuven, Leuven, Belgium
[3]Current address: Department of Farm Animal Health, Utrecht University, Utrecht, The Netherlands

Abstract

Besides business economics other motivations might also play an important role in the adoption of new technology. The objectives of this research were to gain insight in the motivation and background of the farmers who choose for an automatic milking (AM) system in comparison with those who choose for a conventional milking (CM) system. In total 120 randomly selected farmers (60 who recently invested in an AM-system and 60 who recently invested in a CM-system entered the study. The farmers answered a questionnaire with questions on motivation to invest in the milking technology, farm structure, personal circumstances and reduction in labour. The five most important motivations for the farmers to invest in an AM-system instead of a CM-system were: less (heavy) labour, increased flexibility, the possibility to milk cows more than twice a day, the leaving of an employee and the need for a new milking system. The five most important motivations for a farmer to invest in a CM-system instead of an AM-system were: lower costs, being standby for 24 hrs per day, lower operational security of the AM-system, less flexibility with an increasing farm seize and the higher expenses with an increasing farm seize. On average, farmers who invested in an AM-system were of the same farm seize as farmers who invested in a CM-system. There was no difference in educational level between the two groups. In the group that invested in an AM-system, there were more farmers without a successor.

Introduction

One of the key factors influencing the adopting process is the perceived economic gains that producers will reap from new technology. Without knowing the economic consequence of adopting a new technology, managers will be extremely reluctant to employ them. However the economic consequence is not the only factor that will influence the adoption process. Other influential factors include the degree in with the mix of resources (e.g., labour and capital) utilized in the production process is changed, the level of management skills needed to make effective use of the technology, institutional constraints such as government regulations, and the motivation and goals of the producer (Dijkhuizen *et al.*, 1997).

It is suggested that social reasons are more important than economic reasons to adopt an automatic milking (AM)-system. Many farmers delay adoption for more information (less uncertainty) concerning the technology itself as well as the institutional environment. In addition to a trade-off between the direct financial costs and benefits of AM-system adoption, the AM-system also brings about non-financial costs and benefits which are difficult to measure, such as increased flexibility of the farmers working day, impact on animal welfare, etc., but which are important elements in a farmers choice (Meskens and Mathijs, 2002).

Despite the many farmers who adopt an AM-system, in The Netherlands, still more than the half of the farmers, investing in a milking system, choose for a conventional milking (CM)-system (milking parlour). The research of Meskens and Mathijs (2002) does not take this group in consideration. There is no known information of the characteristics of this group of farms and the motivation for a CM-system in relation to farmers who adopt an AM-system. The objectives of this research were to gain insight in the background of the farmers who choose for an AM-system compared to those who choose for a CM-system.

Materials and methods

An random group of 60 farmers who adopted an AM-system and a random group of 60 farmers who adopted a conventional milking system, both in 1998 and 1999 in The Netherlands, have been included in this research. These 120 farmers all have been interviewed by the same person, with the difference that the farmers with an AM-system have been visited and the farmers with a CM-system were interviewed by telephone. This person interviewed the farmer with the help of a questionnaire. Both groups were informed through the robot manufacturers.

Descriptive analyses have been carried out on the collected data. For the non-discrete data, statistical differences between the groups have been calculated using students T-test. To analyse the non-numerical, discrete data, the Fisher-Irwin test, using a hypergeometric distribution is used with PQRS. PQRS is is a tool for calculating probabilities and quantiles associated with a large number of probability distributions. Quantiles and probabilities are displayed and edited in their natural position relative to the probability (density) graph (Knypstra, 2002).

Results and discussion

The response rate of the AM-system group was slightly lower (64%) than of the CM-system group (72%).

Motivations of farmers investing in an AM-system
Of the farmers buy an AM-system, 26% had seriously considered buying a conventional milking parlour. There was a large variety in motivations to invest in an AM-system instead of a CM-system (Table 1).

Not all farmers did have more than 1 important motivation for investing in an AM-system. More than the half of the robot farmers gives less (heavy) labour as a motivation to invest in an AM-system. This is especially true, considering that the motivation less available labour on a farm (due to an employee quitting or a father retiring) has also to do with a decreased need of labour with an AM-system. An improved labour flexibility is many times given as second or third motivation. The possibility to milk more than twice a day and thus improving

Motivation	Reason 1	Reason 2	Reason 3	Total	%
Less (heavy) labour	18	10	5	34	21
Flexibility	7	10	4	21	13
Milking more than twice	7	6	5	18	11
Less labour available	7	5	6	18	11
Need new milking system	9	2	4	15	9
Improved udder health	0	4	5	9	6
Higher milkproduction	0	6	3	9	6
Building new stable	2	4	1	7	4
Future	3	2	1	6	4
Other	7	10	7	25	15
Total	60	59	41	160	100

milk production and/or cow welfare is also an important motivation, while the higher milk production as such was also a motivation for 6% of the farmers. The need for a new milking system was for 9% of the farmers part of the motivation to invest in an AM-system. This indicates that these farmers are under the impression that an AM-system is often bought without the necessity of investing in new milking technology. Improved udder health is connected with the possibility to increase the milking frequency. In particular for high productive animals, it is better for the udder health if the animals are milked more than twice a day. Some farmers gave the necessary building of a new stable as reason for investing in an AM-system. An important aspect in this is the lower space requirements of an AM-system compared to a CM-system, thus reducing building costs. Moreover, when a farm is replaced because of extending cities, these farmers do have a rather large building budget in which a more expensive milking system is less of a problem.

Motivations of farmers investing in a conventional milking parlour
Of the farmers that invested in a CM-system 78% had considered buying an AMS, which is much higher than the other way around (26%). The motivations they had to invest in a CM-system instead of in an AM-system are summarized in table 2.

The most important most important reason for farmers to invest in a CM-system is the cost of the AM-system (depreciation and maintenance). A second important motivation for not investing in an AM-system is the dependency on the system. Although an AM-system gives more flexible working times, failures might happen 24 hours a day and that gives a dependency. A third reason to invest in a CM-system is the uncertainty about the AM-system. Some farmers believe the development of AM-systems is only beginning and that there are still too many weaknesses in AM-systems. AM-systems are still something for the future. Another motivation for investing in a CM-system is the fact that with an AM-system there is a narrow optimal capacity. This means that the farmer is less flexible in expanding the milk production. When the maximum capacity is reached, an increase in milk quota makes investment in an additional milking unit, while in a CM-system the milking times are extended in such a case. Moreover, the farm structure may be such that an optimal capacity

Table 2. Most important motivations to invest in a conventional milking-system.

Motivation	Reason 1	Reason 2	Reason 3	Total	%
Costs AM-system too high	19	7	1	27	29
Dependency on AM-system	6	4	4	14	15
Uncertainty AM-system	1	6	1	8	9
Poor growing possibilities	3	4	0	7	8
2nd milking unit expensive	5	1	1	7	8
Better fit in the stable	4	2	0	6	6
Other	11	9	4	24	25
Total	49	33	11	93	100

of the AM-system is not possible. This fact is resembled in the motivation "2nd milking unit expensive". Some stables are not very suitable for an AM-system, which can also be a motivation for investing in a CM-system.

Farm structure

In table 3 structure of the farms data are given. Farms that invested in an AM-system are, on average, more intensive in terms of kg milk per cow and kg milk per ha, than farms that invested in a CM-system. From the 60 farms with an AM-system, 36 kept their cows inside all year around, compared to 8 of the CM farms. This agrees with the higher intensity of the farms with an AM-system.

Although there was no significant difference in average quota size between the two groups of farms there was a difference in distribution (Figure 1). A larger proportion of the group of farms with an AM-system has a milk quota between 500,000 and 700,000 kg, which

Table 3. Farm structure of the farms with an AM-system and a CM-system.

	Automatic milking		Conventional milking		
	Mean	sd	Mean	sd	P
Cows	87	34	91	36	
Hectares	51	25	55	22	
Quotum (kg)	752,000	285,000	738,000	300,000	
Milk/ha	15.671	4.409	13.867	3.844	0.0093
Milk/cow	8.682	882	8.118	1.101	0.0012
No grazing	33	-	8	-	0.000
Age farmer	44.1	8.7	41.3	10	
Married (nr)	55	-	47	-	0.026
No successor	12		2		0.004
Milking system did not need replacement	25	-	11	-	0.004

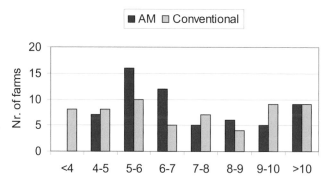

Figure 1. Distribution of milk quotum for the farms that invested in an AM-system and CM-system.

is considered to be "ideal" for a one milking box AM-system. There was a large number of farms with a CM-system and a milk quota under 400,000 kg, whereas there were no farms with an AM-system in this category. It can also be noticed that in the group of farms who invested in an AM-system a larger proportion did make this investment without the direct need of investing in milking technology. When an investment in equipment is made when the old equipment is not completely depreciated, other than pure economical reasons are the reason to do this. There was no difference between both groups of farms with regard to the age of the farmer. The farmers that invested in an AM-system were more often married, but did also have more often no successor. There was no difference in educational level between both groups.

Conclusions

As expected, there was a large difference in motivations between the farmers that invested in an AM-system and in a CM-system. Most important motivations of the robot farmer are the reduction of (heavy) labour by adopting an AM-system, flexibility and the possibility to milk more than twice a day. Most motivations to invest in a CM-system are related to economics (investment and maintenance and costs of a non-optimal herd size). It is clear that for the farmers that invested in an AM-system other factors than pure economical play a major role when adopting the AM-system. Since most of the reasons for CM-system farmers to not invest in an AM-system were economics related, it can be expected that with a (relative decrease) in price of AM-systems when compared to CM-systems in the future a larger proportion of farmers will invest in an AM-system.

Almost the half of the farmers who adopted an AM-system did not need to invest in a new milking system in comparison with 18% of the CM-system farmers. Also robot farmers considered less often the adoption of another milk system. Furthermore farmers with an AM-system are more often married and have more often no successor.

Acknowledgements

The farmers co-operating in this study are gratefully thanked for their hospitality and their time. Also the manufacturers of milking systems (conventional as well as automatic) are acknowledged for their co-operation in contacting the farmers.

References

Dijkhuizen, A.A., R.B.M. Huirne, S.B. Harsh and R.W. Gardner, 1997. Economics in robot application. Computers and Electronics in Agriculture, 17: 111-121.

Meskens, L and E. Mathijs, 2002. Motivation and characteristics of farms investing in automatic milking systems. Deliverable D2 of EU project Implications of the introduction of automatic milking on dairy farms. www.automaticmilking.nl/projectresults/Reports/DeliverableD2.pdf

Knypstra, S., 2002. PQRS, Rijksuniversiteit Groningen. http://www.eco.rug.nl/medewerk/knypstra/pqrs.html

ECONOMIC EFFICIENCY OF AUTOMATIC MILKING SYSTEMS WITH SPECIFIC EMPHASIS ON INCREASES IN MILK PRODUCTION

K.M. Wade[1], M.A.P.M. van Asseldonk[2], P.B.M. Berentsen[2], W. Ouweltjes[3] & H. Hogeveen[2,4]
[1]Department of Animal Science, McGill University, Montreal, Québec, Canada
[2]Business Economics Group, Wageningen UR, Wageningen, The Netherlands
[3]Animal Sciences Group, Wageningen UR, Lelystad, The Netherlands
[4]Current address: Department of Farm Animal Health, Veterinary Faculty, Utrecht University, Utrecht, The Netherlands

Abstract

In order to estimate the effects of automatic milking (AM) on milk production, test-day data from the Dutch DHIA were used, consisting of 2,071,662 test-day milkings from 306 herds. Half of these herds were milking with an AM system and the other half were randomly selected control farms, included to correct for the trend in yearly milk production due to genetic and management progress. The test-day milk production data were corrected for parity, stage of lactation, and season of calving. Milk production increased by, on average, 2% after introduction of an AM-system. Without correction of a year effect, the effect of introduction seemed to be 10-12%. However, a large part of this increase could be explained by a year effect. The milk production data were used in an economic linear programming model of a dairy farm. Using this model, an AM-system was not a viable investment, using the milk production data from this research and other assumptions. This means that other than purely economic factors play important roles when farmers invest in an AM-system.

Introduction

Dairy farming is an economic activity. The cost-efficiency of large investments is therefore an important issue. Several studies have been published on economic aspects of automatic milking (AM) (Arendzen and van Scheppingen, 2000, Armstrong and Daugherty 1997, Cooper and Parsons 1999, Dijkhuizen *et al.*, 1997, Hyde and Engel, 2002 and Pellerin *et al.*, 2001). However, the results of these studies are differ substantially. All of these models were based on normative assumptions. This means that the results of the studies are dependent upon assumptions. For the benefits of an AM-system two factors are important: labour savings and milk production increase. Labour savings are very dependent on farm organisation and attitude of the dairy farmer, see for example van 't Land *et al.*, 2000. Moreover, the economic valuation of labour savings can also differ from farm to farm. The level of milk production increase is much more a subject of debate. An increase in milk yield from 6 to 25% in complete lactations has been shown when increasing the milking frequency from two to three times per day (Hogeveen *et al.*, 2002). These data are often used as input for the economic modelling and a general understanding is reached that a production increase of 5-10% is reasonable. However, it was previously reported that the milk production, in terms of milk production per hour, is dependent on the milking interval (Ouweltjes, 1998). And milking intervals vary much with AM (Hogeveen *et al.*, 2002). For a good economic estimation, unbiased milk production effect estimations are necessary.

The objectives of this study were, therefore, aimed at providing a reliable estimate of the expected change in milk production after introduction of AM in the Dutch context. The resulting solutions would then be implemented in an existing economic model that predicts the changes in labour income after investment in an AM system.

Material and methods

Data selection

The data for this study originated from regular milk recordings of the Dutch DHIA. It is known which farms use an AM-system. Testday records from all farms using an AM-system were selected and combined with a sample from the non-AM-milking farms, whereby the number of control herds was equal with the number of case herds. The dataset contained milk production records covering the period from January, 1990 to March, 2002. The raw data contained 8,332,155 test-day records, which were subjected to a series of edits. The first action was to remove all records from herds that had an average test-day interval of less than 19 days (indicating the possibility of their being a research facility). Subsequent edits were performed on records with inconsistent or seemingly erroneous information. These included animals with unknown or missing birth, calving or test-day dates, missing milk-production values, extreme production values (<2kg milk, and butterfat or protein values that were either <2% or > 9.99%), animals that had greater than 19 test days in any one lactation, and test days were the number of days in milk was greater than 300. This last elimination was carried out not because of any unreliable nature of the data but simply due to the fact that correction factors were not available beyond this point. Records were also eliminated if there was a rejection code indicated on the test day or if the first test day occurred on the date of calving. Finally, since the analysis was carried out on a within-herd basis, "small" herd-test day combinations were also removed (less than 19). This resulted in a data set containing just over 6 million test days from 346,349 cows in 12,420 herds.

The production values in the resulting data set were then corrected for stage of lactation (days in milk), age (months at calving) and season of calving (6 two-monthly periods, starting with December-January) with a procedure comparable to that described by Wilmink (1987). These corrections resulted in standardised production data for the records. In the case of farms that used an AMS at some point in time, an indicator was added to the test day data to show where that observation occurred relative to the introduction of the AMS (days plus/minus the date of introduction). In this way, subsequent analyses were able to account for the transition period and the effect that might have had on the results. Finally, test day records 6 months before or after the introduction of the AM system were removed.

Data analysis

For this study, panel data were used, which included variables in the same temporal interval (cross-sectional data), and annual data from 1990 to 2002 (time series data). Analysing this kind of data makes it possible to compare at the same time "before and after" and "with and without" and hence provides an opportunity to estimate the effect of robot adoption on animal performances, eliminating the influence of trend and herd-specific effects. In addition, the design allows estimation of technology effects for each year after adoption and shows whether the particular technology use causes a gradual response over several years or is immediately apparent at the time of adoption (Mundlack, 1961). The effects of robot adoption on the production of milk, protein, and fat were quantified with the PROC MIX procedure of SAS. The following regression models were used:

$$Y_{ij} = Herd_i + AMS_j + \varepsilon_{ij} \tag{1}$$
$$Y_{ijkl} = Herd_i + AMS_j + Year_k + Parity_l + \varepsilon_{ijk} \tag{2}$$

In model [1], Y (daily milk, fat or protein production) was estimated for herd i (random variable) with only the variable AMS (with j=0 (no AM-system) or 1 (after introduction of the AM-system). In model [2], besides an effect for herd (random) and AM-system, also a year (with k = 1990-2002) and parity (with l = 1-6) efffect were estimated. Estimators of the variance components were obtained with the REML and ML methodology. However, the methods produced almost identical estimates and associated standard errors, therefore, only REML results are presented.

Economic modelling

To evaluate the economics of automatic milking, farming situations with a conventional milking system and with AMS are compared by means of a linerar programming model (Berentsen and Giessen, 1995). modeling. The setup of the basic situation is that of a farm with a milk quota of 800,000 kg and an average yearly cow production of 8,000 kg in the situation with the conventional milking system. The intensity of the farm is varied by using different milk quota/ha (12,000 versus 18,000 kg/ha). This results in an area of 66.7 ha and 44.4 ha for the extensive and the intensive situation respectively. Results from the data analysis in this research were used as the milk production effect of the AMS.The assumed effect of AMS on required labour is a decrease of 5.5 hours per week (Heemskerk, 2002). Investments for conventional milking and AMS are € 78,100 and 195,500 respectively (Hemmer et al., 2003). A final assumption in all calculations is a situation with new buildings. This means that for example the size of the building is adjusted to conventional milking and AMS respectively.

To determine optimum farm results for the different situations a linear programming model of a dairy farm is used (Berentsen and Giesen, 1995). The objective function of this model maximises labour income (i.e. the remuneration for family labour and management that is left after all other costs have been paid). The central element in the model is a dairy cow with a fixed milk production. Young stock is kept for replacement. The land of the farm can be used for growing grass and silage maize. Grass can be used for grazing and for mowing for silage to be fed in the winter period. Feeding is accomplished by matching requirements for energy and protein based on milk production with produced and purchased feed. The latter can be different types of concentrates and, in case of a shortage of roughage, silage maize. The model included Dutch environmental legislation for 2003.

Results and discussion

Introduction of an AM-system gives a production increase of approximately 2% (Table 1). This is lower than expected and lower than reports from farmers. However, the effect is corrected for the yearly production increase because of improving genetics and management. This can be illustrated by looking at the effects estimated by the model without a year effect (model 2). When not taking year into account, the effect of introduction of an AM-system on a farm is 12.4% (kg milk) and approximately 10% for the fat and protein content. Although, the average effect attributable to the introduction of automatic milking was 2%, in the raw data a large variation could be seen, which means that the realized milk production increase on dairy farms is greatly dependent on management.

Table 1. Milk production before and after introduction of an AM-system, estimated with and without year effect.

	Model with year effect AM-system			Model without year effect AM-system		
	no	yes	difference	no	yes	difference
Milk (kg/day)	28.5	29.1	2.1%	27.5	30.9	12.4%
Protein (g/day)	972	991	1.9%	946	1,041	9.9%
Fat (g/day)	1,241	1,266	2.0%	1,220	1,340	9.8%

Economics

Table 2 shows the economic results for the extensive and the intensive farm using conventional milking and AMS respectively. Total returns include the returns of milk, sold animals, and eventually sold silage maize (only in the extensive situation). The small differences in returns between conventional milking and AMS can be explained by the lower number of animals (100 and 98 cows respectively) that affects returns from sold animals and sold silage maize. Major differences between conventional milking and AMS arise on the cost side. Costs of variable labour are almost € 7,000 lower in the situation with AMS due to the lower labour requirement. Other variable costs are lower in the situation with AMS because of the lower number of animals. This advantage is bigger on the intensive farm because of saving on purchased silage maize and because a lower number of animals is very helpfull in complying with environmental policy which is more pressing on intensive farms. Fixed costs of the milking equipment are € 29,507 higher in the situation with AMS due to both the higher investment and the shorter depreciation period of the AMS (7 and 10 years respectively). Other fixed costs are € 5184 lower in the situation with AMS because of the smaller stable. The stable is smaller because of less space required for AMS and the lower number of animals. On balance, the economic situation with AMS is considerably worse. Labour income is some € 16,500 lower.

An important but quite uncertain assumption is the reduction of required labour. The assumption of a reduction by 5.5 hours per week is probably quite conservative. However, in this estimation, not only the labour savings after introduction of an AM-system were

Table 2. *Farm economic results for four situations differing in milking system (conventional vs AM-system) and farming intensity (€/year).*

	Extensive (12,000 kg/ha)			Intensive (18,000 kg/ha)		
	Conv.	AM	Diff.	Conv.	AMS	Diff.
Total returns	312,932	313,168	236	305,865	305,274	-591
Variable labour costs	29,154	22,207	-6947	27,671	20,761	-6910
Other variable costs	109,883	109,404	-479	120,254	118,518	-1736
Fixed costs milk equip.	16,436	45,943	29,507	16,436	45,943	29,507
Other fixed costs	116,476	111,328	-5184	108,053	102,905	-5184
Total costs	271,949	288,882	16,933	272,414	288,127	15,713
Labour income	40,983	24,286	-16,697	33,451	17,147	-16,304

estimated, but corrected for the estimated labour savings with a new conventional milking parlour (Heemskerk, 2002). When the decrease of labour for milking is assumed to be 50% (from 28 to 14 hours per week), the labour income for the situations with AMS by € 8,250. Finally, the effect on milk production is tested for sensitivity. An increase of milk production by 4% instead of 2%, would increase labour income in the situation with AMS by € 2,748 and € 3,135 for the extensive and the intensive farm respectively. This indicates that the economic efficiency of an AM is much more sensitive for labour savings than for milk production increase.

It is important to keep in mind that the presented data are very limited. There is a wide range of farm situations, which have not been described in this paper. However, from these calculations and in line with other publications, it is clear that from a purely business economical point of view, investment in an AM-system is in most farm situations under Dutch circumstances not cost-effective. The farmers that have invested have other reasons to do this investment, e.g., more flexibility in labour, which has been demonstrated by Hogeveen *et al.* (2004).

Conclusions

Milk production (kg milk, fat and protein) increases on average with 2% after introduction of an AM-system. This is lower than expected from practical experiences. From data in this study it is clear that the practical experience does not take a year effect into account, which is probably a reason for an over estimation of the effect of AM. There is a large variation in effect. From a purely economical point of view, investments in an AM-system are not cost-efficient. The labour income on a farm (100 cows with 8,000 kg milk), with conservative assumptions, was approximately € 16,500,- lower in a situation with AM-milking, compared to a situation without AM-milking. This difference could be attributed to higher fixed costs. Other reasons than economical are therefore important for farmers who invested in AM. Labour savings are more important in making AM cost efficient than a milk production increase.

Acknowledgements

We gratefully acknowledge CR Delta for providing the data used in this study.

References

Arendzen, I. and A.T.J. van Scheppingen, 2000. Economical sensitivity of four main parameters defining the room for investment of automatic milking systems on dairy farms. Page 201-211 in: H. Hogeveen and A. Meijering (editors). Robotic milking, Proceedings of the International Symposium, Wageningen Pers, Wageningen, The Netherlands.

Armstrong, D.V. and L.S. Daugherty, 1997. Milking robots in large dairy farms. Computers and Electronics in Agriculture 17: 123-128.

Berentsen, P.B.M. and G.W.J. Giesen, 1995. An environmental-economic model at farm level to analyze institutional and technical change in dairy farming. Agricultural Systems 49: 153-175.

Cooper, K. and D.J. Parsons, 1999. An economic analysis of automatic milking using a simulation model. Journal of Agricultural Engineering Research 73: 311-321.

Dijkhuizen, A.A., R.B.M. Huirne, S.B. Harsh and R.W. Gardner, 1997. Economics of robot application. Computers and Electronics in Agriculture 17: 111-121.

Heemskerk, K., 2002. Differences between farms who invest in an AM-system or in a conventional system. M.Sc. thesis, Farm Management Group, Wageningen University.

Hemmer, H., B. Bosma, A. Everts, and I. Vermeij, 2003. Quantitative Information Animal Husbandry 2003-2004 (in Dutch). Research Institute for Animal Husbandry. Lelystad, The Netherlands.

Hogeveen, H., W. Ouweltjes, C.J.A.M. de Koning, and K. Stelwagen, 2001. Milking interval, milk yield and milk flow rate in an automatic milking system. Livestock Production Science, 72: 157-167.

Hogeveen, H., K. Heemskerk and E. Mathijs, 2004. Motivations of Dutch farmers to invest in an automatic milking system or a conventional milking parlour. In this volume.

Hyde, J. and P. Engel, 2002. Investing in a robotic milking system: A Monte Carlo simulation analysis. Journal of Dairy Science 85: 2207-2214.

Mundlak, Y., 1961. Empirical production free of management bias. Journal of Farm Economics 43: 44-56.

Ouweltjes, W., 1998. The relationship between milk yield and milking interval in dairy cows. Livestock Production Science 56: 193-201.

Pellerin, D., M. Blin, P. Meunier, E. Lapeze and R. LeVallois, 2001. Can you pay a milking robot? (in French). Page 121-131 in: Des défis? Des solutions! 25e Symposium sur les bovins laitiers. Saint Hyacinthe, Québec, Canada.

Van 't Land, A., A.C. van Lenteren, E. van Schooten, C. Bouwmans, D.J. Gravesteyn and P. Hink, 2000. Effects of hus bandry systems on the efficiency and optimisation of robotic milking performance and management. Page 167-176 in: H. Hogeveen and A. Meijering (editors). Robotic milking, Proceedings of the International Symposium, Wageningen Pers, Wageningen, The Netherlands.

Wilmink, J.B.M, 1987. Adjustment of test-day milk, fat and protein yield for age, season and stage of lactation. Livestock Production Science 16: 335-348.

THE ECONOMIC IMPLICATIONS OF AUTOMATIC MILKING: A SIMULATION ANALYSIS FOR BELGIUM, DENMARK, GERMANY AND THE NETHERLANDS

Erwin Wauters & Erik Mathijs
Afdeling landbouw- en milieueconomie, K.U.Leuven, Leuven, België

Abstract

This paper analyzes the economic implications of automatic milking using a standardized simulation analysis in four countries, Belgium, Denmark, Germany and The Netherlands. To ensure comparability, we used the methodology of the International Farm Comparison Network which is based on a set of typical farms. These typical farms are not average farms, but represent a significant number of dairy farms in a region in terms of size, forage and crops grown, livestock systems, labour organisation and production technology used. We consider a 68-cow farm in Germany and an 80-cow farm in Belgium that may invest in a 1-box system, and a 90-cow farm in The Netherlands and a 150 cow-farm in Denmark that may invest in a 2-box system. Herd size increases in The Netherlands and decreases in Belgium and Denmark to be able to better use the capacity of the AM-system. Assuming a 21% reduction in labour use, automatic milking results in higher income per unit of family labour for the Belgian, Danish and Dutch typical farm, but not for the German one. Profit from dairy production increases for the Belgian and Danish typical farm. The results are found to be particularly sensitive to the labour assumption.

Introduction

Usually, the profitability of an automatic milking (AM) system is assessed by calculating its net present value (NPV), i.e., the discounted sum of returns and costs. If NPV>0 the AM-system is profitable. Further, if the NPV of the investment is higher than the NPV of a traditional milking parlour, the former is preferred (Dijkhuizen et al., 1997; Cooper and Parsons, 1999; Hyde and Engel, 2002). Such a simulation can be applied to representative farms (Dijkhuizen et al., 1997; Hyde and Engel, 2002) to make general statements or to real farms (Cooper and Parsons, 1999). The problem with representative farms is that they are usually based on statistical averages and that they usually do not exist in reality. The problem of using real farms is that results are limited to these farms, which may be very specific. Moreover, previous evidence in the formal literature is limited to The Netherlands (Dijkhuizen et al., 1997), the USA (Dijkhuizen et al., 1997; Hyde and Engel, 2002) and the UK (Cooper and Parsons, 1999). Critical factors affecting the profitability of automatic milking compared to conventional milking identified by these studies are milk yield response, changes in labour requirements (and wage rates), quota prices and the lifetime of an AM-system.

In this paper, we adopt a mixed approach by using 'typical' farms. These are virtual farms that are based on existing farms using three sources of data: national statistics, a panel of farmers and expert knowledge. Typical farms are designed to be representative for a certain herd size group and production system. Moreover, they are established following a common methodology coordinated in the International Farm Comparison Network (IFCN). Results only hold for these typical farms and cannot be generalised. This paper starts from four typical farms in four European countries (Belgium, Denmark, Germany and The Netherlands). Changes due to automatic milking are based on data found in the literature and on real farm data from a survey. We refer to Mathijs and Wauters (2004) for more detail.

Methodology and base assumptions

TIPI-CAL (Technology Impact and Policy Impact Calculations), the Excel-based simulation model incorporating physical and economic data of the IFCN, is used to assess the impact of investing in an AM-system. The model simulates typical farms for up to 10 years. Input data include farm data, projections of prices, costs and yields, farm strategies. The model provides a profit and loss account, a cash flow statement and a balance sheet for each year of simulation (dynamic recursive). The major advantage of using this approach is that a common framework is imposed on all countries. The farms selected for simulation include the TIPI-CAL farms DE-68, BE-80, NL-90 and DK-150, where the numbers refer to initial herd size (Table 1). The Belgian and German farms are assumed to purchase a one-box system; the Danish and Dutch farms are assumed to invest in a two-box system. Herd sizes are assumed to adjust to optimize the use of the AM-system. In the base analysis we assume that a farmer must choose between an AM-system and a traditional milking parlour. The assumptions from the base analysis originate from various sources.

Table 1. Characteristics of the initial farm (IF, one year before investment) and the traditional farm (TF, five years after investment).

	Belgium		Denmark		Netherlands		Germany	
	IF	TF	IF	TF	IF	TF	IF	TF
herd size	80	80	150	150	90	128	68	68
land (ha)	50	50	140	140	45.1	64.1	90	90
milk yield (t/cow/year)	7.2	7.6	8.5	9.3	8.3	9.0	7.5	8.9
labour (hours/year)	5500	5500	5450	5450	4810	5850	5200	5200
income (€/hour)	20.92	17.41	8.06	19.87	31.76	38.32	14.12	11.49
cash income (€/hour)	23.98	22.30	30.85	49.54	38.53	43.89	20.99	19.90

Revenues
With an AM system the milking frequency can be higher than with the traditional milking parlours. Milk production per cow is found to increase from 6 to 25% (Hogeveen et al., 2001). We will consider an increase of 6% in the base analysis. This increase will be realised two years after adoption, as the first two years are transition years, with the necessary adaptations to the new technology. The higher milk yield means that an AM farm can milk the same quotum as a conventional farm with less cows. Generally, a reduction in milk price

is assumed following a protein and butterfat percentage decline. However, as the impact on price is relatively small (1-2%) we assume this change to be negligable.

Investments and complementary costs

The investment costs are an important issue to consider in the decision between an AM-system and a traditional parlour. The costs of a traditional parlour depend highly on the degree of automation. We consider a parlour with a rather high degree of automation - since this decision was made in disadvantage of an AM-system - with a price of 50,000 €. The price of an AM-system depends on the way cows are recognised in the system and on the software. We consider a price of 128,000 € for a 1-box system and 168,000 € for a 2-box system. We assume that the old barn does not have to be replaced, but that only some adaptations will be necessary (6,273 € for the conventional parlour and 7,000 € for the AM-system). Furthermore the farmer needs to build a new milking stable, and the AM farmer needs to build a buffer tank. The milking stable for a conventional parlour costs 26,073 € in Belgium and Germany and 36,073 euro in Denmark and The Netherlands. For an AM farmer it is 11,000 € for a 1-box system and 20,000 € for a 2-box system. The buffer tank costs 10,000 €. Both the AM-system and the traditional system are depreciated linearly over 10 years. Interest rates used are specific to country and year.

The more automated a dairy farmer becomes, the more energy will be used on the farm. Klungel *et al.* (2000) found that on an AM farm there is a 42.3% increase in the use of electricity. We will consider an increase of 40%. In the same investigation, Klungel *et al.* found that the use of water on a AM farm was only 1.5% higher than on a comparable traditional farm. In our simulation the use of water on the traditional farm is the same as on the AM farm.

Livestock costs

Breeding, veterinary and medicine costs are assumed to be the same on both farms. Disinfection costs are slightly higher on a conventional farm. This is, among others, due to the fact that a conventional parlour is bigger than an AM-system. We assume that on the traditional farm disinfection costs are 25% higher than on the AM farm. Costs for milk control are also the highest on the traditional farm. A traditional farm needs an external controller, such that we assume a difference of 20%. Finally differences in other livestock costs such as bedding, insecticide, earmarks are negligibly small.

When the milk production increases, there has to be a corresponding increase in energy intake. Cooper and Parsons (1999) use the following formulas to calculate the concentrate intake (K) and the roughage intake (R) - where K and R are expressed as kg dry matter per cow - as a function of the milk yield (Y): $K = 0.339 Y - 340$ and $R = 0.156 Y + 3528$. We only used the exact formulas of Cooper and Parsons on the Dutch and German farms since these were the only TIPI-CAL farms where feed is expressed as kg dry matter per cow per year. In the other two countries feed input in the TIPI-CAL farm was simply expressed as tons per year, including feed not only for dairy cows but also for dry cows, heifers and calves. In this case we interpolated the formulas in the best possible way to indicate the difference in feed input between a traditional and an AM farm.

Labour

The reduction in labour demand is one of the most often mentioned advantages of an AM-system. To what extent labour demand will be reduced is difficult to say and depends on the farm characteristics and on the strategy of the farmer. In this respect, the choice of grazing system and cow traffic are very crucial. We will take a labour reduction of 21%, which is the average labour savings by farms that have kept their herd size constant within a certain range reported in a recent survey (Wauters and Mathijs, 2004). This is lower than the 2.6 hours per day saving assumed by Dijk *et al.* (1997) which approximately equals a 26-28% reduction. Quartile values for the distribution of labour savings (-7% and -33%) are used in the sensitivity analysis. As in Denmark and Germany not all labour is family labour, we assumed that the farmer will reduce hired labour before reducing family labour.

Results and discussion

Base results

The following parameters are investigated and reported in table 2: income, income per unit of family labour, income per kg milk, cash flow, cash flow per unit of family labour, cash flow per kg milk and profit. All parameters are reported five years after investment, for then a relative steady state is reached. Income equals revenues minus costs except family labour and is a measure for the profitability of the farm, independent of how the investment was financed (loan or equity). Cash flow is farm income plus depreciations, changes in inventory and capital gains, and reflects the cash position of the farm. Profit is income minus family labour valued at opportunity cost (country-specific). We also calculated the NPV of income over 7 years assuming an opportunity cost of capital of 5%. We could not calculate the NPV over 10 years as investment was done in the second or third year of the simulation and the software only simulates 10 years. However, it must be noted that the NPV does not take into consideration the potential income of the labour saved.

Investing in AM results in higher income per unit of family labour in Belgium, Denmark and The Netherlands, not in Germany. Expressed per unit of milk produced, income only increases slightly in Denmark. Profit only increases in Belgium and Denmark. However, the NPV is always lower for AM compared to the traditional parlour, which means that after 7 years income decreases in the beginning are not fully compensated by income increases after a transition period. However, these differences are small in Belgium and Denmark, so including more income streams will render the AM-system more profitable.

Sensitivity analysis

We have investigated the sensitivity of our results to three parameters: milk yield (+12%, +18% and +24%), investment costs for an AM-system (10% lower and 10% higher) and labour (-7% and -33%). Note that a sensitivity analysis of 10% higher investment costs is equivalent with a 9 year (instead of 10) depreciation period, while a 10% lower investment cost could be the result of non-zero salvage value. We only report the impact on income per unit of family labour and cash flow per unit of family labour (Table 3).

The sensitivity analysis of yield changes provides somewhat non-intuitive results: farms with yield increases of 12, 18 and 24% have in most of the cases a lower income than the AM farm in the base simulation. According to our assumptions, all farms keep quotum constant, so they adjust their herd size according to the change in milk yield. As a result,

Table 2. Results of the base simulation, five years after investment, in euro.

	Belgium	Denmark	Netherlands	Germany
Traditional parlour				
Income	95731	59637	224174	55157
Income per hour family labour	17.41	19.87	38.32	11.49
Income per kg milk	0.16	0.04	0.19	0.09
Cash flow	122645	148614	256752	95501
Cash flow per hour family labour	22.30	49.54	43.89	19.90
Cash flow per kg milk	0.20	0.11	0.22	0.16
Profit	13640	16671	99152	-23043
NPV(income)*	198843	193782	207070	-114160
AM-system				
Income	81958	62857	187702	34546
Income per hour family labour	18.86	20.95	40.62	8.41
Income per kg milk	0.13	0.05	0.16	0.06
Cash flow	118616	160674	239640	87084
Cash flow per hour family labour	27.30	53.56	51.86	21.20
Cash flow per kg milk	0.19	0.12	0.21	0.14
Profit	17106	19891	88945	-32380
NPV(income)*	196423	191441	131660	-184746

*Net present value of income over a period of 7 years

farms with a higher increase in yield will not produce more, so total farm return will not increase. The change in total farm input is ambiguous. Fixed costs will not change as the change in size is too small. Variable costs have a tendency to decrease because of a lower herd size, but also a tendency to increase because of the need of more feed input and/or input of more expensive feed. In those cases mentioned farm return will be approximately the same while farm input will slightly increase, yielding a lower farm income. Changes in milk yield do not change our results in a significant way.

Also a change in investment costs does not change our results significantly. A ten percent lower investment cost results in a 1-2 percentage point increase of income. The exception is Germany with a 5 percentage point increase. A ten percent higher investment cost results in the same income for AM and the traditional parlour in Denmark.

Assuming a labour reduction of 33% does change our results significantly. Income and cash flow per unit of family labour are now considerably higher for AM compared to a traditional parlour. The exception is Germany, where cash flow is 26% higher, but income is still 14% lower. If labour reduction is only 7%, the AM-system becomes unprofitable in all countries.

Table 3. Results of the base simulation and sensitivity analyses, five years after investment.

	Belgium		Denmark		Netherlands		Germany	
	fi	cf	fi	cf	fi	cf	fi	cf
base	+8.33	+22.42	+5.44	+8.11	+6.00	+18.16	-26.81	+6.53
sensitivity to yield								
ams(y12)	+6.84	+21.26	+2.92	+7.02	+8.46	+20.07	-29.77	+6.28
ams(y18)	+3.96	+19.06	+0.10	+5.81	+9.47	+20.73	-26.46	+7.44
ams(y24)	+4.48	+19.46	+3.22	+6.96	+10.96	+21.80	-16.80	+12.16
sensitivity to investment costs								
ams(i-10)	+10.51	+23.38	+10.87	+9.00	+7.57	+18.50	-21.24	+7.64
ams(i+10)	+6.20	+21.45	0	+7.19	+4.23	+17.79	-32.38	+5.33
sensitivity to labour								
ams(l-7)	-7.98	+3.99	-31.35	-6.52	-9.87	+0.36	-38.29	-9.35
ams(l-33)	+27.74	+44.35	+36.34	+19.88	+25	+39.33	-13.66	+25.63

fi = change (%) in income per hour family labour of AM relative to traditional parlour

cf = change (%) in cash flow per hour family labour of AM relative to traditional parlour

Conclusions

For the typical farms in Belgium, Denmark and The Netherlands investing in an AM-system is profitable if income per hour of family labour is used as an index of profitability. Income only increases for the Danish farm, while profit increases for the Belgian and the Danish farm. In other words, if the farmers of the Belgian and Dutch typical farm do not use the labour saved for other profitable activities, their income will go down. For the German typical farm, investing in an AM-system is not profitable, but it must be noted that also investing in a traditional parlour would be unprofitable given the initial situation of that typical farm (small herd size and low cash flow). However, this does not mean that automatic milking is not profitable in Germany in general. For larger farms that perform better, it is likely that results will be comparable to the other countries. This paper has simulated the investment in AM for typical farms and it is impossible to simulate all types of farms. Obviously, the results will be different for each individual farm as they depend on a variety of factors and each farmer should do his own calculus.

References

Armstrong, D.V. and L.S. Daugherty, 1997. Milking robots in large dairy farms. Computers and Electronics in Agriculture 17: 123-128.

Cooper, K. and D.J. Parsons, 1999. An economic analysis of automatic milking using a simulation model. Journal of Agricultural Engineering Research 73: 311-321.

Dijkhuizen, A.A., Huirne, R.B.M., Harsh, S.B. and R.W. Gardner, 1997. Economics of robot application. Computers and Electronics in Agriculture 17: 111-121.

Hogeveen, H., Ouweltjes, W., de Koning, C.J.A.M. and K. Stelwagen, 2001. Milking interval, milk production and milk flow-rate in an automatic milking system. Livestock Production Science 72: 157-167.

Hyde, J. and P.D. Engel, 2002. Investing in a robotic milking system: A Monte Carlo simulation analysis. Journal of Dairy Science 85: 2207-2214.

Klungel, G., de Koning, K. and I. Arendzen, 2000. Melkrobot verbruikt veel water en stroom, *Veehouderij Techniek*, december.

Wauters, E. and E. Mathijs, 2004. Socio-economic implications of automatic milking on dairy farms, Deliverable 3, EU project, Automatic Milking (QLK5-2000-31006).

ENERGY CONSUMPTION ON FARMS WITH AN AM-SYSTEM

Kees Bos
Applied Research, Animal Sciences Group, Wageningen UR, Lelystad, The Netherlands

Introduction

Changing over to automatic milking, generally means increased costs for energy. To measure energy consumption, the High-tech farm at the Waiboerhoeve was equipped with an energy consumption monitoring system. On two other research farms energy consumption of the Automatic Milking System (AMS) was monitored. Objective was to compare energy consumption of AMS with conventional milking systems and to reduce energy consumption.

Material and methods

One of the targets of the High-tech farm is to achieve a low cost price of milk. An optimal use of the Lely Astronaut milking system necessary but also reduction of energy costs can contribute to a low cost price. Cooling is done with an instant cooling system in combination with a storage tank both from DeLaval. The milk is cooled in two steps in heat exchange: first to 13 °C en next step to 4 °C. Energy consumption is measured on farm level (total energy), the AMS, vacuum pump and boiler within the AMS, cooling(separately on first and second step and bulk tank) and air compressor. Every 15 minutes the energy meter readings are stored in the U1600 (GMC-Instruments) data logger. The program ECSWin is used to download the data from the data logger into a database.

On two other research farms of Applied Research with an AM-system (Nij Bosma Zathe and the Feed&Milk-farm of the Waiboerhoeve) energy consumption of the AMS was also measured. Nij Bosma Zathe is equipped with an Insentec Galaxy and the Feed&Milk-farm with a Gascoigne Melotte Zenith, both are two box systems.

For comparison with conventional milking ,a computer model (WWE, Applied Research) was used to calculate energy consumption for farms with milking parlours.

Comparison

Calculations for conventional milking are done based on 65 cows and 800.000 kg milk quota, comparable to High-tech. Starting point for conventional milking is a milking frequency of 2 times a day and use of heat recovery. On farm level 42% more energy was used on the High-tech farm. Main differences compared to conventional milking were related with the air compressor and the vacuum pump.

Comparing the different AM-systems showed differences up to 25% more energy consumption between brands based on a two-box or two one-box systems.

Energy savings

The energy consumption of the vacuum pump was decreased with 60% by using frequency control. Cooling costs can be lowered by using water for cooling in the first cooling step. The air compressor showed fluctuations in energy consumption which were not related to the number of milkings. It is not clear what the reason of these fluctuations are, but energy saving seem to be possible. Also cooling can be done more efficient on the High-tech farm, which will be done in further research.

EVALUATION OF THE AM-SYSTEM ADOPTING AND USING FARMERS: IN CASE STUDY OF JAPANESE DAIRY FARMS

N. Hatakeyama
Research and Development Center for Dairy Farming, Sapporo, Japan

In recent years, the feature of Japanese dairy farms is the diversification of dairy feeding, breeding and milking technology. In particular, automatic milking system (AM-system) is one of the expectant technologies. However, investment in this technology costs farmers a lot at the initial stage. Dairy farmers determine their investment based on the various expectations under financial situation, herd size, labor load and so on.

The purpose of this study is to investigate farmer's evaluation on AM-system. We asked farmers their evaluation and intension of AM-system focusing on the comparison between expectation and satisfaction. The quantification theory analysis is employed as a statistical method for 37 observations. The feature of this method is to use qualitative data and quantitative data at the same time.

Farmer's quantitative evaluation is from one to ten, and nine qualitative variables were used as dependent and independent variables, respectively. The independent variables include 'investment and economic effect', 'man power improvement effect', 'technology adaptation', 'improvement in production and feeding', 'scale expansion', 'family farm', 'milking type', 'pride' and 'animal welfare'.

The summary of the results is as follows: F-value is 9.8 and adjusted R-square is 0.829. These values imply the appropriateness of this model.

The highest evaluation is 'Manpower improvement effect', and its partial correlation coefficient is 0.722. This is due to the qualitative improvement in labor such as reallocation of surplus labor to other works, for example, feeding management and observing dairy cattle.

The second highest evaluation is 'investment and economic effect', the partial correlation coefficient is 0.544. Farmers who think that AM-system brings the economic effect on their production highly evaluated this technology. It can be considered that the economic effect is the strong factor for the investment decision.

EXPECTATIONS OF AUTOMATIC MILKING AND THE REALIZED SOCIO-ECONOMIC EFFECTS

T. Jensen
Danish Cattle Federation, Danish Agricultural Advisory Center, Udkærsvej 15, DK-8200 Århus N, Denmark

The automatic milking system (AMS) must fulfil the expectations of the farmer to be a successful solution for the investment. This study focuses on the expectations of AMS, the changes in working conditions the effects on the farm economic.

The data material consists of information from 43 Danish cattle farms, which have changed from a conventional milking system to automatic milking between 1st January 1997 and 1st January 2002. Totally 141 Danish farms had AMS in the whole production year 2002. The 43 farms are using a common account system and responded positive on a questionnaire. The farmers completed the questionnaire in the spring of 2003, which focus on the investment reasons, the production system and labour before and after transition as well as some economic figures related to the milking robot. In addition to the questionnaire, the data material consists of the yearly profit and loss account for the last five years from all farms.

AMS is represented by four manufactures, but with one manufacturer representing most of the milking systems. 53% of the farms are having one single-box unit, and the rest are having two or three single-boxes or systems with more boxes and one unit. 42% of the farms changed stall facilities at the same time as installing AMS and half of the farms extended the herd. On average the farms are having 93 cows after transmission.

Looking at the reasons for investing in AMS the wish for less physically hard work is the most important reason, followed by the expectations of less work for the owner himself and flexibility in the daily routine. However, the wishes to have the latest novelties, the expectations of higher yield per cow and profitability in connection with AMS are less important reasons for the investment in AMS.

The change to AMS has not been equally painless on all farms. The running-in period varies from four weeks to three years, with an average on nine month, before the farm manager feels that technique, cows and personnel are working well with the system. Typically problems with the technique are the reason for a special long running-in period. A tendency shows that the running-in period is longer for older farmers than for younger.

After the introduction of AMS and working with the system for at least 16 month, the majority expresses that their work-load has been reduced as well as the daily hours. Further, the majority thinks that the work situation has become less stressful. On average, the reduction in labour was 10 hours per cow yearly according to the farmer's own estimates.

The total economic result on the farm is given as the return on all agricultural assets, after paying own labour but not financing the external capital. On average the result slightly reduces the year after converting to AMS, due to higher depreciations and increase of other costs, which cannot in all situations be covered by reductions and higher returns from milk. There is no connection to the level of investment in the year of converting, indicating that the level of extension is not decisive for the result.

The study shows that parameters about labour are valuated high among the expectations on the AMS. These are fulfilled in most cases as a result of less work-load, less hours and less stress.

FARM AND SYSTEM HYGIENE

FARM HYGIENE AND TEAT CLEANING REQUIREMENTS IN AUTOMATIC MILKING

Karin Knappstein[1], Nele Roth[1], Betsie Slaghuis[2], Reina Ferwerda-Van Zonneveld[2], Hans-Georg Walte[1] & Joachim Reichmuth[1]
[1]Federal Dairy Research Institute, Institute for Hygiene and Food Safety, Kiel, Germany
[2]Applied Research, Animal Sciences Group, Wageningen UR, Lelystad, The Netherlands

Abstract

Within hygiene management in dairying effective teat cleaning before milking is a precondition to ensure high quality of raw milk. The teat cleaning efficiency of Automatic Milking (AM) systems was investigated applying different methods. All brands currently used in practice were included: DeLaval, Insentec, Lely Industries, Fullwood, Prolion/Gascoigne Melotte and Westfalia Landtechnik GmbH. Significant differences in teat cleaning efficiency of different brands were determined (p<0.05), showing the necessity to improve teat cleaning efficiency of at least two brands. Variation between individual farms was of significant influence on teat cleaning efficiency with differences between the 3 farms working with the same AM system (p<0.05). Also of influence was the initial contamination of teats before cleaning. These results indicate that farm management is important to ensure low teat contamination levels.

High coliform counts in bulk tank milk exceeding 100 cfu/ml suggested insufficiencies of teat cleaning on 8 farms, although in some cases these were accompanied by high counts of thermoduric bacteria indicating failures in system cleaning.

On 18 farms with AM systems management aspects with regard to teat cleanliness were studied by means of a questionnaire based interview with farmers and a checklist on the actual hygiene status of barns. The outcome was compared to teat contamination measured by determination of TBC (total bacterial counts) and ATP (adenosine-tri-phosphate) from teat swabs taken before teat cleaning.

AM specific management factors associated with high teat contamination were: replacement of teat cleaning device less than twice per year, moderate/poor status of the teat cleaning device, average milking frequency per day < 2.5 and no selection of cows for robot acceptance (p<0.10). Factors not directly related to AM involved contamination of cubicles: less than one cubicle per cow, cows lying on alleys present in the herd, addition of fresh bedding material less than twice per day, no selection of cows for udder health, moderate/poor status of bedding material and moderate/poor status of claws were significantly related to high teat contamination.

Additional factors like the general impression of the robot, cleaning frequency of the milking box, status of teat cups and the use of cow brushes in the stable were probably more closely related to the general attitude of the farmer towards hygiene than to teat cleanliness.

Introduction

For production of high quality milk bacterial counts in raw milk should be as low as possible. In Europe a threshold of 100 000 colony forming units (cfu)/ml for total bacterial count (TBC) is fixed by Council Directive 92/46/EEC for bulk tank milk at delivery.

Besides mastitis pathogens and bacteria from milk contact surfaces, bacteria from udder and teat surfaces belong to the three main causes of microbial contamination of raw milk (Slaghuis, 1996, Sumner, 1996). Clean udder and teats before milking are demanded by Directive 89/362/EEC on General Conditions of Hygiene in Milk Production Holdings to avoid negative influences on milk quality.

Different systems of teat cleaning are applied by Automatic Milking (AM) systems. Common to all systems is that no control of the teat cleaning effect on individual cows is performed. Limited information is available on teat cleaning efficiency by AM systems. Schuiling (1992) compared udder cleaning with brushes to no cleaning and found removal of 69% of manure due to cleaning after artificial contamination of teats with lithium. Melin et al. (2002) used spores for artificial contamination and found a better reduction of carry over of spores into milk by an automated teat cleaning procedure compared to manual cleaning (98.0 versus 66.5%). In contrast, TenHag and Leslie (2002) could not determine significant differences between effects of automated and manual teat cleaning when a swab method with a simplified determination of bacterial counts was applied. Only one system each was included in these studies with different approaches to determine the teat cleaning effect, making comparisons between studies difficult.

Therefore an investigation including all AM systems currently on the market was performed. The systems were evaluated by a set of different methods.

It was assumed from the beginning that AM systems would have difficulties to deal with high levels of teat contamination. Prolonged teat cleaning on very dirty cows would also limit the availability of the system for milking. Therefore, the hygiene management on the dairy farm should aim for cows with clean udder and teats. In a second part of the investigation factors of farm management were evaluated as regards their association with teat cleanliness.

Materials and methods

Teat cleaning efficiency

Teat cleaning efficiencies of the following brands of AM systems were determined: DeLaval, Insentec, Lely Industries, Fullwood, Prolion/Gascoigne Melotte and Westfalia Landtechnik GmbH. Two different approaches were applied.

Artificial contamination

The first approach was based on artificial contamination of teats with a mixture of poppy seed and manure (20% w/w) and determination of carry over of poppy seed into milk. This approach was applied on 12 farms working with AM systems of 6 different brands (2 farms per brand). Per farm teats of ten cows were contaminated, 5 cows were milked without cleaning and 5 cows after teat cleaning. The number of poppy seeds in milk of each cow was determined by filtering the composite milk through a cotton filter and counting the number of poppy seeds.

Automatic milking – A better understanding

The levels of number of poppy seeds in milk were analysed for the fixed effects of brand and teat cleaning efficiency using a mixed logistic model. In the model the relationship between the probability p ($0<p<1$) of observing the poppy seeds and the explanatory variables was described using the following logit-link function:

$$\text{Logit}(p_{ijk}) = \ln(p_{ijk} /1- p_{ijk})= c + fm_i + br_j + tcleffect_k + br_j * tcleffect_k \qquad [1]$$

where c represents the constant term and is the mean for the combination with all factors at the lowest level, fm_i is the random effect of farm i, considered to be normally distributed with mean 0 and variance equal to σ^2_{farm}, br_j is the fixed effect of brand j and $tcleffect_k$ is the fixed effect of teat cleaning effect k.

The model assumed that the variance of the observed counts Y can be adequately described by variance $(Y(p) = np(1-p)$. Estimates of model parameters and components of variance were obtained using the Generalized Linear Mixed Model (GLMM) Genstat procedure of Welham (2002). Fixed effects were assessed using chi-squares for the Wald statistics.

Natural contamination
The second approach for evaluation was based on the natural contamination of teats. 18 farms (3 farms per AM brand) were included.

On each farm 50 cows were sampled. If less cows were available as many cows as possible were sampled during a period of approximately 8 hours. Teat swabs were taken from two teats per cow before and after the cleaning procedure without interrupting the actions of the robot. For Prolion/Gascoigne Melotte teat swabs after cleaning were taken at the end of the milking procedure to avoid detachment of the milking cluster. Teat swabs were placed into 8.0 ml sterile solution of 0.85% NaCl, 0.1% peptone. ATP (adenosine-tri-phosphate) was measured by the test system HyLite® 2 (Merck, 64293 Darmstadt, DE) for determination of ATP in fluids. Total bacterial counts (TBC) were determined within 24 hours (IDF, 1991). Results were Log_{10} transformed for further analysis and are presented in Log_{10} RLU and Log_{10} cfu/ml swab solution for ATP and TBC respectively.

In order to determine which factors have a systematic influence on teat cleaning efficiency, an analysis of variance was carried out. For this purpose the GLM (General Linear Model) procedure of the statistic package SAS, release 8.01 was used. The linear model had the following equation:

$$Y_{ijkl} = \mu + br_i + fm_{ij} + qu_k + b(X_{ijkl}) + e_{ijkl} \qquad [2]$$

with Y_{ijkl} as the dependent variable teat cleaning efficiency (for TBC: Log_{10} cfu/ml before cleaning minus Log_{10} cfu/ml after cleaning, for ATP: Log_{10} RLU before cleaning minus Log_{10} RLU after cleaning), μ is the overall mean, br_i is the effect of brand i (1, 2, 3, 4, 5, 6), fm_{ij} is the effect of farm j (A to R) of brand i, qu_k is the effect of quarter k (front, hind). The initial teat contamination was used as covariate with b as the slope for TBC (Log_{10} cfu/ml) or ATP (Log_{10} RLU) before cleaning; e_{ijkl} is the random residual error.

Bacteriological quality of bulk tank milk

On each of 18 farms (3 farms per AM brand) a sample of bulk tank milk was taken before starting teat sampling. The sampling time was independent from the collection interval of bulk milk. Differential bacterial counts were determined within 24 hours according to standard procedures: Total bacterial count (TBC) - IDF (1991), coliform bacteria (IDF, 1985) and thermoduric bacteria (Frank *et al.*, 1985).

Farm management

The evaluation of aspects of farm management was performed by means of a questionnaire based interview of 18 farmers (3 farms per AM brand). From the questionnaire 45 questions in total on the categories general housing, AM system, lying area, feeding area and cow management were included into the analysis.

A checklist on hygiene management was applied on all 18 farms. 17 different aspects regarding AM system, housing and cows were scored according to a system of 3 score levels with score 1 = good, 2 = moderate, 3 = poor.

The effect of the different factors on teat contamination before teat cleaning was evaluated by the GLM model of SAS 8.01 according to the following equation:

$$Y_{ij} = \mu + factor_i + e_{ij} \qquad [3]$$

with Y_{ij} representing the dependent variable mean teat contamination per farm before teat cleaning (for TBC: Log_{10} cfu/ml, for ATP: Log_{10} RLU), μ is the overall mean, $factor_i$ is the effect of factor i (factor from a list of management factors based on questionnaire and checklist) and e_{ij} is the random residual error. Statistical significance was assumed at $p<0.10$ and no interactions were tested in the model, because the low number of farms was associated with limited variability.

Results and discussion

Teat cleaning efficiency

In table 1 the teat cleaning efficiency per brand is shown as evaluated by the different approaches based on artificial and natural contamination of teats.

Teat cleaning efficiency varied between different AM systems and results differed dependent on method of evaluation. In general, a similar ranking between brands regarding teat cleaning efficiency was attained by the different approaches. Similar findings have also been found by visual evaluation of teats as well as by sediment tests on teats (data not

Table 1. Teat cleaning efficiency per brand determined by two different methods (artificial contamination - carry over of poppy seeds into milk; natural contamination -TBC and ATP in teat swabs).

Brand	Poppy seed in milk[1] [%]	TBC [Log_{10}cfu/ml] LSQ_M ± se[2]		ATP [Log_{10}RLU] LSQ_M ± se[2]	
		Effect of teat cleaning as calculated from reduction of			
1	88	0.36 [a]	± 0.04	0.51 [bc]	± 0.03
2	56	0.41 [a]	± 0.04	0.55 [b]	± 0.03
3	65	0.15 [b]	± 0.05	0.40 [d]	± 0.04
4	99	0.37 [a]	± 0.04	0.46 [cd]	± 0.03
5	90	0.48 [a]	± 0.04	0.67 [a]	± 0.03
6	51	0.17 [b]	± 0.04	0.48 [cd]	± 0.03

[1]calculated from probabilities; [2] LSQ_M = Least Square Means, se= standard error;
different letters within the same column show significant differences between teat cleaning efficiencies (p<0.05)

shown). It can be concluded that improvements in teat cleaning procedures of certain brands are both necessary and possible.

According to the analysis of variance based on TBC factors of significant influence on teat cleaning efficiency were brand of AM system, individual farm within brand and teat contamination before cleaning ($p<0.05$). The model explained 50% of variance in teat cleaning efficiency.

In figure 1 the cleaning effects are summarized per farm. Contamination levels before teat cleaning varied largely. For all except one brand large differences between the 3 farms per AM brand were determined regarding the average reduction of contamination per teat. On three farms (farms H, K and Q) teat contamination after cleaning was on average higher than before. For five of the six brands significant differences were detected between the three farms working with the same AM system (Figure 1).

These results show clearly that farm management has an influence on teat cleaning efficiency.

The following observed coincidences may give examples of explanations for these differences between farms. Farm B had extremely dirty cows and the teat cleaning device had been in use for about two years explaining the low efficiency. As the only one of the three farms of brand 3 farm H used cold water in the cleaning device which may explain the low teat cleaning efficiency. In a number of cows on farm K wounds were observed at the udder basis, which may add to bacterial contamination of teats during the cleaning procedure. Farm O was only working for 2 months with the AM system. A better condition

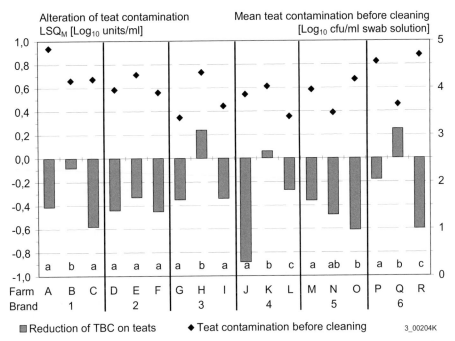

Figure 1. Alteration of TBC per teat [in Log$_{10}$/ml swab solution] by mechanized teat cleaning compared to teat contamination before cleaning - results per farm; LSQ$_M$ = Least Square Mean, different letters (a, b, c) show significant differences in teat cleaning efficiencies between 3 farms of the same AM brand (p<0.05).

of the teat cleaning device is one explanation for the better cleaning results compared to farms M and N. In contrast to farms P and R farm Q did not use any disinfectant for the teat cleaning device which explains the poor teat cleaning efficiency. Obviously bacteria from the brush are smeared over the teat surface during the process intended for cleaning.

Bacteriological quality of bulk tank milk

The results of determination of TBC and coliform counts in bulk tank milk of 18 farms are presented in figure 2.

When comparisons were made between farms with the same AM system (Figures 1 and 2) for all brands except for brand 5 the farms with the lowest teat cleaning efficiency had the highest coliform counts in bulk tank milk (Farms B, E, H, K, Q). The teat cleaning effect seems to be related to the coliform count in bulk tank milk, although other sources for coliforms in bulk milk exist. If teat preparation is sufficient the coliform count of bulk milk should not exceed 100 cfu/ml (Reinemann, 1997). In all farms with coliform counts exceeding 1000 cfu/ml also high counts of thermoduric bacteria (>200 cfu/ml) were detected, which is also an indication for failures of the system cleaning which may contribute to the high coliform counts (Reinemann, 1997). It should be emphasized that several of the farms with high coliform and/or thermoduric counts had TBC of around 10000 cfu/ml. If only TBC had been determined, hints for insufficiencies regarding milk hygiene would have been missed.

Farm management

In addition to brand of AM system differences between individual farms are important for teat cleaning efficiency. Also the initial contamination of teats has a significant influence

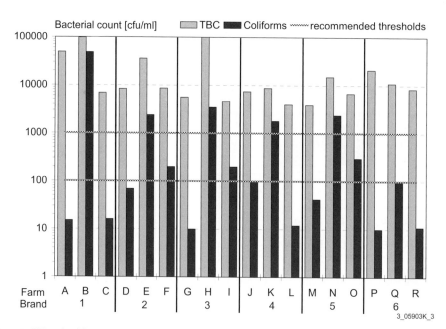

Figure 2. TBC and coliform counts in bulk tank milk of individual farms.

on teat cleaning efficiency, independent of AM brand. Therefore it is necessary to look at management practices on individual farms.

Some of the results of the questionnaire on general aspects of farm management are summarized in table 2.

Table 2. General aspects of management on farms with AM systems, results of questionnaire.

Observation	No. of farms	Average	Min	Max
No. of cows in lactation	18	74	34	158
No of cows per milking box				
single box systems	9	52	34	69
multi-box systems	9	35	26	43
Average milking frequency per cow per day	18	2.6	2.1	3.0
Time until cows are fetched (hours)	15	12.8	10	16
Ratio cubicles/cows	18	1.2	0.8	1.7
Feeding of roughage per day	18	1.5	1	3

As bedding material farms used straw (8 farms), sawdust/shavings (4 farms), specially treated sawdust (3 farms) and sand (one farm). Two farms used no bedding. TBC in fresh bedding material [Log_{10} cfu/g] ranged from 2.7 to 8.0 with an average of 6.3. Lowest bacterial counts were observed in the specially treated sawdust.

Two farms cleaned cubicles once per day, 13 farms twice and one farm three times per day. The frequency of adding new bedding was between once per month and twice per day. The ground in the lying area was slatted floor on 13 farms and concrete floor on 5 farms. In addition to the 5 farms with concrete floor manure scrapers were used on 4 farms with slatted floor. The frequency of scraping was between 4 and 12 times per day. Manual cleaning of floors was performed on 4 farms with scraper and on 4 farms without scraper. 4 farms had cows lying on alleys in the herd. The number of these cows was between one and 8 cows per herd representing 1 to 15% of lactating cows.

In table 3 special aspects of farm management are summarized.

The frequency of replacing the teat cleaning device ranged from every 6 weeks to every second year. None of the farmers cleaned the udders manually. The frequency of claw trimming varied between once per year (7 farms), twice per year (5 farms) and 3 times per year (3 farms), no information was given on 3 farms.

The results of the analysis of variance regarding influence of different factors on teat contamination are shown in table 4. Factors were only included into the analysis when more than 2 farms were positive or negative for the management aspect. Only factors of significant influence are presented, either by evaluation based on determination of TBC or ATP.

Some of the factors are directly related to the teat cleaning process itself. Maintenance of teat cleaning devices is important because devices loose their function with age and may lead to increased teat contamination levels over time. Checking the visual appearance of cleaning devices may give indications for reduced cleaning efficiency. In addition, high levels of teat contamination can be responsible for a poor hygiene status of the teat cleaning device if the device is not cleaned in regular intervals.

Table 3. Special aspects of farm management with regard to teat cleanliness, results of questionnaire.

Variable	No. of farms	yes	no
Disinfection of teat cleaning brushes	9	8	1
Use of teat spray after milking	18	12	6
Separation box in use	18	7	11
Disinfection of cubicles	18	9	9
Presence of cows lying on alleys	18	4	14
Use of cow brushes in the barn	18	15	3
Manual cleaning of udders	18	0	18
Shearing of udders	18	16	2
Shearing of tails	17	12	5
Cutting of tail brush	18	18	0
Selection of cows for			
udder shape	18	9	9
robot acceptance	18	7	11
activity	18	7	11
udder health	18	15	3
claw health	18	11	7

A high average milking frequency is associated with more frequent cleaning of teats and as a consequence reduced teat contamination. For the same reason selection for robot acceptance may be important, because cows with low numbers of voluntary visits to the milking unit have also a lower frequency of teat cleaning.

A number of factors is related to cleanliness of cubicles. If less than one cubicle is available per cow the number of occupied stalls is probably high. Gaworski et al. (2003) determined a positive correlation between stall usage and stall contamination. Opportunities for cows to select clean cubicles are reduced if the number of free cubicles is low. More intensive efforts to clean cubicles may be necessary to compensate for higher stocking densities.

If severe restrictions for lying down or getting up are present cows may prefer not to use the cubicles, but to lie on the alleys which leads to higher teat contamination. An extreme situation was observed on one farm where cows were kept in the former barn for heifers. Cubicles were far too small and cows had serious problems in lying down or getting up. On this farm 15% of cows preferred lying on the floor. This behaviour can be interpreted as an obvious sign for improper cubicle design.

Addition of fresh bedding material to cubicles at least once per day reduces facilities for bacterial growth due to less humidity. Bacterial counts in fresh bedding had no significant influence on contamination of teats, but the quality of bedding material was important. If visual appearance of bedding is scored as moderate or poor it is probably of very bad quality.

Cows with mastitis may contaminate cubicles with high bacterial counts when leaking of milk occurs. Therefore selection of cows for udder health may reduce contamination risk of cubicles and is indirectly associated with lower bacterial counts on teats. Additionally

Table 4. Management factors significantly associated with level of teat contamination.

Management factor	% of variance explained[1]		Higher teat contamination if
	TBC	ATP	
Parameters of questionnaire			
Replacement of teat cleaning device	24.2*	20.5*	< once per year
Average milking frequency	19.7*	10.7	< 2.5/day
Selection of cows for robot acceptance	19.2*	12.1	no
Ratio cubicles/cows	21.7*	16.0*	Ratio <1
Presence of cows lying on floor	16.1*	18.4*	yes
Addition of fresh bedding material	22.7*	10.1	< once per day
Selection of cows for udder health	27.8*	13.7	no
Use of cow brushes in the barn	28.9*	16.9*	no
Cleaning of milking box	5.2	24.5*	< 2x/day
Parameters of hygiene checklist			
Score - teat cleaning device	4.4	17.1*	> 2
Score - quality of bedding	35.4*	19.9*	> 2
Score - claw condition	46.7*	3.0	> 2
Score - impression of robot	8.9	30.2*	> 2
Score - teat cups	7.4	18.7*	> 2

[1]Calculations based on mean teat contamination per farm before cleaning, either based on TBC [$\log_{10}$ cfu/ml] or ATP [$\log_{10}$ RLU];

* significant association with teat cleanliness ($p < 0.10$)

selection of cows could be part of a mastitis control programme in addition to providing a clean environment, which is more directly associated with teat contamination.

The association between use of cow brushes in the barn and teat cleanliness is difficult to explain. If no brushes are used this can be interpreted as a sign that cow comfort plays a minor role for the farmer. Also the cleaning of the milking box has probably no direct influence on teat cleanliness, but more likely reflects the general attitude of the farmer towards hygiene as does the general impression of the robot. The hygiene status of teat cups seems to be a consequence of teat contamination rather than a cause. However, the outer surfaces of teat cups can contribute to teat contamination during attachment of the milking cluster as well.

Striking is that claw condition explained 47% of variance when the evaluation was based on bacterial contamination of teats. One explanation for these findings is that if claw condition is poor, cows contaminate their udders more easily. This could happen by more frequent lying or by difficulties during lying down or getting up as a cause for more contacts between claws and udder. Significant associations have also been found between udder and leg hygiene scores (Schreiner and Ruegg, 2003). Poor claw condition may also negatively influence milking frequency. More likely bad claw conditions are an expression of very unsuitable conditions in the barn thus negatively affecting claw health as well as teat contamination. If management of animal health is good also more attention is paid on cow comfort and cleanliness.

In interpreting the results it should not be forgotten that for some management factors which are expected to be important for teat cleanliness no variability occurred between farms and were thus not included into the analysis. The fact that the majority of farms practised cubicle cleaning twice per day, shearing of udders and cutting of tail brushes shows that these are generally accepted hygiene measures to keep cows clean.

Conclusions

Differences in teat cleaning efficiency between different brands of AM systems exist and although results differed dependent on method of evaluation improvements are necessary and possible. The influence of individual farms and of initial contamination of teats are more important for cleaning efficiency than brand of AM system. Coliform counts exceeding 100 cfu/ml in bulk tank milk indicated insufficient function of teat cleaning on some farms even if TBC was below or close to 10000 cfu/ml.

AM specific management factors associated with high teat contamination were: replacement of teat cleaning device < once per year, moderate/poor status of the teat cleaning device, average milking frequency per day < 2.5 and no selection of cows for robot acceptance ($p<0.10$). Factors not directly related to AM involved contamination of cubicles: less than one cubicle per cow, cows lying on alleys present in the herd, addition of fresh bedding material < once per day, no selection of cows for udder health, moderate/poor status of bedding material and moderate/poor status of claws were significantly related to high teat contamination.

Additional factors like the general impression of the robot, cleaning frequency of the milking box, status of teat cups and the use of cow brushes in the barn were probably more closely related to the general attitude of the farmer towards hygiene than to teat cleanliness.

In interpreting these data it has to be regarded that only a relatively small number of farms could be included into the study. The investigation of management factors was based on farmers´ answers to a questionnaire without any means to check for reliability of these data. In several cases values had to be combined to one value to approximate a more uniform distribution.

Although the cause and effect relationship between parameters used to evaluate hygiene management on farms and teat cleanliness is not always very strong these factors should be considered when improvements of teat cleanliness are intended.

Acknowledgements

The work was funded by the European Commission within the EU project "Implications of the Introduction of Automatic Milking on Dairy Farms (QLK5-2000-31006)" as part of the EU-programme "Quality of Life and Management of Living Resources".

The content of this publication is the sole responsibility of its publisher, and does not necessarily represent the views of the European Commission nor any of the other partners of the project. Neither the European Commission nor any person acting on behalf of the Commission is responsible for the use, which might be made of the information presented above.

The authors thank the farmers for their cooperation during this investigation.

References

Council Directive 89/362/EEC of 26 May 1989 on general conditions of hygiene in milk production holdings, Official Journal of the European Communities No. L156, 08.06.89, pp. 30-32.

Council Directive 92/46/EEC of 16 June 1992 laying down the health rules for the production and placing on the market of raw milk, heat treated milk and milk-based products, Official Journal of the European Communities No. L268, 14.09.1992, pp. 1-32.

Frank, J.F., L. Hankin, J.A. Koburger and E.H. Marth, 1985. Tests for groups of microorganisms. In: Richardson, G.H. (editor), Standard methods for the examination of dairy products, 15th edition, American Public Health Association, Washington D.C., pp 189-191.

Gaworski, M.A., C.B. Tucker, D.M. Weary and M.L. Swift, 2003. Effects of stall design on dairy cattle behaviour. Proceedings of the Fifth Dairy Housing Conference, 29-31 January 2003, Fort Worth, Texas, pp. 139-146.

International Dairy Federation, 1985. Milk and milk products, Enumeration of coliforms - Colony count technique and most probable number technique at 30 °C. IDF Standard 73A:1985.

International Dairy Federation, 1991. Milk and milk products, Enumeration of microorganisms - Colony count technique at 30 °C. IDF Standard 100B:1991.

Melin, M., H. Wiktorsson and A. Christiansson, 2002. Teat cleaning efficiency before milking in DeLaval VMS [TM] versus conventional manual cleaning, using Clostridium tyrobutyricum spores as marker. Proceedings of The First North American Conference on Robotic Milking, March 20-22, 2002, Toronto, Canada, II 60-63.

Reinemann, D., 1997. Troubleshooting high bacteria counts in farm milk. Proceedings 36th Annual Meeting, National Mastitis Council, Inc., Albuquerque, New Mexico, pp. 65-79.

SAS Release 8.01. SAS Institute Inc. Cary, NC, USA

Schreiner, D.A. and P.L. Ruegg, 2003. Relationship between udder and leg hygiene scores and subclinical mastitis. Journal of Dairy Science 86: 3460-3465.

Schuiling, E., 1992. Teat cleaning and stimulation. In: Ipema, A.H. (editor), Prospects for Automatic Milking. Pudoc, Wageningen, Netherlands, pp. 164-168.

Slaghuis, B., 1996. Sources and significance of contaminants on different levels of raw milk production. Proceedings of the IDF Symposium on Bacteriological Quality of Raw Milk. Wolfpassing, Austria, pp. 19-27.

Sumner, J., 1996. Farm production influences on milk hygiene quality. Proceedings of the IDF Symposium on Bacteriological Quality of Raw Milk. Wolfpassing, Austria, pp. 94-102.

Ten Hag, J. and K.E. Leslie, 2002. Preliminary investigation of teat cleaning procedures in a robotic milking system. Proceedings of The First North American Conference on Robotic Milking, March 20-22, 2002, Toronto, Canada, V 55-58.

Welham, S.J., 2002, In: Genstat Release 6.1. 2002. Lawes Agricultural Trust. IACR-Rothamsted, Harpenden, Hertfordshire, UK., pp. 226-228.

THE CLEANING OF AUTOMATIC MILKING SYSTEMS

Erik Schuiling

Applied Research, Animal Sciences Group, Wageningen UR, Lelystad, The Netherlands

Abstract

The cleaning of an automatic milking system is of great importance for the hygienic quality of the milk. From conventional milking several methods and procedures are known and applied to automatic milking systems. The diversity of the cleaning systems of automatic milking systems is given. Important differences do occur in procedure (circulation cleaning versus cleaning with boiling water), in time needed for cleaning and in water and energy consumption. Cleaning time differs between 37 and 130 minutes per day for standard cleaning frequencies, water consumption differs from 284 to 495 l/day.

The effect of the cleaning frequency was investigated by a comparison between 2 and 3 system cleanings per day during 9 weeks each. When three system cleanings were performed, bacterial quality of the milk was significantly better for TBC, coliform, thermoduric and psychrotrofic bacteria. The difference for TBC however was small: 13 versus 10 *10^3 cfu/ml. So, with both frequencies the average TBC is far within the range for first quality milk. It is concluded that two cleanings per day could be sufficient for a good quality of milk, but the cleaning system has to be optimised, well maintained and controlled. Three times cleaning however reduces the risk of increased bacterial growth in the milking system.

To prevent transfer of pathogens by the milking cluster, in AM-systems have a procedure for a cluster flush. The effect of the cluster flush on removal of pathogens is tested on one system, with cold water and with a disinfectant. 98,4 and 98,9% of the pathogens were removed, expressed as log-reduction 1,80 and 2,26. The extra removal of pathogens by using a disinfectant does not compensate for the extra risk for contamination of the milk produced.

In a test on the effect of a cluster flush on the rate of new infections, 46 cows were milked with a deliberately infected cluster. Two cross positioned liners were flushed after infection, the other two liners were not flushed. In none of the quarters an inflammation occurred. The effect of a cluster flush on new infections could not be established.

Because cluster flush does remove most of the bacteria from the liners, does not influence the milking capacity of the AMS and can be performed with small amounts of water, there are no reasons not to perform a cluster flush.

Introduction

Milk quality is equal important with an automatic milking system as with a traditional milking system. The efficiency of the cleaning of the traditional milking system is well known by research, experience and trial and error, resulting in several methods to clean the milking machine. The design of the cleaning system of automatic milking systems is mainly based on this knowledge. For a better understanding of the cleaning of the AM-System the different systems are investigated and compared for consumption of water, detergents and time.

An important difference to traditional milking is the fact that the AM-system is working in a '24-hour economy', where capacity and efficiency are important. Earlier research has shown that cleaning should not be postponed to long, because milk quality tends to decrease after 8 hours of milking without cleaning [Verheij, 1992; Ordolff et al., 1992]. In many EU-countries three times cleaning per day is therefore demanded for AM-systems. However, in order to increase the milking capacity, some systems are cleaned only twice a day, with varying success. More information about the effects of the cleaning frequency of AM-systems is needed.

Mastitis is an important disease in animal production. Large numbers of cows are infected each year. Each infected cow is a potential source for spreading the pathogen, in which the milking machine is a possible vector. The role of the milking machine as a vector is not clear and is, among other factors, depending on the type of pathogen. In traditional milking parlours it is advised to milk infected or suspected cows last, just before the cleaning of the installation. In automatic milking systems it is hardly possible to milk cows in an order corresponding to their udder health status. To minimise the role of the milking machine as a vector, the teat cups can be cleaned after each milking by flushing with water. In this way most of the pathogens, present in the liner after milking an infected cow, can be removed. The effect of the cluster flush on new infection rate however is unknown.

Cleaning systems, procedures and strategies

An overview of the cleaning systems and procedures is made, based on information presented during interviews with the manufacturers of all commercially available systems. In spite of all the differences between systems, the procedures for cleaning are very much alike [Schuiling et al., 2001]. Three procedures can be distinguished:
- *system cleaning*: the whole system, which may consist of one or more units, is rinsed, cleaned and disinfected. Three cycles are needed; 1) pre rinse, 2) cleaning and disinfecting, 3) after rinse.
- *unit flush:* the milking unit (cluster up to the milk line) is rinsed in order to remove milk residues (f.i. contaminated milk) or to prevent adherence of dried milk residues
- *cluster flush:* the cluster is rinsed in order to avoid transfer of (mastitis) pathogens between cows

Table 1 gives an overview of the characteristics of the AM-systems, based on information of the manufacturers. Some systems do have two (or more) milking units per AM-system. In these cases the water consumption is calculated per milking unit of a two-unit-system.

An important difference between systems is the cleaning procedure: circulation cleaning versus boiling water cleaning. Boiling water cleaning takes less time than circulation cleaning, because for circulation cleaning a longer exposure time to the disinfectant is needed. Another factor in the duration of the system cleaning is the time needed to drain the system between the three cleaning cycles.

The water consumption per day ranges from 284 to 495 l. The cluster flush has a large contribution to the differences between brands. The energy consumption is mainly depending on the cleaning procedure: boiling water cleaning takes more energy.

Under practical circumstances the water consumption can differ from the figures given. Especially when the intake of water is based on time instead of volume measurement. In these cases the pressure of the water line does in a large extend influence the volume.

Table 1. Overview of the important characteristics of the cleaning systems of the AM-systems.

AM-system (code)	1	2	3	4	5	6
System cleaning						
Circulation/boiling water	C	C	C	BW	BW	BW
Time needed (min)	35	30	20	17	15	9
Water consumption (l)	75	70	90	70	90	70
Automatic control [1]	d	td	t	t	Td	-
Unit flush						
Time needed (min)	10	1	9	10	2	4
Water consumption (l)	15	0,9	30	12	20	20
Cluster flush						
Time needed (min)	0,25	1	0,4	0,5	0,13	0,17
Water consumption (l)	0,14	0,9	1	0,75	0,5	0,75
Estimated consumption/day [2]						
Time (min)	130	93	83	76	50	37
Water (l)	284	347	495	353	395	373
Energy (kWh)	2,4	1,6	2,4	4,2	4,2	4,2

[1] t=temperature cleaning fluid, d=concentration of detergent

[2] based on 3 system cleanings, 2,5 unit flushes and 150 cluster flushes. For the cluster flush no time consumption is calculated, supposing no milking delay will occur.

The start of the cleaning of AM-systems is automated. However, the milk filter has to be changed manually. It is preferred to remove the dirty filter before cleaning and place the new filter after cleaning. This is not practical, so the standard procedure is changing the filter before cleaning. In several cases however filters will be changed at convenient times for the operator.

Material and methods

Cleaning frequency

The effect of the cleaning frequency, 2 versus 3 system cleanings per day, was studied on commercial farms. The farms were selected on their milk quality in the past 6 month (average TBC<= 15 without large fluctuations) and the willingness to participate in the project under the project conditions. During 9 weeks the system was cleaned 2 times/day and during 9 weeks 3 times per day. The order of the cleaning frequency was random. During both periods of 9 weeks the milk quality was analysed: TBC each delivery of milk, coliform, thermoduric and psychrotrophic bacteria once per two weeks. Also the freezing point of the milk was measured once every two weeks.

On each farm the system cleaning was analysed and the hygienic condition of the AM-system was checked at the start of the project, at change-over of the frequency and at the end of the project. These check consisted of measuring the volumes, concentration and temperatures of the cleaning fluids, the efficiency of the post rinse during one cleaning.

The hygienic condition of the installation was tested by visual checking and ATP-measurements.

Removal of pathogens

The removal of pathogens was tested on an Automatic Milking System of Gascoigne Melotte, installed at the Waiboerhoeve in Lelystad, The Netherlands.

Two tests are performed on different days. Each test consisted of 6 runs, in the first three runs the cluster was flushed with cold water, the last three runs a disinfectant (peracetic acid) was added tot the water. A system cleaning of the AMS was performed before the test.

In each run the liners were infected with a Streptococcus agalactiae suspension in pasteurised milk (approx. 1,5 10^6 cfu/ml). The Str.ag. strain was derived from a virulent udder inflammation. 10 ml of this suspension is flooded at the top end of the barrel of the liner in one circular movement, to provide an equal film of infected milk over the entire interior of the liner. Each time after infection two cross positioned liners were flushed in the standard way of the AMS, the other two liners were not flushed. The number of pathogens in the liner is measured by swabbing a part of the interior surface of the liner. Swabs are taken from the liners by making two circular movements with the swab, approx. 10 cm below the liner mouth. The swabs are analysed for number of Str.ag.

At the end of each run the cluster was disinfected and flushed with cold water.

New infection rate

The effect of a unit flush on the rate new infections was tested on an Automatic Milking System of Gascoigne Melotte in the experimental farm at the Waiboerhoeve, Lelystad [Schuiling et al., 2004]. The cows used for the experiments were accustomed to automatic milking. Cows were selected from a herd of about 50 cows on the basis of good udder health, established by SCC, conductivity and bacteriological examination of foremilk samples. 10 to 12 cows were used in a each challenge test. These cows were separated form the rest of the herd in a part of the cubicle housing system. Twice a day this group of 10 to 12 cows was brought to the AMS to be milked.

The challenge test was done during one evening milking (day 0). All the liners of the AMS were infected with a Streptococcus agalactiae suspension, as described for the removal test.

The cluster was attached by hand to be sure that attachment was conducted without delay and/or retrials. Teat cleaning, as it is usually performed in this AMS, was inactivated, to prevent a flushing of the liners and teats after attachment, which could lead to unwanted removal of pathogens from the liner in this test. The air bleed in the short milk tube was blocked during this milking session, to create sub optimal milking circumstances which will lead to a higher infection risk.

The cows in the experiment were milked in the AMS twice a day, from two days before the start of the infection of the liners until 8 days afterwards. During the evening milking of day -1, 1, 2, 3, 5 and 7 foremilk samples of all quarters of the cows were taken and analysed for SCC and pathogens. Conductivity measured during milking was recorded by the AMS. Foremilk samples of the day before infection of the liners (day -1) were used as a final check on the cows for udder health.

Results

Cleaning frequency

Thirteen farms participated in the project. On these farms the system cleaning was checked before the start of the experiment: in six cases the system cleaning had to be adjusted to improve the concentration of the cleaning fluid or to meet the demands for the system, as set by the manufacturer.

At the change-over of frequency and at the end of the experiment these checks of the system cleaning and of the hygienic condition of the installation were repeated. No defects or problems were found then.

For the analysis the bacterial counts are log 10 transformed, in order to create a normal distribution. The results in table 2 are transformed again to numbers of cfu/ml.

For all bacterial groups there is a significant difference between two and three cleanings per day. The difference for TBC is however small and in average for both frequencies far below the penalty border ($100*10^3$ cfu/ml). In 8 cases on 4 farms TBC however was over 100; the cases were equally divided over both cleaning frequencies.

As aspected, the freezing point of the milk was slightly higher when more system cleanings are performed.

Removal of pathogens

The number of pathogens in the suspension to infect the liners was $1,34*10^6$ cfu/ml. This is a dose comparable to a highly infected quarter. Flushing of the liners resulted in a reduction in the number of Str.ag. found in the swabs, compared with no flushing: 98,4% and 98.9% of the bacteria were removed after flushing with respectively water and the disinfecting fluid. The logreduction was respectively 1,80 and 2,26.

New infection rate

46 cows have been challenged in 4 runs, with respectively 10, 12, 12 and 12 cows. All cows have been checked before the start of the experiment using foremilk samples. None of the cows showed signs of an infection, based on SCC and presence of pathogens. In some cases minor pathogens were found in foremilk samples (aspecific Staphylococci), but no signs of inflammation was found based on SCC. In none of the cases Streptococci were found in the foremilk samples before challenge.

Table 2. Effect of cleaning frequency on milk quality.

Frequency	2 times/day	3 times/day	significance
Quality parameter			
TBC (10^3 cfu/ml)	13	10	<0.001
Coliform bacteria (cfu/ml)	173	13	<0.001
Thermoduric bacteria (cfu/ml)	877	320	<0.001
Psychrotrofic bacteria (cfu/ml)	1047	522	<0.001
Freezing point (°C)	-0.520	-0.519	0.003

After the first run the number of pathogens (Str.ag.) in the suspension used to infect the liners, was increased from 0,5*10^6 to 3*10^6 to increase the risk of infection.

In none of the cases Str. ag. could be found in the quarter foremilk samples on day 1, 2, 3, 5 or 7. None of the cows showed a significant increase in SCC. Also none of the cows were signalled by the AMS to have a change in conductivity.

After challenging 46 cows it was concluded that is was not possible to create new infections with Streptococcus agalactiae by milking with deliberately infected liners, whether the liners were flushed or not flushed after infection.

Discussion

A good quality of milk is not possible without milking equipment, that is well cleaned and disinfected. To keep the equipment in a good hygienic condition, important factors are the frequency of cleaning, the efficiency of cleaning and the sanitary construction of the automatic milking system.

The frequency of cleaning is set on three times per day, according to earlier reseach and the legislation in many EU countries [Verheij 1992; Frost et al., 1999]. In some cases AM-systems are cleaned only twice per day. Research has shown that 3 times cleaning gives better results than two times cleaning, but differences are small. There are indications from this research that the difference between 2 and 3 times cleaning per day is depending on type or brand of the AMS, probably caused by sanitary construction, efficiency of cleaning and or type of cleaning. After installation and during system maintenance the cleaning of AM-systems needs more attention from technicians and operators. Sometimes the software settings in AM-systems are not set according to the instructions, the effect of these settings is not checked and hygienic inspection of the system is not performed often. The complexity of the AM-system compared to a traditional milking machine, makes it difficult to understand how the cleaning is performed and where possible weak points do occur.

When the AM-system is well constructed from a sanitary point of view, the cleaning system is optimised and the cleaning process is controlled, it is possible to produce a good quality milk with 2 times cleaning/day. In average however, milk quality declines. Defects in the construction or in the cleaning procedure will show sooner and have a larger effect. To minimise risk, the farmer should consider three system cleanings per day. The advantage of two times cleaning is less waste of water, energy and detergents and a higher milking capacity. Though, it is possible to choose cleaning times during 'slow milking hours', so the influence on the capacity of the system can be small. Due to the large differences in cleaning time, the influence is depending on brand. In some cases cleaning time can be reduced by optimising the procedure.

It is shown that a cluster flush reduces the number of bacteria in milking liners. In the test conducted with a high dose of a pathogen (Streptococcus agalactiae) 98,4 to 98,9% of the pathogens could be removed, respectively when flushing with cold water or a disinfectant. Log reductions were 1,8 and 2,26 respectively. So, adding a disinfectant increases the reduction of bacteria, but due to the short exposure time which is available, the liners are not fully disinfected. Using a disinfectant has some disadvantages, because of the risk for residues in milk. In this view the small increase in reduction of pathogens does not compensate for extra risk of contamination of food products and/or the extra costs for the more complex flushing system to avoid contamination when using a disinfectant.

The effect of a cluster flush on the rate of new infections could not be established. In total 46 cows were milked with deliberately infected clusters, of which two cross positioned liners were flushed and the other two were not flushed. None of the cows responded in one or more quarters of the udder to the challenge: no pathogens could be found in foremilk samples during one week after challenging and no change in SCC or conductivity was detected during this time.

So the risk for a new infection when milking with contaminated liners is to small to show a difference in flushing and not flushing of liners, even in a sub optimal milking situation as created in the experiment. Pathogens in the liner at the start of milking will be flushed away during milking. The highest risk for a new infection will occur at the end of milking or just after milking, when the teat canal is not fully closed yet and the milk flow has stopped. So the effect of a cluster flush on preventing new infections should not be overestimated. On the other hand it is obvious that reducing the number of pathogens in the liner by flushing the liner will contribute in the decrease of the risk for new infections. As the cluster flush can be performed during changing of cows in the AMS, there will be no effect on capacity of the AMS.

References

British Standard, 1991. Code of practice for equipment end procedures for cleaning and disinfecting of milking machine installations. BS 5226.

Frost, A.R., T.T.Mottram, C.J. Allen & R.P. White, 1999. Influence of milking interval on the total bacterial count in a simulated automatic milking system. Journal of Dairy Research. 66: 125-129

IDF, 1996. General recommendations for the hygienic design of dairy equipment, Bulletin of the IDF 310.

Ordolff, D. & D. Bölling, 1992. Effects of milking intervals on the demand for cleaning the milking system in robotized stations. In: A.H. Ipema, A.C. Lippus, J.H.M. Metz & W. Rossing (Editors), Proceedings for Automatic Milking (EEAP Publication No. 65), Wageningen, The Netherlands, pp. 169-174.

Schuiling, E., J.A.M. Verstappen-Boerekamp K. Knappstein, & C. Benfalk, 2001. Optimal cleaning of equipment: Investigation of systems, procedures and demands. Report D16.

Schuiling, E. & F. Neijenhuis, 2004. Optimal cleaning of equipment: effectiveness optimised teat cup cleaning in the prevention of mastitis pathogens transfer. Report D18.

Verheij, J.G.P., 1992. Cleaning frequency of automatic milking equipment. In: A.H. Ipema, A.C. Lippus, J.H.M. Metz & W. Rossing (Editors), Proceedings for Automatic Milking (EEAP Publication No. 65), Wageningen, The Netherlands, pp. 175-178.

INVESTIGATION OF DIFFERENT CLEANING FREQUENCIES IN AUTOMATIC MILKING SYSTEMS

Christel Benfalk & Mats Gustafsson

JTI - Swedish Institute of Agricultural and Environmental Engineering, Uppsala, Sweden

Abstract

In several European countries there is a demand that milk producers with automatic milking systems (AMS) must clean the milking system three times a day. Cleaning an AMS reduces possible milking time as well as consuming considerable amounts of hot water and detergents. Therefore, it is important to acquire greater knowledge during practical conditions to better understand the consequences of less frequent cleaning.

This study was carried out on 9 Swedish farms during a period of 16 weeks. Three AMS brands were included in the study with three farms of each brand. The two investigated treatments were cleaning with 8 and 12 hours intervals.

Milk quality was analysed for total bacterial count (TBC), coliform count (CC), psychrotrophic count (PC) and thermoduric count (TC). To ensure that milk cooling and cleaning have been done correctly during the sampling period, the temperature was monitored in the milk pipe and in the bulk tank.

The difference for TBC was significant ($p<0.05$) with different cleaning frequencies, but no significant difference was obtained for the other groups of bacteria. Cleaning the system twice a day resulted in a TBC of log 4.15 compared to log 3.98 when cleaning the system three times a day.

The milk in the bulk tank contained an acceptable average level of CC irrespective of farm and cleaning frequency. However, the variation between farms and cleaning frequency was high with the lowest value of 0 and the highest value of log 3.9 CC. With a limit for TC of 200 cfu/ml, all farms regardless of cleaning frequency exceeded that limit for almost all of the milk samples (mean between log 3.02 - log 3.19).

Introduction

In several European countries there is a demand that milk producers with automatic milking systems (AMS) must clean the milking system three times a day. The reason is that when AMS was introduced on farms there was a lack of knowledge regarding the cleaning of this equipment. To be sure to maintain the hygienic quality of the milk, i.e. total bacterial count, the Swedish regulation was set to three cleanings a day. In Sweden and other countries, price premiums exist for high quality milk. For example a total bacterial count below 30 000 cfu/ml will result in the farmer receiving an additional payment.

The cleaning procedure reduces valuable milking time as well as consumes considerable amounts of hot water and detergents. Therefore, it is important to acquire greater knowledge during practical conditions to better understand the consequences of less frequent cleaning. In this study the aim was to investigate the effect of cleaning frequency (twice or three times a day) on the hygienic milk quality during practical conditions in Sweden.

Material and methods

The study was carried out on 9 Swedish farms during a period of 16 weeks (8 weeks per treatment). Three AMS brands were included in the study with three farms of each brand. The treatments were:
- Cleaning with 8h intervals
- Cleaning with 12h intervals

Milk quality was analysed for total bacterial count (TBC; Bactoscan 8000), coliform count (CC; NMKL Method no 44, 5th ed. 2001), psychrotrophic count (PC; IDF 101A:1991) and thermoduric count (TC; Standard Methods for the examination of dairy products" 15th ed. 1985). The sampling frequency was for TBC every bulk milk collection (every second day) and for CC, PC and TC once every week. To ensure that milk cooling and cleaning have been done correctly, the milk temperature was measured in the milk pipe and in the bulk tank. The temperatures were measured with thermistors and logged with Tiny-Tag loggers. One of the thermistors was placed at the end of the pipe just before the tank inlet. The other thermistor measuring the tank temperature was placed close to the tanks own sensor. Both thermistors were mounted and insulated for good heat conduction and to avoid external disturbances.

The cleaning systems on the farms were checked before the milk sampling started. The bacteriologic history of each farm was investigated. During cleaning, conductivity measurements were made to check for the amount of milk or detergent in the last water of the rinse cycles and to check the proper strength of the cleaning solution. Temperature and water flow was measured and logged throughout the cleaning cycle.

Before analyses, all data was checked and corrected for outliers.

Results

The mean TBC, CC, TC and PC for each cleaning frequency are presented in table 1. The difference for TBC was significant ($p<0.05$) with different cleaning frequencies but no significant difference was obtained for the other groups of bacteria. Cleaning the system twice a day resulted in a TBC of log 4.15 compared to log 3.98 when cleaning the system three times a day. The bulk tank milk contained an acceptable average level of CC irrespective of farm and cleaning frequency. However, the variation between farms and cleaning frequency was high with the lowest value of 0 and the highest value of log 3.9. With a limit for TC of 200 cfu/ml, all farms regardless of cleaning frequency exceeded that limit for almost all of the milk samples (mean between log 3.02 - log 3.19).

The first milk entering the tank after tank cleaning was not cooled below 4 °C within 3 hours several times, which can clearly be seen in figure 1. TBC in the milk increased during a period of time and exceeded this farm´s normal TBC.

The water temperature during cleaning was initially too low on one farm (Figure 2) which resulted in an increased TBC. When the starting time of the heater was adjusted so that a sufficient cleaning temperature was reached before start, the TBC decreased to much lower values.

During the initial check of the farms, the cleaning cycle was measured on all farms to make sure that the last cleaning water had a sufficiently high temperature (Figure 3). In the study, all farms obtained a sufficiently high temperature in the last cleaning water.

Conductivity measurements were made on the cleaning solutions and the final rinse water on all farms. There was a sufficient concentration of detergent in the cleaning solution and there were no detergent residues in the final rinse water for all the farms.

Table 1. Mean total bacterial count, coliform count, thermoduric count and psychrotrophic count with cleaning intervals of 12h and 8h.

	Cleaning frequency	
	12h interval	8h interval
Total bacteria count, log	4.15[a]	3.98[b]
Coliform count, log	1.70[a]	1.52[a]
Thermoduric count, log	3.19[a]	3.02[a]
Psychrotrophic count, log	3.02[a]	2.82[a]

[a,b] Means with the same letter are not significantly different (p<0.05)

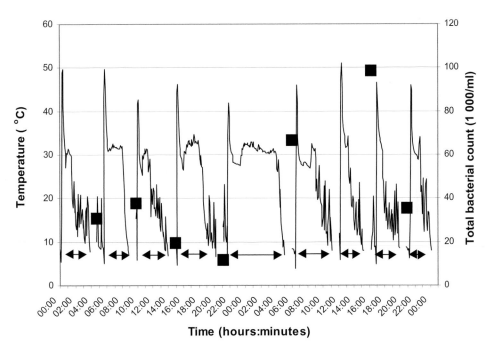

Time (hours:minutes)

Figure 1. Temperature (°C) measured in the bulk tank at one of the farms that had temporary cooling problems. Temperatures measured from collection of milk and cleaning of the bulk tank until the storage temperature is reached (data from July 9 to July 25). TBC (1 000 /ml) in delivered milk during the same period (■).

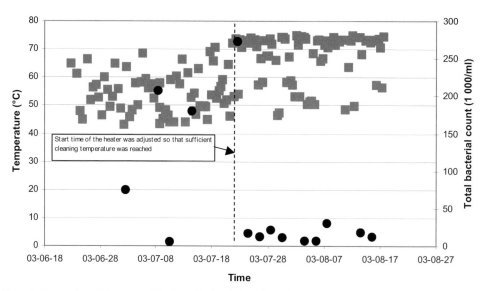

Figure 2. Temperature (ºC) measured in the milk pipe (●) and TBC (1 000/ml) in the bulk tank during the same period (■).

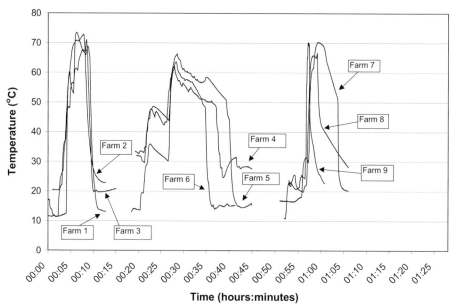

Figure 3. Measured temperature (ºC) during the cleaning cycle for all nine farms. The farms differed in cleaning systems.

Automatic milking – A better understanding

Discussion

In the present study, effects have been noticed on TBC in bulk milk on farms with AMS with cleaning of the system twice a day compared to three times a day. For the cleaning frequency of twice a day a mean level of TBC, log 4.15, is not reaching the level for reduced payment in Sweden. However, the maximum level of TBC on the studied farms during data collection period is exceeding the class limit of extra payment (30 000 cfu/ml) for one or both cleaning frequencies in 7 of 9 farms. For 2 of the 7 farms, 12-56% of the collected milk samples were exceeding 30 000 cfu/ml compared to the other 5 farms that only occasionally exceeded that level. This indicates that these two farms lacked in proper hygiene management.

The effect of cleaning frequency could not be seen for CC, PC and TC and the reason can be both too limited data or that there is no significant difference. The level of CC indicates pre-milking hygiene and equipment cleanliness (Knappstein *et al.*, 2002; Reinemann, 2002). CC between 100 and 1000 are an indication of poor milking hygiene and values exceeding 1 000 cfu/ml indicate that bacterial growth is occurring on milking equipment (Reinemann, 2002). In the present study the CC in most samples in the bulk milk were below 1 000 cfu/ml and 2 out of 9 farms were not exceeding 100 cfu/ml at all during the data collection period.

TC is another bulk milk test that can be of diagnostic value. TC exceeding 200 cfu/ml indicate poor cleaning of the equipment (Knappstein *et al.*, 2002; Reinemann, 2002). All the investigated farms in the present study exceeded the above recommended limit. Since the farms and especially the 2 farms with high TBC have not been investigated further, the reason can not be fully explained. However, there can be two possible explanations. One is that for the 2 farms with high levels of bacteria, all investigated groups of bacteria were high, and therefore it can be a lack in the cleanliness of the environment as well as the cows and the milking equipment. The other is that the milk samples from these farms were transported for a longer time than milk samples from other farms before analysis.

There is a significant difference in TBC between cleaning frequencies and therefore recording the temperature in the cooling and the cleaning procedure might be a useful tool in the management system. This can clearly be shown in the present study.

References

Knappstein, K.; Reichmuth, J. & Suhren, G. 2002. Influence on bacteriological quality of milk in herds using automatic milking systems and experiences from selected German farms. Proc. First North American Conference on Robotic Milking, March 20-22, 2002, Toronto, Ontario, Canada. pp V13-24.

Reinemann, D. 2002. Application of cleaning and cooling principles to robotic milking. DeLaval Hygiene Technology Center. In: Inaugural Symposium (Eds. Hemling, T. & Ingalls, W.) May 15-16, 2002, Kansas City, Missouri, USA. pp 112-120.

INDUCTION OF MILK EJECTION DURING TEAT CLEANING IN ROBOTIC MILKING SYSTEMS

Rupert M. Bruckmaier & Heinrich H.D. Meyer
Physiology Weihenstephan, Technical University Munich, Freising, Germany

Abstract

In a series of investigations we have tested the stimulatory effect of the teat cleaning devices of the different AMS types and of subsequent teat cup attachment. Teat preparation induces the release of oxytocin (OT) and hence alveolar milk ejection already before the milking vacuum is applied to the teats. If milking is performed at short intervals from previous milking or in late lactational stages, i.e. at low degrees of udder filling, small amounts of cisternal milk and a long latency period until milk ejection in response to a pre-stimulation (up to 3 min) occur concomitantly. Due to the frequent occurrence of short intervals between milkings, an adequate pre-milking stimulation adapted to the expected actual degree of udder filling is crucial for a successful milking process. In a multi-box AMS udder cleaning caused immediate release of OT and after cows have moved to the milking box and teat cups were attached alveolar milk was available for milking. Several types of single stall AMS provide sequential cleaning of the teats one by one via two rolling brushes before the attachment of teat cups. This type of udder cleaning induced release of OT and milk ejection, too. Thus, alveolar milk is available immediately after attachment of the teat cups. Also in AMS with teat cleaning via warm water in a special teat cup OT is released immediately after the start of cleaning. It could be shown that even water temperatures as low as 12 to 15 degrees Celsius induce normal OT release and milk ejection. In some of the tested AMS an individual adaptation of the cleaning time to the actual udder filling was possible. At very low udder filling a prolonged teat cleaning reduced the total milking time whereas a prolonged teat cleaning in well filled udders just consumed additional time. In conclusion, all tested AMS teat cleaning systems are suitable for pre-stimulation to induce milk ejection already before the start of milking.

Pre-stimulation and milk ejection

Usually a maximum of 20% of the milk stored in the udder, the cisternal milk, is immediately available for milking. Most of the milk, the alveolar fraction, must actively be shifted into the cisternal cavities by the milk ejection reflex before being available for milk removal. In late lactation and shortly after previous milking, the cisternal fraction can be close to zero, i.e. the alveolar fraction amounts to up to 100% of the milk (Knight *et al.*, 1994; Pfeilsticker *et al.*, 1996) and no milk can be removed before milk ejection has commenced.

Tactile stimulation of the teats and the release of oxytocin (OT) are essential for milk ejection (Bruckmaier & Blum, 1998). Increased concentrations of OT cause contraction of the myoepithelial cells surrounding the alveoli. During milk ejection, alveolar milk is shifted into the cistern, which causes an increase of intracisternal pressure (Bruckmaier *et al.*,

1994). Only slightly elevated OT concentrations above baseline are sufficient to elicite a maximum milk ejection (Schams *et al.*, 1984; Bruckmaier *et al.*, 1994). At the start of milking, stimulatory effects of machine milking without pre-stimulation or with a manual pre-stimulation and subsequent machine milking cause the release of comparable amounts of OT and hence milk ejection (Bruckmaier & Blum, 1996). Pre-stimulation, i.e. a stimulation without simultaneous removal of milk, is however crucial for the timing of alveolar milk ejection to avoid milking on empty teats already at the start of milking. Surprisingly, the intensity and the way of stimulation is less important as long as it does not cause pain to the teats. Thus, the stimulation of only one teat caused an OT release similar to the stimulation of all four teats (Bruckmaier *et al.*, 2001). Even a teat stimulation at extremely low intensity, i.e. the application of the milking vacuum without pulsation in liner-closed position after teat cup attachment caused OT release, albeit on a lower level than that caused by standard milking pulsation (Weiss *et al.*, 2003). However, also this moderate OT release was sufficient to induce milk ejection.

Udder filling and milk ejection

The release of OT in response to a tactile stimulation occurs always immediately within 30 to 60 s of the start of a tactile stimulation. However, the commencement of milk ejection in response to the released OT varies considerably. It was shown that the start of milk ejection is delayed at short intervals from previous milking and in late lactation, i.e. in all situations when the degree of udder filling is low (Bruckmaier & Hilger, 2001). The start of milk ejection after the start of teat stimulation as a function of udder filling follows a negative linear regression and accounts to less than 1 min in well-filled udders and up to 3 min in little filled udders. It needs to be considered that the lag time until occurrence of milk ejection does not depend of the amount of stored milk per se. Milk ejection occurred after a similar lag time in animals of different production levels at the same stage of lactation although the obtained milk was much lower in the low than in the high producing animals (Wellnitz *et al.*, 1999). In this case the degree of filling of the individual udder was similar, because lower producing udders had also lower storage capacity. We assume that in partially filled alveoli more contraction of the myoepithelial cells and therefore more time is needed until milk is ejected in milk ducts and cistern. Only if the alveolus is well filled, myoepithelial contraction results immediately in the ejection of milk. Myoepithelial cells surrounding a partially filled alveolus must more contract before milk is pressed into the milk ducts.

Concomitantly with late milk ejection, the amount of cisternal milk, i.e. milk available for milking before milk ejection, is small at low degrees of udder filling (Pfeilsticker *et al.*, 1996). This phenomenon is due to the fact that only after alveoli are considerably filled milk is shifted into the cisternal cavities by the secretory pressure (Knight *et al.*, 1994). For practical machine milking this means that at low udder filling there is not sufficient cisternal milk to bridge the time until milk ejection occurs. Therefore, a long-lasting pre-stimulation is of particular importance at low udder filling whereas in full udders the milking on huge amounts of cisternal milk is not finished until alveolar milk is available (Weiss & Bruckmaier, unpublished). In robotic systems, the intervals between milkings and hence udder filling may vary considerably. Therefore, an efficient pre-stimulation, if possible adapted to the expected udder filling at each individual milking, is crucial for complete and fast milk removal.

The actual degree of udder filling at each milking does not only affect milk ejection but also the composition of the removed milk. The highest variation is observed in the milk fat content. It is high at a low degree of udder filling and low in well filled udders (Weiss *et al.*, 2002). A short-term variation of milk composition due to variable milking intervals in AMS is egalized in bulk milk but needs to be considered if milk samples are taken to determine the milk composition of individual cows.

Attachment delay after pre-stimulation

Due to technical reasons the start of milking in AMS may not be immediately after teat cleaning. Furthermore, teat cup attachment sequentially one by one is performed much slower than in conventional milking. Depending on the system a lag time of up to several minutes may occur between teat cleaning and start of teat cup attachment.

The situation in AMS was simulated in a series of experiments (Bruckmaier *et al.*, 2001). Sequential attachment of teat cups one by one every 20 s or every 60 s after a 1 min manual pre-stimulation and start of milking in each quarter immediately after attachment did not have adverse effects on OT release or milk removal. The pulsation of the liner of the first attached teat cup was able to maintain elevated OT concentrations and milk ejection.

In contrast, total delay of teat cup attachment after pre-stimulation for 120 s caused a transient numerical reduction of OT concentrations during the period from 2-4 min after the start of milking. If milking was thereafter started wihout a new pre-stimulation milk removal was reduced (Bruckmaier *et al.*, 2001).

A delay of 60 s after the end of a 1-min pre-stimulation until start of teat cup attachment did not reduce OT concentrations and milk removal even if teat cups were in addition attached sequentially every 20 s. Surprisingly, numerically highest OT concentrations and optimal udder emptiing was observed after a 30 s pre-stimulation, 30 s delay, and thereafter sequential teat cup attachment every 20 s. This clearly demonstrates that the duration of teat stimulation is less crucial for milking performance than the total interruption of an established milk ejection (Bruckmaier *et al.*, 2001).

Milk ejection during udder cleaning and milking in automatic milking systems

While the main purpose of teat/udder brushing before milking in AMS is to clean the teats for milking, this mechanical treatment is assumed to have also stimulatory effects on milk ejection. Several studies in various AMS were conducted to investigate the stimulatory effect on milk ejection of the respective automatic pre-milking udder cleaning devices.

The stimulatory effect of teat cleaning by rolling brushes or towels or by water in a special cleaning teat cup was investigated in various multi-box and one-box AMS types (Macuhova *et al.*, 2003; Dzidic *et al.*, 2004a, b). All investigated cleaning devices induced release of OT and hence alveolar milk ejection comparable to manual or mechanical pre-stimulation in conventional milking. Furthermore, the attachment of teat cups induced OT release and milk ejection. Therefore, even during (experimental) milkings without pre-milking udder cleaning, milk ejection occurred before the start of milking in AMS types where milk removal was only started after attachment of all four teat cups (Macuhova *et al.*, 2003). In AMS where milk removal started in each quarter directly after attachment, milking without pre-milking teat cleaning frequently caused delayed milk ejection after the start of milking, i.e. a transient decrease of milk flow after the removal of the cisternal milk or a milking on empty teats until milk ejection if no cisternal milk was present (Dzidic *et al.*, 2004a, b).

During the attachment process, already attached teat cups induced OT release and milk ejection. Therefore, the frequency of occurrence of delayed milk ejection at the start of milking decreased with the number of attached teat cup.

Elevated levels of OT persisted throughout the entire milking in all AMS under study (Macuhova *et al.*, 2003; Dzidic *et al.*, 2004a, b). In a multi-box AMS the interruption of tactile udder stimulation between udder cleaning and teat cup attachment due to the cow´s move from the prep box to the milking box had no negative effect on milk ejection and milk removal (Macuhova *et al.*, 2003). This may be mainly due to the fact that milk removal in this system started only after attachment of all four teat cups, i.e. after a renewed udder stimulation. Pre-milking teat cleaning with warm water (30-32 °C) was as effective as cleaning with brushes. Most surprisingly, cleaning with water at 12-15 °C had a similar stimulatory effect as cleaning with warm water (Dzidic *et al.*, 2004b).

Duration of teat cleaning in automatic milking systems

In some AMS, the number of cleaning cycles, each lasting for less than 20 s can be determined. Given that the cleaning effect is sufficient this means the possibility for an individual adaptation of cleaning duration based on the actual udder filling estimated from stage of lactation and interval from the previous milking. It was demonstrated that in well filled udders a cleaning for 20 s is sufficient whereas one or two additional cleaning cycles improved milk removal in udders with low degree of filling (Dzidic *et al.*, 2004a). In other systems, udder cleaning lasts for about 60 s in any case. The stimulatory effect of this cleaning was sufficient at all degrees of udder filling and a prolonged cleaning had no beneficial effect on milk removal (Macuhova *et al.*, 2003; Dzidic *et al.*, 2004b).

Extremely delayed and long-lasting teat cup attachment

The time needed for teat cup attachment in an AMS is usually longer than manual cluster attachment in conventional milking. As shown above, this has no negative effects on OT release, milk ejection and milk removal as compared to conventional milking. In some milkings, however, the time from udder cleaning until attachment of all teat cups can be extremely prolonged to more than 5 min or can even fail totally if several attempts are necessary to attach teat cups. Despite attachment delay after udder cleaning OT release during subsequent milking was normal and cows did not show any disturbance of OT release during long-lasting teat cup attachment, obviously because teat cup attachment provided a new stimulation before the start of milking (Macuhova *et al.*, 2004). Continuously elevated OT levels where observed throughout the entire attachment process during long-lasting teat cup attachment and the entire subsequent milking. Attachment delay and long-lasting teat cup attachment in AMS had no negative effect on OT release and subsequent milk removal (Macuhova *et al.*, 2004). This underlines the stimulatory function for the induction and maintenance of milk ejection by almost any mechanical action on the teats during milking.

Conclusions

Pre-milking udder cleaning devices in all tested AMS adequately induced OT release and milk ejection as a prerequisite for optimal udder emptiing during milking. An adaptation of duration of udder cleaning to the expected degree of udder filling could help to reduce the time required for the milking process, provided that the cleaning effect on the teat is sufficient after a short cleaning time.

References

Bruckmaier, R.M. and J.W. Blum, 1996. Simultaneous recording of oxytocin release milk ejection and milk flow during milking of dairy cows with and without prestimulation. Journal of Dairy Research 63: 201-208.

Bruckmaier, R.M. and J.W. Blum, 1998. Oxytocin release and milk removal in ruminants. Journal of Dairy Science 81: 939-949.

Bruckmaier, R.M. and M. Hilger, 001. Milk ejection in dairy cows at different degrees of udder filling. Journal of Dairy Research 68: 69-376.

Bruckmaier, R.M., J. Macuhova and H.H.D. Meyer, 2001. Specific aspects of milk ejection in robotic milking: a review. Livestock Production Science 72: 169-176.

Bruckmaier, R.M., D. Schams and J.W. Blum, 1994. Continuously elevated concentrations of oxytocin during milking are necessary for complete milk removal in dairy cows. Journal of Dairy Research 61: 323-334.

Dzidic, A., D. Weiss and R.M. Bruckmaier, 2004a. Oxytocin release, milk ejection and milking characteristics in a single stall automatic milking system. Livestock Production Science, in press.

Dzidic, A., J. Macuhova and R.M. Bruckmaier, 2004b. Effects of cleaning duration and water temperature on oxytocin release and milk removal in an automatic milking system. Journal of Dairy Science, in press.

Knight, C.H., D. Hirst and R.J. Dewhurst, 1994. Milk accumulation and distribution in the bovine udder during the interval between milkings. Journal of Dairy Research 61: 167-177.

Macuhova, J., V. Tancin and R.M. Bruckmaier, 2003. Oxytocin release, milk ejection and milk removal in a multi-box automatic milking system. Livestock Production Science 81: 139-147.

Macuhova, J. , V. Tancin and R.M. Bruckmaier, 2004. Oxytocin release and milk removal after delayed or long-lasting teat cup attachment during automatic milking. Livestock Production Science, in press.

Pfeilsticker, H.U., R.M. Bruckmaier and J.W. Blum, 1996. Cisternal milk in the dairy cow during lactation and after preceding teat stimulation. Journal of Dairy Research 63: 509-515.

Schams, D., H. Mayer, A. Prokopp and H. Worstorff, 1984. Oxytocin secretion during milking in dairy cows with regard to the variation and importance of a threshold level for milk removal. Journal of Endocrinology 102: 337-343.

Weiss, D., A. Dzidic and R.M. Bruckmaier, 2003. Effect of stimulation intensity on oxytocin release before, during and after machine milking. Journal of Dairy Research 70: 349-354.

Weiss, D., Hilger, M., Meyer, H.H.D., Bruckmaier, R.M., 2002. Variable milking intervals and milk composition. Milchwissenschaft 57, 246-249.

Wellnitz, O., Bruckmaier, R.M., Blum, J.W., 1999. Milk ejection and milk removal of single quarters in high yielding dairy cows. Milchwissenschaft 54, 303-306.

EVALUATING CLEANLINESS OF UDDERS WITH AN IMAGE PROCESSING SYSTEM

D.W. Ordolff
Federal Dairy Research Centre, Kiel, Germany / Federal Agricultural Research Centre, Braunschweig, Germany

Abstract

In an initial investigation an industrial image processing system was used to evaluate optical parameters for indicating the cleanliness of udders. The condition of surfaces were described by luminance and by the balances of colours red-cyan and yellow-blue. The amount of pixels found at unclean and clean surfaces differed in a highly significant way (P<1%). It was possible to identify the status of surfaces using limits for the amount of pixels not to be exceeded by clean surfaces. Further investigations are required to set more reliable limits and to avoid problems by allowing more stable cow position and optimised illumination of udders.

Introduction

Automatic milking systems actually are not able to evaluate the status of cleanliness of udders and to detect lesions of teats to manage cleaning of udder and teats according to the demands of actual regulations.

Results of basic research on application of optical parameters to fulfil these demands have been presented by Bull *et al.*(1995). Problems were mainly found with respect to pigmented surfaces. In a further step Bull *et al.*(1996) used a CCD-colour-camera to evaluate the cleanliness of teat surfaces. Correct recognition of dirty teats was possible by connecting type and intensity of colours of all pixels.

Analysis of spectroscopic parameters according to an industrial standard to evaluate the efficiency of cleaning udders and teats indicated that manual cleaning mainly caused modifications of luminance of surfaces (Ordolff, 2002). The parameters red/green and yellow/blue were not useful to indicate cleaning efficiency at white surfaces, but a significant reduction of the level of "yellow" due to cleaning was observed at black surfaces. The parameter red/green was most efficient to detect bloodstained surfaces. It was concluded that for practical application a remote sensing system based on video cameras would be more useful than the device used in this investigation requiring direct contact with the surface to be evaluated.

Material and methods

In a first experiment in two recording sessions an industrial image processing system (Manufacturer: ISRA-Vision systems, Karlsruhe, D) was used to collect information about optical parameters to indicate the cleanliness of udder surfaces of ten cows, housed in an experimental stanchion barn at the Federal Agricultural Research Centre (FAL) at

Braunschweig. In each session two sequences of images were produced. The initial sequence represented clean surfaces, the second sequence was recorded after controlled application of faecal material.

The data recording system produced three parameters, Y, U, V, to describe the optical condition of the evaluated surface, using the numeric range from 1 to 255, corresponding to an 8 bit data transmission system (Schwarz et al., 1996). The parameter Y indicates the luminance. The chrominance signals U and V indicate the balances of red-cyan (parameter U) and yellow-blue (parameter V). A white object is represented by the values of 255 (Y), 127 (U) and 127 (V). The Y-U-V colour model used here is written down in the standard CCIR-601, dealing with conditions for transmission of colour video signals.

The image processing system applied offers two ways for setting the critical limits for the parameters. The more practical proposal is to select critical points at the image shown at the screen, using the pointing device of the computer ("mouse"). The range of parameters to characterise corresponding areas then is set automatically by the software. Another option is to enter the limits numerically by the keyboard which can be useful to reduce "false positive" or "false negative" indications.

Images of the rear part of udders were recorded by a CCD-camera which was placed on a trolley, also equipped with two 55 W halogen-lights to obtain stable illumination of the respective udder surface. Due to their triangular arrangement the light beams indicated the central area of the image to be recorded, simplifying a correct placement of the camera with a distance of about 1.5 m behind the cow.

The evaluation of images was based on four conditions of surfaces to be found: dirt, white, black, shadow/contour. The last mentioned situation was included into the analysis because initial tests have shown that the signals corresponding to unclean surfaces to a certain amount may also be found at clean areas due to poor illumination, caused by the movement of the cow or at some parts of outlines of udder or legs. Similar observations were mentioned by Bull et al. (1996).

Ten images of unclean udders recorded in the first session were used to define the range of the optical parameters to be used for evaluation of the status of surfaces (Table 1), corresponding to the averages of lowest and highest values resulting from the teaching procedure.

Evaluation of surfaces was done in two ways, using the most frequent values of all parameters, based on visual inspection of images (Setting A) and using the averages of the parameter Y in combination with the most frequent values for U and V (Setting B).

Table 1. Ranges of optical parameters.

Surface	Average values					
	Ymin	Ymax	Umin	Umax	Vmin	Vmax
dirt	46.4	78.3	111.7	123.3	133.7	139.1
white	92.6	150.8	112.2	123.4	136.3	151.2
black	25.1	51.5	126	131	128.7	133.1
shadow	55.7	74.8	122.8	131.5	132.6	143.3

For statistical treatments F-test and range-test according to Newman-Keuls (e.g. Haiger, 1966) were used.

Results and discussion

The total surface evaluated included 391554 pixel per image. In table 2 and 3 the amounts of pixels are given, obtained at different types of surfaces in unclean and clean condition. Both settings for evaluating the parameter "dirt" produced corresponding results. The amount of pixels found at unclean and clean surfaces differed in a highly significant way (P<1%). A significant difference between unclean and clean surfaces also was found for the parameter "white", while for "black" an "shadow" no difference between unclean and clean surfaces were to be seen.

It was possible, as shown in figure 1, to identify unclean and clean surfaces with both settings for the parameter "dirt", using limits for the amount of pixels representing this parameter not to be exceeded by clean surfaces. Using setting A the critical limit could be set at 10000 pixels, with setting B a limit of 15000 pixels seemed to be more appropriate. The limits corresponding to setting A produced less false results than setting B.

Similar to results presented by Ordolff (2002) also in this investigation, luminance (Y) was the most variable parameter. However, the figures representing the structure of colours at surfaces evaluated indicate that only the application of all parameters can lead to a

Table 2. Dirt (pixel) found at evaluated surfaces.

	dirt (setting A)		dirt (setting B)	
	unclean	clean	unclean	clean
Average	15890	4127	24011	8372
Std. dev.	7866	2522	10253	6400
F	38,52		31,81	

Table 3. Other optical conditions (pixel) of evaluated surfaces.

	Type of surface					
	white		black		shadow	
	unclean	clean	unclean	clean	unclean	clean
Average	104769	143622	79028	70772	12218	11712
Std. dev.	35460	55422	41876	49741	7722	7148
F	6,63		0,31		0,04	

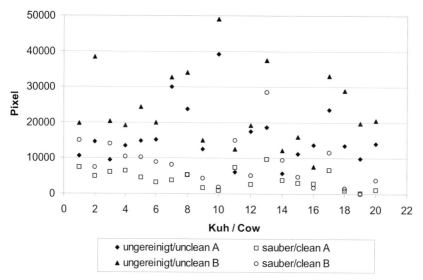

Figure 1. Presentation of unclean and clean surfaces with setings Aand B.

reliable decision to what extend cleaning the udder is necessary and whether it was done efficiently.

Based on only 20 recordings it would be too early to set general limits for clean and unclean surfaces, especially with respect to the questions about shading effects as described by Bull *et al.* (1996).

Visual inspection of images indicated, that not all cows were in identical position when clean and unclean udders were to be recorded. This situation, already mentioned by Bull *et al.* (1996), may explain some of the irregular results to be seen at figure 1. For practical application of the system the relative position of the camera and udder therefore is to be stabilised, e.g. by using the signals to be obtained by sensors for monitoring the cow position, included in most automatic milking systems.

While shading, found to be a problem by Bull *et al.* (1996), in the investigation described here did not affect the efficiency of classification of unclean and clean surfaces, it should be avoided by optimised illumination of udders. Since evaluation of the total udder surface requires at least two cameras this problem may be solved by adapting illumination individually to the surface inspected by the respective camera. This also could be the way to avoid irregular classification of contours.

Summary

The analysis of images of unclean and clean surfaces of udders, recorded in two sessions with an industrial image processing system, indicated, that the combination of luminance and chrominance allowed setting limits to recognise with some certainty unclean and clean surfaces. For practical application, however, more investigation is required to analyse additional aspects like positions of cows, adding cameras for evaluation of full udder surfaces and optimisation of illumination of surfaces to be checked.

References

Bull, C., T. Mottram and H. Wheeler (1995): Optical teat inspection for automatic milking systems. Computers and electronics in agriculture 12 (2) 121 - 130

Bull, C.R., N.J.B McFarlane, R. Zwiggelaar, C.J.Allen, T.T. Mottram (1996): Inspection of teats by colour image analysis for automatic milking systems. Computers and electronics in agriculture 15 (1) 15-26

Haiger, A. (1966): Biometrische Methoden in der Tierproduktion. (Biometric methods in animal production) Österr. Agrarverlag, Wien, ISBN: 3-7040-0744-7

Ordolff, D. (2002): Farbparameter zur Bewertung der Eutersauberkeit. Landtechnik 57, (6) 328 - 329

Schwarz, J., G. Sörmann (1996): Kompressionsalgorithmen. Seminararbeit WS95/96, ZTT, FH Worms, www.ztt.fh-worms.de/de/sem/

DIFFERENT LOCATIONS OF INSTANT COOLING IN THE AUTOMATIC MILKING SYSTEM AND THE EFFECT ON MILK QUALITY

Mats Gustafsson & Christel Benfalk
JTI - Swedish Institute of Agricultural and Environmental Engineering, Uppsala, Sweden

An automatic milking system will be used almost all day, which means that warm milk will stay in the transportation line longer than in a conventional milking system. However, to cool milk as quickly as possible will reduce problems with bacterial growth. Therefore present study was to evaluate the effect on milk quality with different locations of instant cooling system. An automatic milking system on an experimental farm was used. In the experiment, three treatments were studied: Treatment A - milking with no instant cooling; Treatment B - milking with instant cooling located at the AMS and Treatment C - milking with instant cooling located at the bulk tank. The milk temperature was measured at the end unit, before the milk reached the bulk tank, in the bulk tank and before and after the instant cooling system (treatments B and C). The bulk milk was analysed for FFA (Free fatty acid), Total Bacterial Count (TBC) and Psychrotrofic Count (PC) both the same day as sampling and when stored for 72 hours in a refrigerator (4°C).

The data clearly show that when the instant cooling system was placed at the AMS the milk was cooled to 3-4°C but during transportation to the bulk tank the temperature of the milk was increased to 10-15 °C. However, when the instant cooling system was placed at the bulk tank the milk was cooled down to 3-4 °C when enter the bulk tank without increasing the temperature of the stored milk in the bulk tank. The level of FFA was significantly higher (0,94 meqv/l) when the instant cooling system was placed at the AMS in comparison with the treatments with no instant cooling system (0,61 meqv/l) and with instant cooling system placed at the bulk tank (0,66 meqv/l). The TBC in the bulk tank showed no significant difference irrespective of treatment if the milk was analysed the same day as sampling day. However, when the milk samples were stored for 72 hours before analysis the TBC was significantly ($p < 0,001$) increased for the treatments with instant cooling (Treatment B - 7,73 Log_{10} CFU/ml; Treatment C - 7,47 Log_{10} CFU/ml) compared to Treatment A (3,95 Log_{10} CFU/ml). The pattern was the same for PC.

The conclusion of the study is that if the instant cooling system is placed at the AMS the transportation line must be isolated so the milk is kept cooled during transportation to the bulk tank. Also the most likely reason for the increased level of FFA in the milk when the instant cooling system was placed at the AMS was a long transportation line to the bulk tank and at the same time was the milk passing a throttle to get the right flow trough the instant cooling system. Furthermore the instant cooling system was placed 0,5 m below floor level which increased the back presser even more.

TEAT CLEANING EFFICIENCY BEFORE MILKING IN DELAVAL VMS™ VERSUS CONVENTIONAL MANUAL CLEANING

Martin Melin[1], Hans Wiktorsson[1] & Anders Christiansson[2]
[1]Department of Animal Nutrition and Management, Swedish University of Agricultural Sciences, Uppsala, Sweden
[2]Swedish Dairy Association R&D, Lund, Sweden

With automatic milking, teat cleaning before milking is performed without visual inspection of teats and foremilk. Teat cleaning is a crucial step in automatic milking systems. The aim of the present study was to investigate the efficiency of the removal of *Clostridium tyrobutyricum* spores and manure slurry from teats using the DeLaval VMS™ teat preparation module, and to compare the results with conventional manual teat cleaning.

For standardisation of dirtiness, twenty minutes before milking of a cow, each teat was dipped into a sterilised manure-water slurry contaminated with 200,000 *Clostridium tyrobutyricum* spores/ml. After contamination the cows waited outside the milking unit and were prevented from laying down. The study included the following treatments: DeLaval VMS™ cleaning and fore-milking; conventional manual cleaning and fore-milking, and no cleaning and no fore-milking (control). In VMS™ the separate teat preparation module, handled by the robotic arm, uses lukewarm water and compressed air for cleaning. The vacuum transports the water/ air mixture and the dirt to the water and foremilk collector. During the fore-milking step, the water and the compressed air supply stop, and the foremilk is sucked from the teat by vacuum. The conventional manual cleaning and fore-milking was accomplished by cleaning the teats with a fresh moist paper tissue for each cow and then by hand stripping of 2 streams of milk. This procedure took 20-24 seconds in total and was similar to the regular cleaning procedure in a milking parlour. Milking was performed in the VMS™ for all treatments.

Of the total of 12 cows included in the study, 9 were randomly divided into a Latin Square, with 3 treatments and 3 periods. The remaining 3 cows were used as cleaning control cows before and between the milking of the cows on the different treatments. All milk from each cow was collected and sampled. The mean daily milk yield was 31.1 kg and mean somatic cell count 49,700 cells/ml. Spores were collected by filtration and enumerated anaerobically on RCM agar. The logarithm of the total number of spores was calculated for all observations and used in the statistical analysis by means of the general linear models procedure of SAS. The mean values were tested in the model, where test day, cow, treatment and teat were fixed effects.

The mean total numbers of spores recovered in the milk for the control, manual and VMS™ cleaning were 32,246 cfu, 10,805 cfu and 616 cfu, respectively ($p < 0.01$). The conventional cleaning and the VMS™ cleaning removed 65% and 98% of the spores, respectively. The results after conventional cleaning were in line with other similar studies and the VMS™ was comparable to very careful manual double cleaning.

CLEANING EFFECTS OF TEAT CLEANING IN ROBOTIC MILKING SYSTEMS BY USE OF POPPY SEED AS TRACER

B.A. Slaghuis, R.T. Ferwerda-Van Zonneveld & A. Mooiweer
Applied Research, Animal Sciences Group, Wageningen UR, Lelystad, The Netherlands

Teat cleaning is necessary to prevent milk from contamination with manure, dirt and included bacteria. A lot of research has already been done to investigate the effect of teat cleaning on milk quality. Robotic milking also includes automatic teat cleaning, being performed either by brushes or by water and air in a (separate) teat cup within a system. All teats are cleaned in the same way on a farm; the dirtiness of the teats is not taken in consideration.

The aim of this study was asses the cleaning effects of different types of cleaning devices in robotic milking systems.

Therefore a protocol was developed testing the efficiency of teat cleaning by applying a mixture of sterilised manure and poppy seed (20% w/w) on teats of ten cows per farm. Five cows were milked after teat cleaning and five cows were milked without a previous teat cleaning. Milk per cow was separated and filtered through a cotton filter; poppy seed remained on the filter. The filters were dried and the poppy seed on the filter was counted.

Two farms per brand were selected by the producers/suppliers. Six brands and two conventional farms were tested, so results of 14 farms were obtained.

Results showed significant effects of teat cleaning and brands. Differences between brands in teat cleaning efficiency could be concluded; from three brands teat cleaning was more effective (85-99% reduction in poppy seed in milk) than from three other brands (50-70% reduction in poppy seed in milk). Conventional teat cleaning gave a reduction of 99%. One of the reasons for less effective cleaning was the maintenance of the system. In some cases brushes, teat cups or liners were not replaced in time.

Teat cleaning in practice might be less effective compared to this experiment, because the waiting time after application was less than one and a half hour. The adhesion of poppy seed to the teats is probably stronger in case of dried manure and/or bacterial spores.

Teat cleaning determined by the poppy seed method showed to be effective for all brands. However, good maintenance of the teat cleaning system is very important for a good effect of teat cleaning.

Nevertheless, the effect of teat cleaning on milk quality might not be overestimated, because hygiene in the stable is the first concern in regard to the effect on milk quality.

COBALT AS A TRACER FOR ESTIMATION OF RESIDUES IN MILK AFTER APPLICATION OF AUTOMATIC TEAT CLEANING SYSTEMS

B.A. Slaghuis[1], R.T. Ferwerda-Van Zonneveld[1], M.C. te Giffel[2] & G. Ellen[2]
[1]Applied research, Animal Sciences Group, Wageningen UR, Lelystad, The Netherlands
[2]NIZO food research, Ede, The Netherlands.

Teat cleaning is necessary to prevent contamination of milk with manure, dirt and included bacteria. A lot of research has already been done to investigate the effect of teat cleaning on milk quality. Automatic Milking (AM) also includes automatic teat cleaning, being performed either by brushes or by water and air in a (separate) teat cup. Teats of all cows in a herd are cleaned in the same way within an AM-system. Different teat cleaning systems may give different cleaning effects. In some systems brushes or teat cups are disinfected after teat cleaning to prevent contamination of milk cross-infection of cows with udder diseases. AM-systems of three manufacturers were included in the study. The aim of this study was:

1. To develop a protocol for the evaluation of teat cleaning systems with regard to risk of milk contamination in automatic milking systems.
2. To estimate the level of residues in milk after teat cleaning making use of a with cobalt solution treated brush or teat cup.

1. A mixture (mean 2,5 g/cow) of sterilised manure, spiked with a Co-solution (5 g Co/kg of manure) was put on teats of ten cows. Five cows were milked without teat cleaning and five cows were milked after automatic teat cleaning. Milk from all ten cows was sampled and Co-contents were measured.
2. AM-systems with disinfection of the teat cleaning system were tested by replacing the disinfectant by a cobalt solution (5g/liter; to avoid the need to measure the levels of the various individual chemicals in the disinfectants) and milking three consecutive cows afterwards. Milk of these cows was sampled and Co-contents were measured. Milk samples were dry-ashed and Co was determined by Inductively Coupled Plasma Atomic Emission Spectrometry (ICP-AES).

Results

1. The effect of teat and udder cleaning via cobalt levels in the milk could not be determined because no significant difference was found between samples from Co-manure treated cows with and without teat cleaning. The amount of cobalt added to the manure (5 g/kg) was probably not high enough.
   Recoveries of more than 100% were calculated in milk, probably due to inhomogeneous distribution of Co in manure.

2. Disinfection of teat cleaning systems showed Co-residues in milk, but the levels were rather low (4-80 µg/kg). Differences between AM-systems were not significant, but only two farms per brand were studied.

To test the efficiency of automatic teat cleaning systems another tracer (poppy seed) was used; see abstract on poppy seed.

Using cobalt as a tracer combined with manure or other materials is not appropriate, but application of cobalt in carry over experiments without mixing with other materials might be useful. However, more research on adhesion characteristics of cobalt in comparison with regular chemicals in disinfectants is necessary.

ASSESSMENT OF TEATS CLEANING SYSTEM EFFICIENCY OF A MILKING ROBOT

F.M. Tangorra[1], V. Bronzo[2], A. Casula[2], M. Cattaneo[1], C. Martinazzi[2], P. Moroni[2], M. Zaninelli[1] & A.G. Cavalchini[1]
[1]Dept. of Veterinary Sciences and Technology for Food Safety, University of Milan, Milan, Italy
[2]Dept. of Animal Pathology, Hygiene and Public Health, University of Milan, Milan, Italy

Many studies led in different European Countries have emphasized an increase of the total bacterial count (TBC) in bulk tank milk after the introduction of the Automatic Milking Systems (Klungel, 2000; Billon & Tournaire, 2002; Everitt et al., 2002; Van der Vorst et al., 2002). In order to evaluate the efficiency of the cleaning system of a milking robot (Fullwood-Merlin) installed in an Italian farm, field-tests have been run in order to evaluate milk TBC of different cow groups using various sets up of the automatic teats cleaning system. Field-tests have been run in two following steps. First one involved 12 cows divided in 3 groups (G1, G2 and G3): G1, no teats cleaning; G2, one cleaning cycle; G3, two cleaning cycles. Second step has been carried out following a Latin Square Outline on 16 cows divided in 4 groups (G1, G2, G3 and G4) subjected to the following treatments: manual cleaning by a disposable disinfectant towel, one cleaning cycle, two cleaning cycles and no teats cleaning. All the cows involved in both the experimental designs were chosen at random among all those milked by the robot and with less than 150 days of lactation in order to have animals with approximately the same day in milking. Foremilk samples were taken weekly on first experimental design and daily on second one by sampler fitted on milk pipeline. Milk samples were subjected to TBC, sown samples by inclusion in Plate Count Agar Petri dishes. Data observed in the first experimental step show the automatic teats cleaning system guarantees better results than no teats cleaning ($4,42 \pm 1,05$ Log_{10} CFU/ml), as observed by Schuiling (1992), although the results do not show statistically significant differences, and one cleaning cycle ($4,19 \pm 0,54$ Log_{10} CFU/ml) is better than a double one ($4,20 \pm 0,75$ Log_{10} CFU/ml). The second experimental phase has shown statistically significant differences in groups subjected to no teats cleaning and to a double teats cleaning cycle, emphasizing a total bacterial count of the milk higher than the reference group subjected to a manual cleaning of each teat by a disposable disinfectant towel. The TBC trend of the milk is growing starting from the reference group ($2,74 \pm 0,35$ Log_{10} CFU/ml) to the groups subjected respectively to manual cleaning ($2,98 \pm 0,52$ Log_{10} CFU/ml), one cleaning cycle ($3,02 \pm 0,31$ Log_{10} CFU/ml), two cleaning cycles ($3,19 \pm 0,33$ Log_{10} CFU/ml), and no teats cleaning ($3,29 \pm 0,44$ Log_{10} CFU/ml). In both phases we have observed that a single teats cleaning cycle is more efficient in the removal of organic residues from the teats surface than a double one. The most likely hypothesis is that an extended action of the cleaning rollers of the robot do not increase the removal of organic residues from the teats but, on the contrary, favor its redistribution on the teats surface. Overall and as regards the test carried out, the teats cleaning unit of tested automatic milking system, guarantees a satisfactory cleaning of the teats surface.

References

Billion P. and Tournaire F., 2002. Impact of automatic milking systems on milk quality and farm management: the French experience. Proceedings of the First North American Conference on Robotic Milking, March 20-22, Toronto, Canada, V 59-63.

Everitt B., Ekman T., Gyllensward M., 2002. Monitoring milk quality and udder health in Swedish AMS herds. Proceedings of the First North American Conference on Robotic Milking, March 20-22, Toronto, Canada, V 72-75.

Klungel G. H., Slaghuis B. A., Hogeveen H., 2000. The Effect of the introduction of Automatic Milking Systems on Milk Quality. Journal of Dairy Science. 83, 1998-2003.

Schuiling E. (1992) Teat cleaning and stimulation. In: A.H. Ipema (Editor), Prospects for Automatic Milking. Pudoc, Wageningen, Netherlands, pp. 164-168.

Van der Vorst Y., Knappstein K., Rasmussen M.D. (2002) Milk Quality on farms with an automatic milking system. Report within EU project "implications of the introduction of automatic milking in dairy farms", Deliverable D8, http://www.automaticmilking.nl.

ANIMAL HEALTH

IMPACT OF AUTOMATIC MILKING ON ANIMAL HEALTH

J.E. Hillerton[1], J. Dearing[1], J. Dale[1], J.J. Poelarends[2], F. Neijenhuis[2], O.C. Sampimon[3],
J.D.H.M. Miltenburg[3] & C. Fossing[4]
[1]Institute for Animal Health, Compton, United Kingdom
[2]Animal Sciences Group, Wageningen UR, Lelystad, The Netherlands
[3]Animal Health Service, Deventer, The Netherlands
[4]Institute for Animal Sciences, Foulum, Denmark

Abstract

Assessments of cow health on the transition of the herd from conventional to automated milking (AM), by use of a milking robot, have been made on up to fifteen farms in each of three European countries. No adverse effects of the transition have been found for body condition, lameness or teat condition. A potential risk is that fertility of the herd may decline faster than the current trend for conventional dairy farms. An increase in lameness after 12-months is possible and must be managed.

The only obvious change was that milk cell count, often an indicator of the prevalence of mastitis increased overall. However, this response to AM is inconsistent between herds, countries and studies such that the risk factors are not known, or consistent. In any herd the deleterious effects appear confined to a portion of the herd only. Other biological and technical factors including the frequency and periodicity of milking, and milk sampling techniques may also be involved and require examination.

Overall few changes have been found in the health of dairy cows milked by an automated system compared to a conventional parlour system. The quality of most current management is at least adequate for successful conversion although not all farms do convert successfully. On the contrary no particular benefits for the health of the cows have been found so it is likely that automated milking systems are far from optimised.

Introduction

The quality of milk and the efficiency of its production depend on the interactions between the cow, the farmer and the equipment used set upon the environmental background of manageable aspects such as feeding and housing, and uncontrollable factors including weather, milk price, production policies etc. No justification for any change in the overall model should exist whatever the farming system employed. However, a plethora of experience suggests that major dynamic components of the farm system, especially staffing or structural changes, affect the response of the farming performance. The response is usually measured either financially at the herd level or how the individual cows change.

The major EU project on Automatic Milking includes a study, carried out in Denmark, The Netherlands and the United Kingdom, on "Health of dairy cows milked by an automatic milking system". The aim has been to measure the effects of converting from conventional parlour milking to milking all cows by an automated milking system (AMS). The responses of the cows and the herd in terms of health have been measured using baseline data from up to 6-months prior to installation and for up to 12-months after installation of an AMS.

Parallel studies have been undertaken in the three countries gathering information on the same general factors but with each country focussing in much more detail on particular hazards of animal health posing measurable risks to successful milk production.

The project has used a range of farms that have chosen to install one or more AMS units for whatever reason. Within the study, data have been collected on the status of the farms prior to the commissioning of the AMS. Detailed analyses of the farms choosing to adopt AMS will follow but it is clear that generally interest in AMS occurs from farms and farmers where facilities and management require replacement or refreshment. The conventional milking systems on the study farms included many varying from obsolete to poorly effective, especially towards the time of AMS commissioning. Staff time for husbandry and maintenance was increasingly restricted as efforts were directed to the planning or actually development of the new facilities. On many farms the herd was also being restructured with abrupt changes including aggressive culling to remove health problems and cows likely to be unsuitable for AMS due to temperament and conformation. This did not occur on all farms. Some were significantly restricted in opportunities to change the herd as a common underestimation of cost and implications meant that the final development and the herd available for automatic milking (AM) did not meet all of the intentions and expectations.

The majority of farms convert to AM in less than ideal circumstances. Perhaps only those moving to purpose-built dairy units were close to optimum conditions. Most were adapting buildings, accommodation, yards, passageways, feed-ways etc but often with little forward planning and often in the absence of adequate or any experience and useful advice.

Disconcertingly many farmers were ill-prepared in their facilities and in most cases lacking in much advance training on how to manage cows for AM or operate the AMS. It was not uncommon for training to start when the first cow was being milking and any sustained programme to be offered three or more months after the first cow was milked by the AMS.

Rarely was AM installation easy for the farmer. Exceptions were on large, purpose-built units where the extra staffing allowed dedication of people and time to more detailed planning and training. The self-employed farmer did not usually have these opportunities.

These background conditions may help to explain why some farms fail to succeed and continue with an AMS. Fortunately the cows usually do much better. In several areas the AMS appears to improve performance overall especially in the longer term, despite transitional problems. This overview presents a synopsis of the preliminary findings of the EU project data on the impact on animal health of AM.

All of the project data are available in an Oracle database that can be made available for further analyses.

Materials and methods

In each of Denmark, The Netherlands and the UK farms that planned to install an AMS were asked if they wanted to cooperate in the study with the intention of recruiting fifteen farms in each country. In each country farms already being monitored withdrew at some time in the study either from use of the AMS or from dairy farming entirely. Not all manufacturers sell into all countries so the recruitment of farms by type of AMS or breed of cow was not controllable. The distribution of farms and AMS are shown in table 1.

Table 1. Number of farms studied by country and manufacturer of AMS.

Country	Recruited	Withdrew	Manufacturer				
			DeLaval	Fullwood	Insentec	Lely	Prolion
Denmark	15	1	2	0	0	9	3
Netherlands	17	2	6	1	2	5	1
UK	16	2	Not sold	6	Not sold	8	0

Each herd was selected to have 45-120 cows, varying between the countries, allowing 1000 cows or more to be studied in each country. Most animals in most herds were black and white, a mixed Friesian-Holstein population, but other breeds were included because of limited availability of herds (Table 2). The mixture of breeds varied between countries and will be considered more fully in later analyses.

Table 2. Distribution of breeds in herds studied in each country.

Country	Breed				
	Holstein/Friesian	Cross	Jersey	Danish Red	MRIJ
Denmark	10	1	2	1	0
Netherlands	6	0	0	0	2
UK	10	1	3	0	0

All farms were visited to assess the interest and commitment of the farmers to complete the study. An initial questionnaire on the type of farms, their structure, management and the philosophy of the farmers was completed. Subsequent questionnaires were completed on installation of the AMS and up to twice further to the end of the study. These document any changes. They will be reported elsewhere.

Each farm was visited at least twice before installation of the AMS and a minimum of twice, but often up to six times, after installation. The timing varied with the installation of the AMS, problem solving in running of AMS and the intensity of the study on any one farm. On these visits assessments were made of at least half of the cows or fifty animals on body condition and locomotion, and forty cows for teat condition (on some farms in The Netherlands and UK only). Farm data including milk production, milk quality, animal records on individual cow cell count, fertility, animal treatments, animal movements, veterinary purchases etc. Downloads of data collected automatically by the management systems or AMS were obtained.

Common methods of assessing locomotion, body condition and teats were applied. Those making these assessments made joint visits to compare and agreed standardisation between their methods. Body condition was scored on a scale 0 to 5 with increments of one half (Dearing et al., 2004). Locomotion scored 1 to 4. Teat condition was in considerable detail in The Netherlands and the UK using the Teat Club International method (Mein *et al.*, 2001).

Milk quality data, principally bulk milk cell count and individual cow cell counts, were obtained where possible to inform on the prevalence of mastitis. DHIA data are available for 14 farms in The Netherlands. In the UK data are only available for 8 farms, two others stopped individual cow cell counting on installation of the AMS because of technical problems. Four different recording systems were used in the UK so only limited comparisons can be made. On selected farms mastitic cows were sampled to indicate pathogens, also records of all treatments were collected and veterinary invoices obtained, to describe clinical mastitis. These results will be reported separately.

Additionally data were collected on all other infections, diseases and conditions likely to affect milk quality, production, attendance at the automated milking system and involve special staff time. Data have been collected such that the management and staff inputs into animal husbandry and health management can be calculated. Similarly for costs on preventive treatments e.g. teat disinfectants, udder preparation costs, insecticides etc.

Data on individual cow and herd fertility are available principally from National Milk Records in the UK, and from DHIA in The Netherlands.

A series of databases have been constructed using an Access framework subsequently migrated to an Oracle system. They comprise the herd descriptions, all data on individual cows, teat condition, body condition, locomotion, fertility, milk production, milk quality including cow cell count, veterinary involvement and purchases, and the farm questionnaires. The databases will be made available for further study.

Results

Fifteen herds were recruited in each country initially. Additional herds were required to replace several that either did not progress to install an AMS or ceased milking sufficiently soon after installation to provide insufficient data. The herds recruited represented the types of AMS marketed in each country (Table 1). A variety of breeds were used with the predominant breed for each herd shown in table 2. The effect of breed has not been considered fully to date.

The results considered here will be confined to overall descriptions comparing the three countries in the effects of introducing AMS on body condition, lameness assessed as locomotion score, overall milk cell count changes and herd fertility. More detailed consideration of body condition and fertility (Dearing *et al.*, 2004), teat condition (Neijenhuis *et al.*, 2004) and individual cow cell count (Poelarends *et al.*, 2004) recorded in this study but specific to study populations in The Netherlands or the UK are reported elsewhere in these proceedings.

Body condition

Substantial data are available from The Netherlands and the UK and data from six herds in Denmark. The body conditions varied more between countries than in response to the introduction of AM (Table 3). The UK cows averaged a half point more in condition than the Dutch cows. This is unlikely to be operator difference as they trained together. The cows were at the same stage of lactation (201 DIM at installation in The Netherlands, 198 DIM in the UK). Both were slightly earlier in lactation for the post installation scores

Table 3. Average body condition score (mean±SEM, range) before and after installation.

Country	Before	DIM	After	DIM	No. herds
Denmark	3.02±0.42 (2.38-3.53)		3.00±0.40 (2.32-3.42)		6
Netherlands	2.82±0.34 (2.06-3.41)	201	2.71±0.32 (2.27-3.25)	176	15
UK	3.32±0.22 (2.99-3.80)	198	3.34±0.30 (2.83-4.04)	187	14

reflecting the introduction of newly calved cows after the AM installation. Most herds were reduced in size to help in coping with the milking system transition. The smaller Danish dataset produced an average body condition between the Dutch and UK averages. In Denmark and the UK there was no change in body condition between 3-6 months prior to AM installation and 6 months post installation. A slight but not significant drop occurred with the Dutch cows.

More Dutch farms had a decrease in body condition score than an increase reflecting the change in the average and the range. In the UK the same number of farms showed an increase in average body condition score as showed a decrease. On the Dutch farms the range of body condition narrowed significantly from 1.35 to 0.98 points score suggesting that the farms are managing body condition, probably feeding, better. However, most of the change was due to the farm with the thinnest cows increasing condition by three quarters of a point. In the UK the range increased mostly because the thinnest cows became thinner and the fattest cows became fatter suggesting an exacerbation of the lack of adequate feeding management on two of the poorer units.

Locomotion

Data on scoring of locomotion are available for The Netherlands and the UK. Small but insignificant differences were found between the countries with a lower score in the UK (Table 4). No change had occurred by one month after AM installation. The scores had increased slightly, but not a significant difference, in both countries by 3-months after AM installation. The ranges also increased and the average score increased on seven farms whilst

Table 4. Average lameness score (mean±SEM, range) before and after installation.

Country	Before	After one month	After three months	After twelve months	No. herds
Netherlands	1.88±0.20 (1.54-2.04)	1.84± 0.21 (1.56-2.11)	1.91±0.34 (1.61-2.43)	–	12
UK	1.67±0.12 (1.35-1.93)	1.67±0.15 (1.40-1.90)	1.72±0.15 (1.50-2.00)	1.88±0.11 (1.70-2.05) n=12	14

unchanged on 6 farms in the UK. Scoring was continued on 12 of the UK farms. Twelve months after installation of AMS the lameness has increased significantly (t=5.52, P<0.001). Prior to installation eleven of fourteen UK herds were grazed but only six after installation. The poorer locomotion may reflect the increase in constant housing.

Milk cell count

The detailed effects on individual cow cell count were determined for Dutch herds and are reported in a companion paper (Poelarends *et al.*, 2004).

Overall production changes

The overall impact of conversion to AM may be assessed by comparing how each individual farm has coped in terms of the main indicators of animal health and welfare related to the changes in production methods. This has been attempted for the Dutch and the UK data.

Comparing 12 Dutch farms (Table 5) it appears that only one farm (number 11) improved in all aspects measured. This farm had a significant improvement in locomotion score from a poor pre-installation standard (>2.0). It was one of only two farms that reduced the cell count and the proportion of cows with a high cell count. However, the cows were already some of the thinnest in the study and became thinner so body condition may not truly have improved. One herd (number 1) deteriorated in all three parameters but this may also be an anomaly as the cows were particularly thin prior to installation (score =2.06) and although body condition score increased, the score only regressed towards the mean. Overall, little change was apparent. Locomotion improved in five herds and deteriorated in five herds. Body condition score decreased in eight herds but only by a small amount. It increased in two herds but not making the cows any fatter, just more typical. The only major deterioration was in average milk cell count and the proportion of cows with a cell count above a threshold (Poelarends *et al.*, 2004) where only two of the herds produced better quality milk. Average milk yield in the Dutch herds decreased in continuation of a trend starting up to 12-months prior to installation of the AMS and the cows became thinner with only a small reduction in DIM. This might suggest additional stresses including in management, especially as cell count increased, or inadequate feeding.

The same comparison is shown for the UK farms in table 6. Similarly to the Dutch farms there were no overall significant improvements, no farm improved in all parameters. One farm (Number 9) revealed significant problems, cows became fatter (BCS>4.0), they became more lame (Score >2.0) and cell count increased to more than 300,000 cells/ml. However, this AMS was being used to milk an increasing proportion of 'hospital' cows and so this is

Table 5. Overall changes in locomotion and body condition score, and bulk milk cell count for Dutch farms from before to after installation of the AMS. + = increase in score, usually poorer; o = same or very similar score; - = lower score, usually better.

Farm	1	2	3	4	7	9	11	12	13	14	15	16
Locomotion	+	-	+	+	-	o	-	-	-	-	+	o
Body condition	+	-	-	+	o	o	-	-	-	-	-	-
Cell count	+	+	o	-	+	+	-	+	+	+	+	+

Automatic milking – A better understanding

Table 6. Overall changes in locomotion and body condition score, and bulk milk cell count for UK farms from before to after AMS installation. + = increase in score, usually poorer; o = same or very similar score; - = lower score, usually better.

Farm	1	2	3	5	6	7	8	9	10	11	12	13	14	16
Locomotion	+	+	o	o	o	+	+	+	+	o	o	o	+	-
Body condition	+	+	+	-	+	-	-	+	+	-	-	-	-	+
Cell count	-		+	-	o				+		o		o	o

a not a reflection in any way of the AMS but a peculiar herd. Contrarily one herd (number 5) improved in all production aspects, marginally in lameness but reducing bulk mill cell count from more than 300,000 to fewer than 200,000 cells/ml. Some cell count data in some herds were confounded as a particular problem was encountered when the testing service had no understanding of the implications of variable milking frequencies and intervals, and proportioned the sample between milkings wrongly. In one herd the average cell count was reported as greater than 400,000 cells/ml when it was less than 200,000 cells/ml. This farm stopped using this service.

Fertility

Data on herd fertility are only available, so far, from the UK and then only from 6 herds. They are considered in detail in a companion paper (Dearing et al., 2004). Overall there is little evidence of major changes occurring in the common measures of fertility. The data come from the five UK herds that showed an overall deterioration in performance and one herd that had a small improvement. Any bias this increases is unknown. One month after AM installation services to conception had increased slightly and the conception rate had declined from 59 to 49%, both remained slightly poorer after AM installation. Consequences of these changes are shown in a 10% increase (average 13 days) in days to conception three-months after installation and a 14 days increase in calving index to 416 days by 12-months after installation. None of the changes were statistically significant but all suggestive of poorer fertility, at least in the transition period from conventional milking to AM.

Discussion

Body condition

Body condition scores are used as an indicator that the balance between intake of food and output of milk and reproduction is in some form of balance. It is necessary to maintain the energy balance of the cows so that they become neither too thin nor to fat. Within any herd individual cows vary according to production level, stage of lactation and the quality of the food supply. Wide variation was found between herds (Dearing et al., 2004), with no obvious effect of breed, and between the three countries in this study but relatively little within herds. Despite a minor short-term increase in body condition in some herds soon after installation of the AMS no longer term changes were found other than in an abnormal herd and a herd of thin cows regressing to the mean. Either the conversion to an AMS, often including zero-grazing, on any farm was of little or no consequence or the farms adopting AMS already are alert to proper feeding management. However, no significant increase in milk yield (Dearing *et al.*, 2004; Poelarends *et al.*, 2004), as might be expected on conversion to more frequent milking were found so no particular production stresses were imposed on the cows. Study of body condition on high producing cows responding by a yield increase of 10-15% may be required to identify any risk. It should be noted that neither in this study nor in many others from Kremer & Ordolff (1992) to Wagner-Storch & Palmer (2003) has the promise from more frequent milking of much higher yields (Hillerton & Winter, 1992) been achieved.

Locomotion

Locomotion scores were determined as an indicator of lameness. An increase in lameness might be predicted in AMS herds if cows become less mobile when zero-grazed, when confined to new concrete or persistent walking on concrete, become over-fat if fed excessively (not found here), stand for excessive lengths of time if access to the AMS is inadequate or if cows are poorly trained. The transition period showed little change but hoof wear is a process of attrition and the finding that a significant deterioration may be occurring after 12-months suggests that a lameness management programme is an essential part of adapting to an AMS.

Udder health

The detailed study on Dutch farms (Poelarends *et al.*, 2004) showed that milk cell count increased significantly after installation of the AMS particularly for newly calved cows and cows previously used to conventional milking. This may indicate problems in the adaptation of cows to the AMS or training of cows. The Dutch data, and the limited UK data, show that the problem did not occur on all farms. Previous studies have shown that the cell count response is inconsistent. Earlier work in this EU project (v d Vorst *et al.*, 2002) found an initial impairment in milk quality in three countries whilst a Swedish experimental study (Berglund *et al.*, 2002) found no effect on cell count from AM. This indication of a potential problem in the transition period to AM is a clear indicator that the quality of milk is at risk. How some farms avoid the problem, whether by design or accidentally, is not yet described although the Danish self-monitoring programme resolved the problem (Rasmussen *et al.*, 2002) suggesting it is partly due to management.

Teat condition in properly controlled AMS was improved and this occurred within a short time (Neijenhuis *et al.*, 2004). Infection was not necessarily the sole cause of the increase cell count especially as the increase was relatively small for many cows (Poelarends *et al.*, 2004). Hillerton & Winter (1992) showed that the initial cell count increase occurring when milking frequency is changed take several days to resolve. In the transition period to AM the milking frequency and the periodicity of milking may both be highly variable, especially for certain cows. It may be that the increased cell count in certain cows is a sign of their slow adaptation to AM. Poelarends *et al.* (2004) commented that all Dutch farms investigated had an increased work load related to the farm redevelopment. The higher cell counts may results from the consequences of poorer management mostly from lack of time to attend fully to the cows. Further analyses of the data in this project may be able to identify any such risk factor.

The technical problems in determining cell count in AM cows should not be overlooked. Peeters and Galesloot (2002) demonstrated, in estimation of milk fat from single daily samples, problems analogous to those in two UK herds in estimating individual cow cell counts in AM.

Fertility

Slight signs of lower fertility were found in the transition period to AM both before and after installation suggesting a lack of attention to oestrus. Although no significant changes were found the rate of deterioration of fertility in AMS could be faster that in conventional milking. Any fertility problems have a large time penalty to remedy so impaired fertility, still unproven, remains an important risk of conversion to AM. If significant yield affects were achieved this risk might be greater although a powerful argument exists that sustained duration to lactation is more beneficial if losses related to poorer fertility and health problems connected with parturition are reduced.

Conclusions

No major problems in converting from conventional milking to AM have been identified but equally none of the 44 farms has been found to achieve a substantial improvement in any aspect of cow health. Given that most of the farms studied were confronted by many and varied problems in conversion, with no clear advice readily available, they have all achieved a remarkable success. The transition period to AMS comprises a period of higher risk to health that extends from weeks before installation when resources start to be diverted from cow management. The length of the transition will vary on individual farms related to many unique factors. Several potential problems may develop in the longer term and anticipation of these is necessary. Clearly AMS succeeds but its longer-term promises for animal welfare and milk quality are unfulfilled to date.

Acknowledgements

This work would not have been possible without the immense support and goodwill of the cooperating farmers. The industrial partners gave huge support. Many staff in all institutes are also gratefully acknowledged.

This study was performed within the EU project Implications of the introduction of automatic milking on dairy farms (QLK5 2000-31006) as part of the EU-program ' Quality of Life and Management of Living Resources'. The contents of this publication are the sole responsibility of the authors, and do not necessarily represent the views of the European Commission. Neither the European Commission not any person acting on behalf of the European Commission is responsible for the use of the information.

References

Berglund, I., G. Pettersson and K. Svennersten-Sjauna, 2002. Automatic milking: effects on cell count and teat-end quality. Livestock Production Science 78: 115-124.

Dearing, J., J. E. Hillerton, J. J Poelarends, F. Neijenhuis, O. C. Sampimon and C. Fossing, 2004. Effects of automatic milking on body condition score and fertility of dairy cows. In: Automatic milking; A better understanding. Editors: A. Meijering, H. Hogeveen and C.J.A.M. de Koning, Wageningen Academic Publishers, Wageningen, The Netherlands.

Hillerton, J. E. and A. Winter, 1992. The effects of frequent milking on udder physiology and health. . In: A. H. Ipema, A. C. Lippus, J. H. M. Metz and W. Rossing (editors), Prospects for automatic milking, Pudoc Scientific Publishers, Wageningen.

Kremer, J. H. and D. Ordolff, 1992. Experiences with continuous robot milking with regard to milk yield, milk composition and behaviour of cows. In: A. H. Ipema, A. C. Lippus, J. H. M. Metz and W. Rossing (editors), Prospects for automatic milking, Pudoc Scientific Publishers, Wageningen.

Mein, G. A., F. Neijenhuis, W. F. Morgan, D. J. Reinemann, J. E. Hillerton, J. R. Baines, I. Ohnstad, M. D. Rasmussen, L. Timms, J. S. Britt, R. Farnsworth, N. Cook and T. Hemling, 2001. Evaluation of bovine teat condition in commercial dairy herds: 1. Non-infectious factors. In: Proceedings of the 2nd International Symposium on Mastitis and Milk Quality, NMC/AABP, Vancouver, p374-351.

Neijenhuis, J., K. Bos, O. Sampimon, J. J. Poelarends, J. E. Hillerton, C. Fossing and J. Dearing, 2004. Changes in teat condition in Dutch herds converting from conventional to automated milking. In: Automatic milking; A better understanding. Editors: A. Meijering, H. Hogeveen and C.J.A.M. de Koning, Wageningen Academic Publishers, Wageningen, The Netherlands.

Peeters, R. and P. J. B. Galesloot, 2002. Estimating daily fat yield from a single milking on test day for herds with a robotic milking system. Journal of Dairy Science 85: 682-688.

Poelarends, J. J., O. C. Sampimon, F. Neijenhuis, J. D. H. M. Miltenburg, J. E. Hillerton, J. Dearing and C. Fossing, 2004. Cow factors related to the increase of somatic cell count after introduction of automatic milking. In: Automatic milking; A better understanding. Editors: A. Meijering, H. Hogeveen and C.J.A.M. de Koning, Wageningen Academic Publishers, Wageningen, The Netherlands.

Rasmussen, M. D., M. Bjerring, P. Justesen and L. Jepsen, 2002. Milk quality on Danish farms with automatic milking systems. Journal of Dairy Science 85: 2869-2878.

Van der Vorst, Y., K. Knappstein and M. D. Rasmussen, 2002. Milk quality on farms with an automatic milking system. Deliverable 8, EU project Implications of the introduction of automatic milking on dairy farms, pp22.

Wagner-Storch, A. M. and R. W. Palmer, 2003. Feeding behaviour, milking behaviour, and milk yields of cows milked in parlor versus an automated milking system. Journal of Dairy Science 86: 1494-1512.

EFFECTS OF AUTOMATIC MILKING ON BODY CONDITION SCORE AND FERTILITY OF DAIRY COWS

J. Dearing[1], J.E. Hillerton[1], J.J Poelarends[2], F. Neijenhuis[2], O.C. Sampimon[3] & C. Fossing[4]
[1]Institute for Animal Health, Compton, United Kingdom
[2]Applied Research, Animal Sciences Group, Wageningen UR, Lelystad, The Netherlands
[3]Animal Health Service, Deventer, The Netherlands
[4]Institute for Animal Sciences, Foulum, Denmark

Abstract

Body condition and fertility were studied on farms in the UK changing from conventional to automatic milking (AM). No significant effect was seen on body condition or fertility as a result of AM although trends were seen to suggest that fertility was slightly poorer after installation and may become a problem in the longer term. Body condition was not shown to affect fertility during this study. Daily production was shown to be higher after installation of an automatic milking system which may have accounted for the slight changes in fertility, although neither were proven to be statistically significant. Further statistical analyses remain to be completed as these results do not support suggestions elsewhere that body condition influences fertility.

Introduction

As part of the EU project on Automatic Milking a study was carried out on "Health of dairy cows milked by an automatic milking system". In Denmark, The Netherlands and the UK farms were recruited that were changing from conventional to automatic milking (AM). These farms were studied intensively and general aspects of herd health were monitored from a few months before until one year after installation of the automatic milking system (AMS). Particular attention was paid to udder health, teat condition, lameness, body condition and fertility. The UK study on body condition and fertility is reported here.

Body condition scoring is potentially of enormous value in assessing the success of management of feeding in a dairy herd (Ward, 2003); with, in recent years, a growing realisation of the importance of correct management and feeding during lactation and the dry period, to achieve better milk production and fertility in the subsequent lactation (Chamberlain, 2003b). This study also examines the effect body condition has on fertility as there is a large amount of data to support the hypothesis that fertility is affected by body condition. Problems can occur both with cows calving when too fat or too thin. Chamberlain (2003a) and Pushpakumara et al. (2003) contend that cows in severe negative energy balance after calving have reduced fertility and that condition loss and negative energy balance are common in freshly calved cows. This can impair fertility by a number of mechanisms. Over-fat cows may store excessive amounts of fat in the pelvic area resulting in a reduction in the size of the pelvic canal, leading to difficult calvings (Chamberlain, 2003b).

Methods and materials

Data collection
Farms planning to install an AMS were identified. They were monitored for up to six months prior to installation and up to 12 months post installation by regular visits and access to farm data. Protocols for body condition scoring were discussed and practised with all partners of the study in the three cooperating countries. This was to achieve consistency throughout therefore making the assessments between farms and between countries comparable.

On each farm, 50% of cows were scored for body condition taking 10-15 seconds per cow (Ward 2003). Hady *et al.* (1994) found this method gave a significant estimate of the mean score of the whole herd. Cows were scored on average every eight weeks from 3-6 months prior to installation until a maximum of one year after installation, so allowing for seasonal effects and stage of lactation. The scoring system allocated a score of 1-5, from one indicating virtually no fat reserves to five indicating excessive fat (ADAS, 1978). Average body condition scores and days in milk (DIM) from 14 farms in the UK were useable and compared in a preliminary analysis using paired t-tests to examine the effect of installation of the AMS on body condition.

Fertility data from six farms were taken from three different milk-recording services depending on the farm. The preliminary analysis compared, using paired t-tests, 12-monthly rolling averages from 12 months prior to installation to 12 months after installation, to examine any effect of AMS and to allow seasonal effects. Fertility determinants, services to conception and conception rate, were compared at one month prior compared with one month after and 12 months after installation of the AMS. Days to conception was compared for one month prior with three months after and 12 months after installation. Calving index was compared for one month prior with 12 months after installation. Any correlation between body condition score and fertility was examined for farms with full data.

Daily production was compared for up to 12 months before and 12 months after installation to determine if changes might affect fertility. This was compared statistically using paired t-tests and a Mann-Whitney test, excluding the transitory period of one month either side of installation.

Results

Body condition
No significant difference was found between the overall average body condition score of cows for all 14 farms from before AMS installation (mean±SEM = 3.32±0.06) to after installation (mean±SEM = 3.34± 0.08). On some seven farms the average condition score increased after installation whilst on the other seven farms it decreased although none were statistically significant changes. Three general trends were seen between the farms.

Trend 1 (Figure 1) was found for three herds. Condition score was rising and then dropped sharply after installation, gradually increasing again for the 12 months after installation. When season was considered little or no difference was found in condition, this pattern describes a herd with a synchronised calving pattern and shows no effect of AMS.

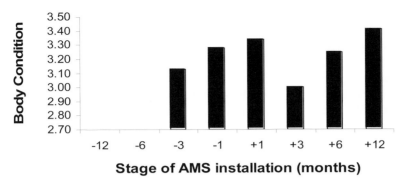

Figure 1. Average body condition scores on farm 1 (Trend 1).

Trend 2 was found for four herds. Condition score was rising until installation followed by a plateau or slight decrease for the 12 months after installation. Again time of year had little or no effect on condition of the cows.

Trend 3 was found for five herds. Condition score fluctuated throughout the study within half a score on each farm but overall no changes were seen. Season showed no effect.

Stage of lactation
The effect of days in milk of cows on body condition score was further examined to determine any effect of AM. No significant effects were found with DIM (before installation mean±SEM = 197±10 and after installation mean±SEM = 182±14).

Breed
Most of the farms milked only Holstein-Friesian cows but two herds milked only Jerseys and one was a mixture of Brown Swiss and Holstein-Fresian cows. No obvious difference in body condition score between breeds was found.

Fertility
For services to conception and conception rate no differences were found to be significant although small variations occurred for both parameters at one month post installation. Both also suggested fertility to have decreased a little by 12 months after installation (Table 1).

Table 1. Average fertility performance for the six farms installing AMS.

	Time relative to AMS installation (months)							
	-12	-6	-3	-1	+1	+3	+6	+12
Services/ conception	1.75	1.73	1.74	1.77	1.86	1.70	1.77	1.81
Conception rate (%)	59.0	57.3	52.0	51.8	48.7	55.8	51.2	52.6
Days to conception	121	132	135	133	135	151	140	144
Calving interval (d)	393	403	405	405	404	415	413	416

No significant differences were found for days to conception and calving index although overall days to conception was slightly longer after installation of the AMS and calving index was also lengthening (Table 1).

Daily production

No significant difference was seen between daily production immediately before installation (mean±SEM = 24.7±1.0 kg/day) and after installation (mean±SEM = 25.6±1.2 kg/day) and although an increase in milk produced after installation was seen, there was also a trend whereby the volume decreased up until installation. No significant increase in daily production yield was seen (P=0.08) when accounting for any changes during the transition to the new AM system.

Body condition and fertility

No correlations were found between body condition and fertility.

Discussion

In this UK study, body condition was not affected significantly by the installation and subsequent use of an AMS. Some farms did experience a marginal increase in body condition score for a short period after installation. This was a period when milk yield did not respond to AM, when there was general disruption as cows and staff, were training to the AMS and the nature and organisation of labour were altered (Meijering *et al.*, 2002). Any of a number of factors may have affected food intake and milk production including an increase in concentrates given in the box to facilitate training (Koning and Vorst, 2002). On these farms body condition score generally reduced to the pre installation level once the AMS routine was established, indicating that training and management, including that of feeding, posed no obvious problems.

No difference in body condition score was observed between Holstein-Fresian, Jersey or Brown Swiss cows. Body condition scores for Jerseys or Swiss should be higher than for Holstein-Friesians (Schwager-Suter *et al.*, 2000; Ward, 2003) and they should maintain the same condition when dry rather than gaining condition (Koenen *et al.*, 2001; Ward, 2003), however the difference between body condition during lactation and throughout the dry period was not examined in our study. No reason for this lack of difference was discerned.

On all of the farms with fertility data a small increase in days to conception occurred resulting in a later increase in calving index. On all farms the breeding cows in the period immediately after installation of the AMS required a slight increase in services to achieve conception resulting in a slight decrease in conception rate. Some recovery of the conception rate occurred but it did not reach the levels of six and 12 months prior to installation. Compared to conventional dairy farms a rate of 52% is poor when the aim is to exceed 60%. A slight decline in fertility is symptomatic of UK dairying in general but the 12 months pre installation data are typical of the average for the UK e.g. 121 days to conception (Barrett & Boyd, 2003). The increase of more than 20 days in this parameter appears at a faster rate than would be expected and suggests that further study is required. This could include more data from these UK herds and the herds studied in The Netherlands. A previous Dutch study (Kruip *et al.*, 2002) concluded that there were no negative effects from AM on fertility but the data considered were limited to the non return rate at 56 days and days to first service. No indication of effect with time after installation of the AMS was given.

The slight impairments in fertility were not significant and, therefore, not proven to be a problem as a result of AM. The variations in fertility could be consequences of any of a number of management alterations, active or passive, occurring around installation including time to observe oestrus, changes in routine for and opportunity for cows to display oestrus and dedication of staff time for breeding.

An initial aim was to consider if any changes in body condition during the change from conventional milking to AM might exacerbate any problems with fertility. Body condition is known to affect fertility. Cows that dry off and calve in lean condition tend to eat more and lose less condition than cows calving in fatter condition (Ward, 2003), and they have a lower risk of disease and a higher fertility than cows calving fat. Pryce *et al.*, (2001) reported that body condition score could be used as management and selection tools to improve reproductive performance. Cows with a low body condition score also have a longer calving index, exacerbated by high levels of milk production (Pryce *et al.*, 2002). However, due to the lack of significant changes in body condition score or fertility and only a small improvement in daily yield, no inferences can be drawn to link the two in AM.

In our study no effects on body condition score were seen suggesting that maybe there were other factors affecting fertility. Daily production for the study period was compared to determine if sufficient changes might explain the small perturbations in fertility rather than management reasons. There was no significant difference, although a trend was seen whereby yield increased from the preceding 12 month period (mean 24.7kg/day) to the post installation 12 month period (mean 25.6kg/day). On re-assessing the results without the immediate transitory period, avoiding effects such as training, stress etc., a significant difference in the daily yield was still not found although the increase was approaching significance (P=0.08). These results contrast with those from The Netherlands where there was a significant decrease (P=0.03) in daily production from before installation (26.6kg/day) to after installation (24.0kg/day) (Poelarends *et al.*, 2004).

Our fertility study was only on a few of the farms, in comparison to the body condition scoring, and this may account for only slight changes seen in fertility. For the full consequence of any changes in fertility to be studied, given that they may take more than 9 months to occur, a longer trial period may be required. This would allow other factors such as changes in management, body condition and any production increases, as may be expected from more frequent milking in an AMS, to have stablised (Kruip *et al.*, 2002).

Conclusions

This study suggests that, overall, cow management in the AMS herds was good as both body condition and fertility of the herd were relatively similar. Short term variations occurred but no particular problems were identified in any herd. Some evidence in the trends suggests that fertility could worsen in the longer term but a larger and longer study will be required to determine if this is true and identify risk factors other than simply AM. So far, the results suggest that changes in body condition or fertility on introduction of an AMS are not likely to pose a problem to the health of the dairy cows.

Acknowledgements

The authors would like to thank the farmers for their co-operation, and John Dale for his database construction and management. This study was performed within the EU project Implications of the introduction of automatic milking on dairy farms (QLK5 2000-31006) as part of the EU-program ' Quality of Life and Management of Living Resources'. The contents of this publication are the sole responsibility of the authors, and do not necessarily represent the views of the European Commission. Neither the European Commission not any person acting on behalf of the European Commission is responsible for the use of the information.

References

ADAS, 1978. Condition scoring of dairy cows. P612. (Leaflet no longer on sale).

Barrett, D.C. and H. Boyd, 2003. Herd fertility management (b) dairy herds Chapter 41 In: A.H. Andrews, R.W. Blowey, H. Boyd and R.G. Eddy (editors), Bovine Medicine 2nd edition, Blackwell Science Ltd, Oxford.

Chamberlain, A.T., 2003a. Feeding the fresh calver. Cattle Practice 11: 69-74.

Chamberlain, A.T., 2003b. Dry cow feeding. Cattle Practice 11: 93-99.

Koenen, E.P.C., R.F. Veerkamp, P. Dobbelaar, and G. De Jong, 2001. Genetic analysis of body condition score of lactating Dutch Holstein and Red-and-White heifers. Journal of Dairy Science 84: 1265-1270.

de Koning, C.J.A.M. and Y. v. d. Vorst, 2002. Automatic milking-changes and chances. Proceedings of the 15th.British Mastitis Conference, Gloucester, Institute for Animal Health/Milk Development Council, p68-80.

Kruip, T.A.M., H. Morice, M. Robert and W. Ouweltjes, 2002. Robotic milking and its effect on fertility and cell counts. Journal of Dairy Science 85: 2576-2581.

Meijering, A., Y. v. d. Vorst and C.J.A.M. de Koning, 2002. Implications of the introduction of automatic milking on dairy farms, an extended integrated EU project (www.automaticmilking.nl). Proceedings from the first North American conference on Robotic milking, Plenary I-29, March 20-22, Toronto, Canada. Wageningen Academic Publishers, Wageningen, The Netherlands.

Poelarends, J.J., O.C. Sampimon, F. Neijenhuis, J.D.H.M. Miltenburg, J.E. Hillerton, J. Dearing and C. Fossing, 2004. Cow factors related to the increase of somatic cell count after introduction of automatic milking. In Automatic milking; A better understanding. Editors: A. Meijering, H. Hogeveen and C.J.A.M. de Koning, Wageningen Academic Publishers, Wageningen, The Netherlands.

Pryce, J.E., M.P. Coffey & G. Simm, 2001. The relationship between body condition score and reproductive performance. Journal of Dairy Science 84: 1508-1515.

Pryce, J.E., M.P. Coffey, S.H. Brotherstone & J.A. Woolliams, 2002. Genetic relationship between calving interval and body condition score conditional on milk yield. Journal of Dairy Science 85: 1590-1595.

Pushpakumara, P.G.A., N.H. Gardner, C.K. Reynolds, D.E. Beever and D.C. Wathes, 2003. Relationships between transition period diet, metabolic parameters and fertility in lactating dairy cows. Theriogenology 60: 1165-1185.

Schwager-Suter, R., C. Stricker, D. Erdin and N. Kunzi, 2000. Relatioship between body condition scores and ultrasound measurements of subcutaneous fat and *m. longissimus dorsi* in dairy cows differing in size and type. Animal Science 71: 465-470.

Ward, W.R., 2003. Body condition scoring - technique and application. Cattle Practice 11: 111-115.

CHANGES IN TEAT CONDITION IN DUTCH HERDS CONVERTING FROM CONVENTIONAL TO AUTOMATED MILKING

F. Neijenhuis[1], K. Bos[1], O.C. Sampimon[2], J. Poelarends[1], J.E. Hillerton[3], C. Fossing[4] & J. Dearing[3]

[1]Applied Research, Animal Sciences Group, Wageningen UR, Lelystad, The Netherlands
[2]Animal Health Service, Deventer, The Netherlands
[3]Institute for Animal Health, Compton, United Kingdom
[4]Institute for Animal Sciences, Foulum, Denmark

Abstract

Teat condition was studied on 15 farms changing from conventional to automatic milking. Data from the farm visits before, one to two months after installation, and a year after installation of the automatic milking system were used. Teat end callosity decreased, the amount of rough callosity was less with automatic milking especially on front teats. Short-term changes in teat condition, like swelling and wedged teats, were seen less frequent with automatic milking. The conclusion is that the overall teat condition improved after changing to an automatic milking system.

Introduction

As part of the EU project on Automatic Milking a study was carried out in The Netherlands, United Kingdom and Denmark on "Health of dairy cows milked by an automatic milking system". In each country 15 farms were recruited that were changing from conventional to automatic milking General aspects of herd health were studied on these farms from a few months before until one year after installation of the automatic milking system (AMS). More detailed attention was paid to udder health, teat condition, lameness, body condition and fertility.

Review of the literature on conventional milking and appraisal of AMS suggests that automated milking could result in good teat condition but that not all hazards may be known and that the level of any risk will only be reduced with proper operation and good management (Neijenhuis and Hillerton, 2003). It appears important to monitor teat condition in the transition from conventional to automated milking in any herd under conversion as many other changes in operation and management will occur by default or by design. The objective of the present study is to describe the changes in teat condition in fifteen Dutch farms changing from conventional to automatic milking.

Material and methods

Farms

Companies selling automatic milking systems (AMS) in The Netherlands were asked to provide addresses of farms who decided to install an AMS. Farmers were asked if they were willing to cooperate in this research. Installation date of the AMS was between December 2001 and June 2002. The fifteen farms selected had 5 different brands of AMS. The farms were visited to collect data on udder health, lameness, body condition, fertility and management.

Teat condition

During three visits teat condition was studied, just before changing to automated milking, then two months and one year after changing to automated milking. On five of the farms teat condition was studied more intensively with an extra 5 visits, one every two to three months, resulting in two visits before and six after changing to automated milking. On the intensively studied farms, photographs of the teats were taken just after milking. On the 10 farms, which were visited three times to assess teat condition, teats were scored directly after milking. On average 33 randomly selected cows were scored for teat condition at each visit.

Teat condition was scored on presence of warts (yes/no), gross damage (yes/no), infections (yes/no), teat skin (normal, dry, open lesions), and the more short term machine related condition in terms of vascular damage (yes, no), wedged teats (yes/no), open teat canals (yes/no), teat base swelling (yes/no), teat end swelling (yes/no), and teat colour (normal/red or blue). Long term changes on teat end condition, teat end callosity (TEC), was scored in terms of TEC thickness (TECT) and TEC roughness (TECR) using a standardized classification system. In this system, TECR is scored as smooth / no callosity ring, or rough callosity ring. TECT is scored as no callosity ring (N), thin (A), moderate thick (B), thick (C) to an extreme thick callosity ring (D). TEC is a combination of both TECR and TECT, e.g. TEC of 1B is a smooth callosity ring of moderate thickness and TEC of 2C is a rough and thick callosity ring (Neijenhuis *et al.*, 2000).

Data analysis

Data from the 3 visits of all 15 farms was used to examine the changes in teat condition from just before the AMS installation compared to two months and a year after installation (2908,1925, and 1680 scored teats respectively). All the models analysed differences in data with respect to two scoring dates (before versus directly after or one year after installation of an AMS), and a complete random structure of the data was taken into account: scoring date within quarter within cow within farm.

TECT was tested using random regression model using the REML method ($P<0.05$, Genstat 6, version 6.1.0.200, 2002). Data on TECR are binomial and were tested using generalized linear mixed regression with link logit using the procedure GLMM ($P<0.05$, Genstat 6, version 6.1.0.200, 2002). The effects of installation of AMS (BA = before or after), teat position (TP = front or rear), parity (PAR = first, second, third, fourth or older parity cows), and days in milk (DIM = 0-60, 61-120, 121-180, 181-240, 241-300, or >300 days in milk) was tested. Interactions between BA with TP, PAR or DIM were tested.

Differences before versus after installation in other teat condition parameters were tested using Chi-square statistics (P<0.05).

Results

General descriptive results

In total 3462 quarter observations in the conventional milking parlour were made and 5599 observations after change-over to automated milking. The results showed on general a decrease in thickness and roughness of the callosity ring after change over to automated milking (TECT 2.65 to 2.51 and TECR 19% to 15%). A decline in the proportion of teats in TEC categories 2C and 2D and an increase in teat scored in category 1A is obvious (Figure 1).

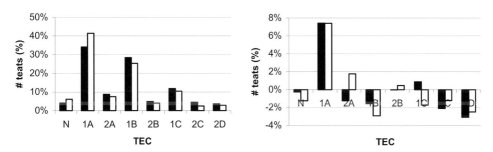

Figure 1. Teat end callosity (TEC) before AM installation for front (■) and rear (□) quarters (left) and the differences in percentages of teat end callosity categories from before to after installation of the AMS (right).

Model of callosity

The amount of TECT found was lower just after installation of the AMS compared to during conventional milking especially for mid lactation cows (Figure 2). However, the difference in TECT before and a year after installation of the AM was less. Front teats had more TECT than rear teats in both milking systems (Figure 3).

After installation of the AMS TECR was lower, especially in cows at the end of lactation (Figure 2). Before AMS milking, front teats had more TECR than rear teats. This difference in TECR between front and rear teats was not found with AMS milking (Figure 3).

Not all farms showed the same trend in decreasing TECT and TECR after installation of the AMS (Figure 4).

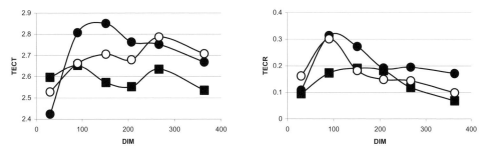

Figure 2. TECR and TECT before (●), just after (■) and a year after installation (○) of the AMS (fitted values).

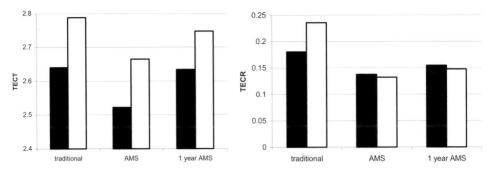

Figure 3. TECR and TECT for front (□) and rear (■) quarters before AMS, directly after and one year after AMS installation as calculated by the model (fitted values).

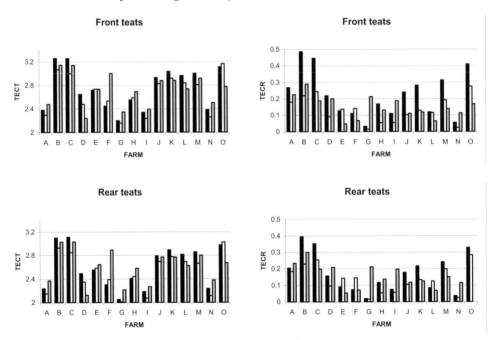

Figure 4. Fitted values per farm and teat position for front and rear teats before AMS (black), directly after (white) and one year after (grey) AMS installation as calculated by the model (fitted values).

Other teat condition parameters

Few teats were seen with short term machine related teat condition problems (Table 1). The swelling at teat base and teat end decreased after installation of the AMS. Teats were less often red after installation, however the percentage blue teats increased. This was also found when cows calved within 14 days were excluded. Percentage wedged teat ends decreased directly after installation, but this effect was not significant one year after. Teats with open teat canals was decreased a year after installation. No differences in vascular damage was found when milking changed from conventional to AMS.

Percentage of teats with dryness of the skin decreased directly after installation. Open lesions increased directly after installation but this effect was disappeared one year after

Automatic milking – A better understanding

installation. The percentage of teats with warts was more than 20% in conventional milking and decreased after installation of AMS. There is no evidence that the percentage of teats with gross damage or infections are different before or after installation of AMS.

Table 1. Percentages of teats differing from normal before, just after and one year after installation of AMS (* is significant different from before installation with P< 0.05).

Teat parameter		Conventional	Directly after AMS	One year after AMS
Swelling teat base		2	1.2*	0.6*
Swelling teat end		2.8	0.5*	0*
Teat colour	Red	3	1.7*	1.2*
(ref. Normal)	Blue	0.1	0.8*	1.8*
Wedged teats		2.5	1.4*	1.8
Openness teat canal		2.8	2.5	1.1*
Vascular damage		1.3	0.9	1.2
Teat skin	Dryness	0.6	0.1*	0.2
(ref. Normal)	Open lesions	0.1	0.4*	0
Warts		21.5	14.3*	14.3*
Gross damage		0.8	0.8	1.0
Infections		0.2	0.3	0

Discussion

The amount of teat end callosity decreased with AM. A higher proportion of teats only had a smooth thin callosity ring. Likewise, Berglund et al. (2000) found that teat condition, a parameter described by them as extended teat canals, improved after the move from a herringbone parlour to an AM. The most obvious improvement was the decrease in rough callosity of the front teats after installation of the AMS. The reason for less TEC may be a shorter machine-on time with quarter milking especially on front quarters that appear to be over milked in whole udder milking (Berglund et al., 2000). Moreover, the one farm (F in figure 4) that showed the most increase in TECT on front quarters after installing an AMS practised whole udder take-off. On the 15 farms, milk yield did not increase after installation of AMS (Poelarends et al., 2004) which may also have led to shorter machine-on time and better teat condition. In the period before installation (200 days and less) of the AMS an increase in SCC and a decrease in milk yield was found (Poelarends et al., 2004). Reasons for this are unknown but the farmers all had a significant increase in workload particularly because of building activities around the milking parlour and on many farms a milking machine with problems.

According to the Teat Club International (Mein et al., 2001) a herd should have fewer than 20% of the teats with roughened rings extending 1-3 mm from the orifice (R = category 1C & 2B) and fewer than 10% of the teats with rings extending more than 4 mm (VR = 2C&2D). With conventional milking, 7 farms were above this critical line. This number decreased to 6 and 5 respectively immediately and a year after installing the AMS. The influence of AM might be slight but encouraging especially as the degree of genetic influence on this condition is unknown.

The number of teats with short-term condition changes because of milking in terms of vascular damage, wedged teats, open teat canals, teat base and end swelling, and teat colour decreased from 13% of the teats to 9% after installation of the AMS and 7% one year after installation. Milking with an AMS may adversely affect teat condition less than conventional milking. Less redness of the skin with AMS milked cows was also found by Berglund *et al.* (2002). Fewer warts were found after installation of the AMS probably because of better teat hygiene.

Dryness of teat skin with conventional milking was low and decreased after installation of AMS, which is in contrast to finding of Berglund *et al.* (2000). It is known that the dryness of the skin can be managed in automated milking (Hovinen & Pyörälä, 2002; Rasmussen *et al.*, 2002), and kept low by use of a post milking disinfectant with an appropriate amount of emollient.

Conclusions

Teat condition improved when milking changed from conventional to AMS.

Acknowledgements

This study is performed within EU project Implications of the introduction of automatic milking on dairy farms (QLK5 2000-31006) as part of the EU-program ' Quality of Life and Management of Living Resources'. The content of this publication is the sole responsibility of the authors, and does not necessarily represent the views of the European Commission. Neither the European Commission not any person acting on behalf of the European Commission is responsible for the use of the following information.

The authors gratefully thank the farmers for their cooperation, without whom the data could not have been collected. Reina Ferwerda, Hans Miltenburg, Hans van den Heuvel and several students are acknowledged for their assistance during the study.

References

Berglund, I., G. Pettersson and K. Svennersten-Sjaunja. 2002. Automatic milking: effects on somatic cell count and teat end-quality. Livestock Production Science 78: 115-124.

Genstat Release 6.1. 2002. Lawes Agricultural Trust. IACR-Rothamsted, Harpenden, Hertfordshire, UK.

Hovinen, M. and S. Pyörälä. 2002. Observations on udder health of automatically milked cows in Finland. First North American Conference on Automatic Milking, Wageningen Pers, The Netherlands, IV: 71-74.

Mein, G.A., F. Neijenhuis, W.F. Morgan, D.J. Reinemann, J.E. Hillerton, J.R. Baines, I. Ohnstad, M.D. Rasmussen, L. Timms, J.S. Britt, R. Farnsworth, N. Cook and T. Hemling (2001) Evaluation of bovine teat condition in commercial dairy herds: 1. Non-infectious factors. In: Proceedings of the 2nd International Symposium on Mastitis and Milk Quality, NMC/AABP, Vancouver: 374-351.

Neijenhuis, F. & J.E. Hillerton. 2003. Health of dairy cows milked by an automatic milking system : Review of potential effects of automatic milking conditions on the teat. Deliverable D21. EU project Implications of the introduction of automatic milking on dairy farms (QLK5 -2000-31006) as part of the EU-program 'Quality of Life and Management of Living resources': 24 pp.

Poelarends, J.J., O.C. Sampimon, F. Neijenhuis, J.D.H.M. Miltenburg, E. Hillerton, J. Dearing and C. Fossing· 2004. Cow factors related to the increase of somatic cell count after introduction of automatic milking. In: Proceeding International Symposium Automatic Milking, a better understanding. Lelystad, The Netherlands.

Rasmussen, M.D., L. Foldager and T.C. Hemling (2002) Teat conditioning, udder health and frequency of milking: impact of teat dip composition. In: Proceeding of the DeLaval Hygiene Technology Centre Inaugural Symposium, Kansas City MO: 16-23.

Svennersten-Sjauna, K., I. Berglund and G. Petterson. 2000. The milking process in an automatic milking system, evaluation of milk yield, teat condition and udder health. In: Robotic milking. Eds. H. Hogeveen & A. Meijering, Wageningen Pers, Wageningen: 277-288.

COW FACTORS RELATED TO THE INCREASE OF SOMATIC CELL COUNT AFTER INTRODUCTION OF AUTOMATIC MILKING.

J.J. Poelarends[1], O.C. Sampimon[2], F. Neijenhuis[1], J.D.H.M. Miltenburg[2], J.E. Hillerton[3], J. Dearing[3] & C. Fossing[4]
[1]Applied Research, Animal Sciences Group, Wageningen UR in Lelystad, The Netherlands
[2]Animal Health Service, Deventer, The Netherlands
[3]Institute for Animal Health, Compton, United Kingdom
[4]Danish Institute for Agricultural Science, Foulum, Denmark

Abstract

Cow somatic cell count (SCC) was studied on 15 farms changing from conventional to automatic milking. SCC was significantly higher and milk yield was significantly lower after installation of the AM-system. Fresh cows (<60 days in milk) and second and third parity cows particularly showed a significant increase in SCC. Milk yield decrease was shown in fresh cows (<60 days in milk), late lactation cows (>300 days in milk) and in cows of second, third and higher than fourth parity. It is concluded that fresh cows should receive more attention when changing to automatic milking.

Introduction

As part of the EU project on Automatic Milking a study was carried out in The Netherlands, United Kingdom and Denmark on "Health of dairy cows milked by an automatic milking system". In each country 15 farms were recruited that were changing from conventional to automatic milking (AM). These farms were studied intensively. General aspects of herd health were studied on these farms from a few months before until one year after installation of the automatic milking system (AM-system). Attention was paid to udder health, teat condition, lameness, body condition and fertility. In The Netherlands additional attention was paid to individual cow somatic cell counts (SCC).

From other studies it is known that the bulk milk somatic cell count (BMSCC) increases on farms after installation of an AM-system (Rasmussen *et al.*, 2002; Vorst & Hogeveen, 2000). Cow SCC has also been found to increase on farms that installed an AM-system (Kruip *et al.*, 2002; Rasmussen *et al.*, 2001). Rasmussen *et al.* (2001) found a peak in SCC at about two months after installation and also concluded that the mean cow SCC in the year after installation was higher compared to the year before. However, it is not described in either study which specific cows had an increase in SCC. It is important to know if particular cows should receive extra attention after changing over to AM. Therefore, the objective of the present study is to characterise the cows that show an increase in SCC after the introduction of AM.

Material and methods

Data

In The Netherlands 15 farms that planned to install an AM-system were asked if they wanted to cooperate in the study. Before installation of the AM-system they were visited once or twice and after installation they were visited at least four times to collect data. Most farms were participants in monthly milk recording schemes including cow SCC measurement. Milk recording data were retrieved for cows that were present on the farm at the time of installation. These data were retrieved electronically from the DHIA from about one year before installation until one year after installation. Data consisted of daily milk yield, cow SCC, days in milk and parity.

Statistical analysis

Before statistical analysis raw data were examined to look at cow SCC patterns before and after installation. It appeared that there was a peak in cow SCC around 2 to 3 months after installation. Based on this it was decided to analyse the SCC records from the last milk recording before installation and the second milk recording after installation (between 50 and 100 days after installation, from two farms the third milk recording was used). SCC was log transformed in order to achieve a normal distribution of the data.

The effects of installation of the AM-system, days in milk and parity on LOG(SCC) were examined using the residual maximum likelihood method (REML) (Genstat 6, version 6.1.0.200, 2002). The model to analyse LOG(SCC) used was:

$$LOG(SCC) = \mu + herd + herd(cow) + herd(cow(BA)) + BA + DIM + PAR + (BA*DIM*PAR) + e$$

Where:

μ = overall mean; herd = random effect of herd; herd(cows) = random effect of cows within herd; herd(cow(BA)) = random effect of installation of AM-system (before / after) within cows within herd; BA = fixed effect of installation of AM-system (before / after); DIM = fixed effect of days in milk; PAR = fixed effect of parity; (BA*DIM*PAR) = fixed effect of the interactions between BA, DIM and PAR; e = residual random error. Daily milk yield was analysed with the same model.

Results

The number of cows per farm varied from 41 to 106, on average 62 cows per farm. On average the cows were 201 days in milk before and 176 days in milk after installation. Parity of cows averaged 2.5 lactations.

Descriptive statistics for cell count

For every cow the difference between the two SCC values (last milk recording before AM and the second milk recording after) was calculated. Some 59% of the cows had an increase in SCC, with 24% the increase was less than 50.000 cells/ml.

The percentage of cows with a SCC above a critical threshold (250.000 cells/ml for cows and 150.000 cells/ml for heifers) was examined for every period of 50 days in the year before and after start of AM. All available milk recordings with SCC data were used, approximately

20.500. The results are presented in figure 1. It appears that after installation of the AM-system an increase in the percentage of cows with high SCC occurred, with a peak during the period 50-150 days after installation. It can also be seen that the percentage of high SCC cows was already increasing before installation. On average the percentage of high SCC cows during the year before installation was 21.6% and during the year after installation it was 26.8%.

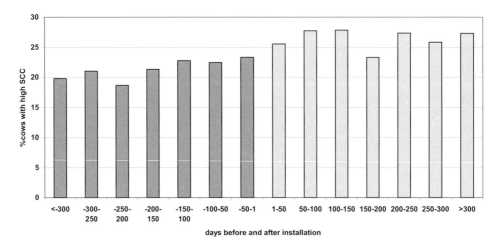

Figure 1. Percentage of cows with SCC above the critical threshold (150.000 cells/ml for heifers and 250.000 cells/ml for cows).

Descriptive statistics of daily milk yield

For every period of 50 days in the year before and after the start of AM the mean daily milk yield of all cows was calculated (in total 22.000 milk recordings, on average 1558 records per 50 day period) and are presented in figure 2. The mean daily milk yield in the year before start of AM was 26.6 kg/day and the year after it was 24.0 kg/day. It appears that milk yield was already decreasing in the period before installation.

Statistical analysis SCC

For the statistical analysis of SCC some 1659 milk recording records were used from the last milk recording before and the second milk recording after installation of the AM-system. It appeared that there was a significant effect of AM-system installation (BA) on cow SCC (P<0,001). This effect varied with parity and stage of lactation of the cows, because the interactions BA*PAR and BA*DIM were significant (P=0,05 and P=0,002 respectively) (Table 1 and 2).

The estimated LOG(SCC) for all cows before installation was 4.71 (last milk recording before installation) and after installation it was estimated to be 4.85 (2nd milk recording after installation), this difference was significant (P<0,05). The biggest effect occurred with the newly calved cows. Cows <60 days in milk in the period before AM had an estimated LOG(SCC) of 4.18 and the cows <60 days in milk milked automatically had an estimated LOG(SCC) of 4.82. This difference is statistically significant. These are not the same cows. The group of cows that was 60-120 days in milk before AM had an estimated LOG(SCC) of

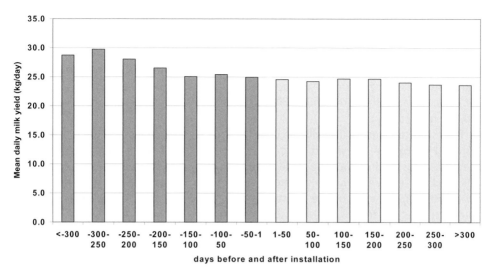

Figure 2. Mean daily milk yield for all cows in the year before and after installation of the AM-system.

4.17 and the analogous group after AM installation had an estimated LOG(SCC) of 4.24 (difference not significant). No significant differences have been found for the other stages of lactation (Table 1).

The effect of AM-system installation on cow SCC also varied according to the parity of the cows. table 2 shows the estimated LOG(SCC) for the different parities before and after installation of the AM-system. Some cows changed parity because they calved between the two milk recordings used. The results show that the cows in second and third parity had a significant increase in estimated LOG(SCC). Heifers and older cows showed no significant change in SCC.

Table 1. Estimated LOG(SCC) for different DIM groups before and after installation of an AM-system.

	<60 DIM	60-120	120-180	180-240	240-300	>300
Before	**4.18[a]**	4.17	4.71	4.81	5.15	5.22
After	**4.82[b]**	4.42	4.72	4.93	4.91	5.32

[a,b] different superscripts within a column represent significant differences ($P<0.05$).

Table 2. Estimated LOG(SCC) for different parities before and after installation of an AM-system.

	1st parity	2nd parity	3rd parity	4th parity	≥5th parity
Before	4.33	**4.45[a]**	**4.70[a]**	5.02	5.04
After	4.37	**4.74[b]**	**5.04[b]**	4.92	5.20

[a,b] different superscripts within a column represent significant differences ($P<0.05$).

Statistical analysis of daily milk yield

For the statistical analysis of daily milk yield 1674 milk recordings were used from the last milk recording before and the second milk recording after installation of the AM-system. It appeared that there was a significant effect of AM-system installation (BA) on daily milk yield (P=0,03). This effect also significantly depended on parity and stage of lactation of the cows, because the interactions BA*DIM and BA*PAR were significant (P<0,001 and P=0,014 respectively).

The estimated mean daily milk yield during the last milk recording before the start of AM was 25.7 kg/day and was 24.3 kg/day after the start of AM, a significant drop. The effect of AM-system installation also depended on DIM and parity. The estimated mean daily milk yields for every DIM group are presented in table 3. It appears that recently calved cows (<60 DIM) and cows in late lactation (>300 DIM) had a significantly lower daily milk yield after AM compared to the same DIM group before (but other cows).

Table 3. The estimated mean daily milk yields for every DIM group before and after AM.

	<60 DIM	60-120	120-180	180-240	240-300	>300
Before	34.1[a]	32.5	27.6	24.2	19.7	16.3[a]
After	30.3[b]	31.4	27.4	23.1	20.1	13.6[b]

[a,b] different superscripts within a column represent significant differences (P<0.05).

The estimated mean daily milk yields per parity before and after AM are presented in table 4. The results show that heifers and fourth parity cows did not produce less after installation. Cows of parity 2, 3 and ≥5 had significantly lower milk yields after the start of AM.

Table 4. The estimated mean daily milk yield for every parity before and after the start of AM.

	1st parity	2nd parity	3rd parity	4th parity	≥5th parity
Before	22.0	25.7[a]	27.2[a]	27.7	26.1[a]
After	21.6	23.8[b]	25.8[b]	26.9	23.3[b]

[a,b] different superscripts within a column represent significant differences (P<0.05).

Discussion

Overall an increase in cow SCC and a decrease in milk yield occurred after start of automated milking. However, in an experimental study of Berglund et al. (2002) an increase in milk yield and a decrease in quarter SCC was found when AM was compared to conventional milking under the same circumstances in the same herd. The remarkable difference between experimental studies and observational studies in practice like Rasmussen et al. (2001) and the present study indicate that introduction of AM on a farm implicates more than just a change in milking technique. The management of the farm is completely changed and the farmer has to get used to the system, the attention lists and monitoring his cows, that also have to get used to the new routine.

Figure 1 gives an indication of the increase in cow SCC after installation. Rasmussen *et al.* (2001) also found an increase in cow SCC after the start of AM. However, in the present study it appeared that before installation of AM the percentage of infected cows (heifers > 150.000 cells/ml and cows >250.000 cells/ml) was already increasing and the mean daily milk yield was already decreasing. The reasons for the increase in SCC and decrease in milk yield in the period before installation are unknown. However, during the farm visits it appeared that the farmers all had a significant increase in workload particularly because of building activities around the milking parlour and on many farms there was a milking machine with problems. The higher percentage of high cell count cows, may be indicative of more infection related to poorer management and/or a poor milking system.

In the present study the effect of parity was examined and it appeared that heifers did not seem to have problems with udder health and milk production after installation of the AM-system. Farmers participating in the study believed that cows that were used to conventional milking had more difficulties in getting used to the AM-system. Therefore, heifers should be the group with the fewest problems to adjust. The second and third parity cows showed an increase in SCC, but the older cows (parity ≥4) did not show significant changes in SCC. In addition, daily milk yield of the second and third parity cows after installation was significantly lower compared to before installation. In general it is expected that milk yield will increase with an AM-system because of the higher milking frequency. In the present study milk yield declined. Analysis of daily milk yield revealed that 28% of the variance in milk yield was explained by differences between farms, compared to only 5% of the variance in SCC. This indicates that farm management is very important to maintain milk yield. In addition milk yield may also decline because an increase in udder infections.

Analysis of the effect of DIM revealed that recently calved cows in particular (<60 DIM) had a higher SCC after the start of AM. The reason for this is not exactly clear. It may be that after installation farmers and cows are not yet used to the system or there may be some start-up problems with the system. For example, it is possible that in the beginning of AM the milking frequency and milking intervals are not optimal and cows are milked too late or milkings are incomplete (one or more quarters) due to start-up problems or due to farmers that have to get used to the AM-system and the attention lists. All these factors may be critical for the high producing fresh cows and give them more (physiological) stress.

In addition, daily milk yield analysis showed that fresh cows after installation produced less compared to the fresh cows before installation. This indicates that management of fresh cows after installation is not optimal. Therefore fresh cows should receive even more attention when changing to automatic milking.

Acknowledgements

The authors gratefully thank the farmers for their cooperation. Reina Ferwerda, Kees Bos, Hans van den Heuvel and several students are acknowledged for their assistance during the study.

This study is performed within EU project Implications of the introduction of automatic milking on dairy farms (QLK5 2000-31006) as part of the EU-program 'Quality of Life and

Management of Living Resources'. The content of this publication is the sole responsibility of the authors, and does not necessarily represent the views of the European Commission. Neither the European Commission not any person acting on behalf of the European Commission is responsible for the use of the following information.

References

Berglund, I., G. Petterson and K. Svennersten-Sjaunja. 2002. Automatic milking: effects on somatic cell count and teat-end quality. Livestock Production Science, 78, 115-124.

Genstat Release 6.1. 2002. Lawes Agricultural Trust. IACR-Rothamsted, Harpenden, Hertfordshire, UK.

Kruip, T.A.M., H. Morice, M. Robert and W. Ouweltjes. 2002. Robotic milking and its effect on fertility and cell counts. Journal of Dairy Science, 85, 2576-2581.

Rasmussen, M.D., J.Y. Blom, L.A.H. Nielsen and P. Justesen. 2001. Udder health of cows milked automatically. Livestock Production Science, 72, 147-156.

Rasmussen, M.D., M. Bjerring, P. Justesen and L. Jepsen. 2002. Milk quality on Danish farms with automatic milking systems. Journal of Dairy Science, 85, 2869-2878.

Vorst Y.v.d. and H. Hogeveen. 2000. Automatic milking systems and milk quality in The Netherlands. In: Hogeveen, H. and A. Meijering (eds.), Proceedings International Symposium Robotic Milking. Lelystad, The Netherlands. Wageningen Pers, pp. 73-82.

THE INFLUENCE OF AN AUTOMATIC MILKING SYSTEM ON CLAW HEALTH AND LAMENESS OF DAIRY COWS

B. Vosika[1], D. Lexer[1], Ch. Stanek[2], J. Troxler[1] & S. Waiblinger[1]
[1]Institute of Animal Husbandry and Animal Welfare, University of Veterinary Medicine Vienna, Vienna, Austria
[2]Clinic of Orthopaedics in Ungulates, University of Veterinary Medicine Vienna, Vienna, Austria

Abstract

Lameness and claw diseases in dairy farms became increasingly important over the recent years. Studies of possible effects of automatic milking systems (AMS) on claw lesions as well as on occurrence of lameness do not yet exist. In the present study we compared occurrence and pre-valence of claw lesions and lameness of cows milked in an automatic milking system (Astronout®, Lely Industries NV, NL; AMS-group) and cows milked in a 2x6 herringbone parlour (Happel Ltd., Germany ; HP-group). A herd of 60 cows split into the two trial groups (15 Austrian Simmental, 15 Brown Swiss cows each) was housed in the same building under identical conditions (loose housing system with cubicles and slatted floor), except the milking system. The time of data collection included periods with partially forced cow traffic as well as free cow traffic.

Locomotion scores (Manson and Leaver, 1988, modified) were recorded for 36 times over $1\frac{1}{2}$ years. Twice yearly the cows` claws were examined in connection with the regular claw trimming (in total 4 times). The condition of the claws was recorded using the scoring system of Boosman et al. (1989) modified.

In this study neither a positive nor a negative effect of an automatic milking system on claw health and lameness was found under the investigated conditions. However, the relatively high replacement percentage of cows due to various reasons might cover possible long-term effects of the milking system on claw health and lameness.

Introduction

Lameness in dairy cattle is a major cause of suffering as well as a serious economical loss (direct and indirect), and it is one of the most important health and welfare issues on today`s dairy farms. It is known, that lameness in cattle has a multifactorial causation (Bargai, 2002, Galindo et al., 2000). Among the various problems of the locomotor system in high yielding dairy cattle, diseases of the digit including the claws dominate. Foot lesions cause approximately 90% of cattle lameness (O`Callaghan et al., 2002).

Pain is a serious component of lameness. Galindo and Broom (2002) observed, that a lame cow is less able to cope successfully with her environment, as pain might seriously affect walking and other movements. Singh et al. (1993) found, that cows that stood for longer presented more haemorrhages in the sole horn, which could predispose them to lameness. Standing still in passageways has been related to low- ranking cows which are blocked in their movements and subsequently have to spent more time waiting for available space to move (Miller and Wood-Gush, 1991).

Robotic milking changes the daily life of dairy cows and may affect their welfare. The waiting time in front of the automatic milking system (AMS), combined with prolonged standing and reduced human supervision also in the region of the claws should be considered as possible negative factors.

Studies of possible effects of AMS on claw lesions as well as on occurrence of lameness do not yet exist. Therefore, the aim of this study was to compare occurrence and pre-valence of claw lesions and lameness of cows milked in an automatic milking system (Astronout®, Lely, NL; AMS-group) and cows milked in a conventional milking system (2x6 herringbone parlour, Happel, Germany ; HP-group).

Animals, materials and methods

Animals and housing

A herd of 60 cows, split into two groups (15 Austrian Simmental, 15 Brown Swiss cows each group), was housed in the same uninsulated building under identical conditions, with exception of the milking system. The animals were housed in separate loose-housing pens with 30 cubicles each group (cubicles: width 120 cm, length 185 cm; height of the floor 22.5 cm, soft rubber mattresses, small amounts of straw) and slatted floor. Food and water was given ad libitum. The AMS-group was milked in a single box robot (Astronout®, Lely Industries NV, NL). The HP-group was milked in a 2x6 herringbone parlour (Happel Ltd., Germany) twice a day. The average daily milk yield during the recording period was 20.8 l in the AMS-group and 21.9 l in the HP-group. The time of data collection included periods with partially forced cow traffic as well as free cow traffic.

Data collection and statistical analysis

Claw health was systematically checked at intervals of 6 month (in total four times) and recorded using the Boosman et al. (1989) scoring system, modified by Stanek (1994). An extra score was built for heel horn erosion. The condition of the claws was recorded at the beginning of the claw trimming procedure. The claw trimming was carried out on a tipping table.

Locomotion scores were recorded fortnightly for 36 times within 11/2 years. The lameness scoring system used was adapted from that described by Manson and Leaver (1988). Each cow was observed walking away from the observer and a score on a scale of 0 to 6, including half-scores, was given. Higher scores indicated poorer locomotion. The scoring was always carried out by the same person. In total, 97 animals including replacements were examined. During the recording period from March 2001 until October 2002 37 cows had to be replaced.

For further analysis the 12 different grades of lameness were aggregated to five new groups (0=0; 1=1-2; 2=2.5-3; 3=3.5-4.5; 4=5-6). Score groups of 0 and 1 were considered as no lameness (0) or uneven gait (1). Score groups from 2 to 4 were considered as clinically lame.

For the statistical analysis of the claw score data we considered only cows staying at least for two months in the trial groups. For the comparison of the trial groups with regard to lameness only cows with at least 18 consecutive recording dates were used. For each of these cows the percentage of lameness score groups over the observation period was used in the statistical analysis.

For group comparisons Mann-Whitney-U tests were used. All tests were performed at alpha=0.05 (two-tailed).

Results

The median claw scores of the AMS-group were 5, 6, 6 and 7, respectively. The median claw scores of the HP-group were 4, 8, 4 and 7 (Table 1). The groups did not differ significantly in claw scores at any of the four scoring dates.

The AMS-group showed a higher occurrence of heel horn erosions in the early phase of the observation period, whereas the HP-group had a higher frequency of erosions towards the end of the period.

At any time for both groups the percentage of clinically lame cows was lower than 30% (Figure 1 and 2). The groups did not differ significantly with regard to lameness (Table 2).

Table 1. Descriptive statistics of claw scores in AMS-group and HP-group at the 4 different scoring dates (Score 1 to 4); AMS-group: score 1: N = 29, score 2: N = 29, score 3: N = 28, score 4: N = 30; HP-group: score 1: N = 26, score 2: N = 30, score 3: N = 30, score 4: N = 29, (SD=standard deviation).

	AMS-group				HP-group			
	score 1	score 2	score 3	score 4	score 1	score 2	score 3	score 4
minimum	0	0	0	0	0	0	0	0
maximum	22	33	32	66	64	33	49	92
median	5	6	6	7	4	8	4	7
mean	6	8	7	10	10	9	9	13
SD	6	7	7	13	15	8	14	18

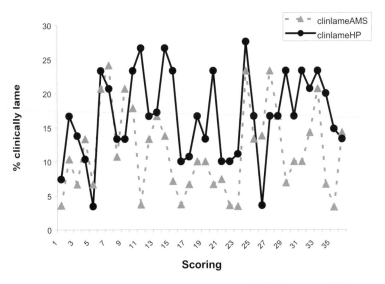

Figure 1. Comparison of clinically lame cows (lameness score groups 2-4) of AMS-group and HP-group over 36 recording terms from March 2001 to October 2002 (all cows present during the recording terms).

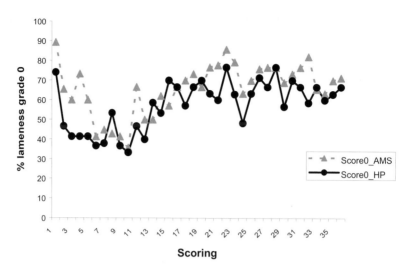

Figure 2. Comparison of cows with no lameness (lameness score group 0) of AMS-group and HP-group over 36 recording terms from March 2001 to October 2002 (all cows present during the recording terms).

Table 2. Comparison of AMS-group and HP-group with regard to score groups (mean ± standard deviation; Mann-Whitney-U tests, data: cows with at least 18 consecutive recording dates; AMS group: N=26, HP-group: N=30).

	score group 0 [%]	score group 1 [%]	score group 2 [%]	score group 3 [%]	score group 4 [%]
AMS-group	62 ± 26	22 ± 12	6 ± 8	3 ± 6	1 ± 2
HP-group	57 ± 27	25 ± 13	10 ± 12	6 ± 10	1 ± 2
P (MWU)	0.527	0.360	0.355	0.391	0.787

Discussion

In cubicle buildings, the level of competition for eating and lying places increases, and lame cows are less likely to be successful under these conditions (Wierenga and Metz, 1986). Lame cows are probably not strong enough to engage in attempts to displace another cow from a cubicle, a feeder or the AMS. Hypothesis of the presented study was, that the adaptation to the AMS system might cause situations and stress that induce a worsening of claw condition. This hypothesis was not confirmed by the present study.

At 28 out of 36 recording dates the percentage of clinically lame cows was higher in the HP-group compared to the AMS-group. However, this might have been caused by the higher replacement rate in the AMS-group (23 cows versus 14 cows in the HP-group) reducing the average residence time of a cow in the trial. Generally, the replacement rate was relatively high, but overall (both groups) only 2 cows had to be replaced due to claw lesions. A further reason for the result of our study might be that the single box robot used in this study is recommended by Lely for about 60 dairy cows. As there have been just 30 cows in the AMS-group, stress in order to cope with this system might have been not as high as in a fully occupied AMS and might not differ to the HP-system. There, one possible source of stress and cause of lameness is the regular handling by the stockperson (both moving to and interactions in the milking parlour). When cows are crowded together, and the stockman continues to push them, tussling can cause unplanned foot placement as cows are unable to see the ground with their heads above the backs of those in front (Chesterton *et al.*, 1989). Lame cows most often walked at the back of the herd and therefore were more affected by an impatient stockman in the investigation of Clackson and Ward (1991). Additional to this problem the waiting times in front of the 2x6 herringbone parlour could likewise lead to prolonged standing. As well, the crowded situation in the waiting area and the milking parlour might cause social stress and sudden, possible injurious movements.

To conclude, neither a positive nor a negative effect of an automatic milking system on claw health and lameness was found under the investigated conditions.

Acknowledgements

This study was part of a project in cooperation with the *Institute of Animal Husbandry and Animal Welfare of the University of Veterinary Medicine Vienna*, the *Landwirtschafliche Bundesversuchswirtschaften GmbH (BVW)* in *Wieselburg* and the *Österreichische Agentur für Gesundheit und Ernährungssicherheit GmbH Milchwirtschaft Wolfpassing*, and financed by the Austrian Federal Ministry of Agriculture, Forestry, Environment and Water Management (Grant 1206 sub). All experimental work was performed at the *BVW Wieselburg*, Austria.

References

Bargai, U., 2002. Herd examination for lameness - guidelines for the practitioner. In: 12th International Symposium on lameness in Ruminants, 9-13 Jan. 2002, Orlando, Florida: pp. 318-320.

Boosman, R., F. Nemeth, E. Gruys, A. Klarenbeek, 1989. Arterio-graphical and pathological changes in chronic laminitis in dairy cattle. The Veterinary Quarterly 11: 144-155.

Chesterton, R.N., D.U. Pfeiffer, R.S. Morris, C.M. Tanner, 1989. Environmental and behavioural factors affecting the prevalence of foot lameness in New Zealand dairy herds - a case-control study. New Zealand Veterinary Journal 37: 135-142.

Clackson, D.A., W.R. Ward, 1991. Farm tracks, stockman`s herding and lameness in dairy cattle. The Veterinary Record 129: 511-512.

Galindo, F., D.M. Broom, 2002. The effects of lameness on social and individual behavior of dairy cows. Applied Animal Welfare Science 5: 193-201.

Galindo, F., D.M. Broom, P.G.G. Jackson, 2000. A note on possible link between behaviour and the occurence of lameness in dairy cows. Applied Animal Behaviour Science 67: 335 -341.

Manson, F.J., J.D. Leaver, 1988. The influence of concentrate amount on locomotion and clinical lameness in dairy cattle. Animal Production 47: 185-190.

Miller, R., D.G.M. Wood-Gush, 1991. Some effects of housing on the social behaviour of dairy cows. Animal Production 53: 271-278.

O`Callaghan, K., R. Murray, P.J. Cripps, 2002. Behavioural indicators of pain associated with lameness in dairy cattle. In: 12th International Symposium on lameness in Ruminants, 9-13 Jan. 2002, Orlando, Florida: pp. 309-312.

Singh, S.S., W.R. Ward, R. Murray, 1993. Aetiology and pathogenesis of sole lesions causing lameness in cattle: a review. Veterinary Bulletin 63: 303-315.

Stanek, C., 1994. In: Brandejsky, F., C. Stanek, M. Schuh,1994. Zur Pathogenese der subklinischen Klauenrehe beim Milchrind: Untersuchungen von Klauenstatus, Pansenstatus und Blutgerinnungsfaktoren. Deutsche Tierärztliche Wochenschrift 101: 68-71.

Wierenga, H.K., J.H.M. Metz, 1986. Lying behaviour of dairy cows influenced by crowding. In: Nichelmann, M. (Ed.): Ethology of Domestic Animals. Privat I.E.C., Toulouse, pp. 61-66.

INTRODUCTION OF AMS IN ITALIAN DAIRY HERDS: EFFECTS ON TEAT TISSUES, INTRAMAMMARY INFECTION RISK, AND SPREAD OF CONTAGIOUS PATHOGENS

A. Zecconi[1], R. Piccinini[1], G. Casirani[1], E. Binda[1] & L. Migliorati[2]
[1]Dipartimento di Patologia Animale Igiene & Sanità Pubblica Veterinaria, Milano, Italy
[2]Istituto Sperimentale per la Zootecnia, Sez. Vacca da Latte, Cremona, Italy

Abstract

This paper reports the results of a follow-up study undertaken in 2 herds focused on the impact of AMS on teat tissues, teak skin and orifice, the monitoring the dynamic of intramammary infections, the relationship between milking parameters, milk biochemical parameters, cows and IMI. The introduction of AMS in herd free from contagious pathogens didn't influence the frequency of IMI and the SCC. In herd characterized by the presence of *Staph.aureus* IMI, the frequency of IMI showed to increase progressively, very likely as a consequence of the spread of infections during milking. The presence of a significant number of herds with *Staph.aureus* IMI in different countries emphasizes the importance to address this aspect of AMS milking. Teat thickness changes showed different patterns in the two herds, probably because of the different type of AMS. In general terms the decrease in thickness after milking was relatively small in herd A and in the range -2% and -9% in herd B, The peak/average ratio (PAR) is the ratio between the peak flow and the average flow and it highly influenced NAGase mainly in infected animals. These results suggest that milking of infected quarters could increase the level of udder inflammation probably by prolonged milking.

The results of this field trial confirm that AMS have no negative impact on IMI incidence, SCC and teat tissue conditions, when the initial cow health status and overall herd management are good. In presence of IMI, and when cows have more than 300 DIM the frequency of negative outcomes (increased SCC, increased NAGase activity, decrease in teat apex conditions) significantly increase and therefore improvement in AMS management and functionality are needed.

Introduction

The application of automatic milking systems (AMS) in Italian dairy herds should cope with peculiar environmental conditions with hot and humid climate for 4-5 months/year and the presence of herds with contagious pathogens (*Staph.aureus*). A research project was planned to assess the impact of this new technology, involving both commercial and experimental dairy farms with different AMS and different managements.

This paper reports the results of a follow-up study on focused on:
- the impact of AMS on teat tissues, teak skin and orifice
- the monitoring the dynamic of intramammary infections
- the relationship between milking parameters, milk biochemical parameters, cows and intramammary infections (IMI).

Material and methods

Herds
The herds considered in this study were a commercial dairy herd (herd A) and an experimental herd (herd B). Herd A is milking 120 cows with 2 Lely Astronaut™ AMS. Herd B includes 50 cows selected among the original herd of 80 cows previously milked with a herringbone milking parlour. The AMS machine is a single-stall robot (VMS™, DeLaval).

Cows and sampling
The study was divided in 2 parts. In the first part, heifers were included after calving and sampled for at least 12 months. In the commercial dairy herd, samples were taken 1 week after calving from all the freshening heifers with a weekly frequency in the first month and then monthly. In experimental dairy herd, cows were sampled twice in the first month and then monthly. In the second part, focused on milk and milking parameters, both cows and heifers of herd B were included, applying the same sampling procedure described.

Teat thickness and skin and orifice conditions
Teat thickness was assessed by a cutimeter as described by Hamann and Mein, (1990). Teat conditions were assessed by a modification of the method proposed by Neijenhuis *et al.* (2000). Practically a digital picture was taken, then transferred on PC and scored by a panel of 3 people using reference images (Casirani *et al.*, 2002).

Milk samples & analysis
Quarter milk samples (QMS) were collected before milking by an aseptic procedure. At the laboratory, samples (0.01 ml) of each QMS were spread on blood agar plate Colonies were isolated and identified by appropriate methods according to National Mastitis Council (N.M.C., 1999). Somatic cells were counted on a Bentley Somacount 150 (Bentley USA). The quarter status was defined as follows: a bacteriologically positive sample with SCC $\geq$200.000 cells/ml was defined as *infected*; a bacteriologically positive sample with SCC <200.000 cells/ml as *latent IMI*; a bacteriologically negative sample was defined as *healthy* if SCC $\leq$100.000 cells/ml and as *inflamed* when SCC >100.000 cells/ml.

NAGase and lysozyme
Lysozyme was assessed in duplicate by the procedure described by Metcalf *et al.* (1986) based on the lysis of *Micrococcus lysodeycticus* measured by changes of optical density at 450 nm after 2 min on a microplate spectrophotometer (Spectramax 340, Molecular Devices, USA). N-Acetyl-_-glucosaminidase (NAG) was assessed in duplicate by the procedure described by Kitchen *et al.* (1978), and expressed as units (pmol of 4-methylubelliferon released per min at 25°C catalyzed by 1 µl of milk) on a microplate fluorimeter at 355 exc and 460 em (Ascent, ThermoLabsystem, FL).

Statistical analysis

Data from laboratory analysis and from AMS software were recorded in a database and analysed by the means of ANOVA for repeated measurements and χ^2 statistics on SPSS 11.5 (SPSS, 2002), while IMI prevalence was calculated by software specifically developed to analyse IMI epidemiology (Sala and Zecconi, 2002).

Results

The first part of the study included 28 cows in herd A and 27 in herd B with an overall number of samples of 1528 and 816, respectively. Cows were sampled from the beginning of the lactation up to drying-off, therefore the number of samples was different for each cow, depending on the length of lactation.

Intramammary infections and somatic cell counts

The two herds considered showed a different epidemiological pattern. Herd B was free from contagious pathogens, while herd A had *Staph.aureus* intramammary infections (Table 1). The IMI prevalence were below 20% during the whole lactation in herd B, except for the period between 210-270 DIM. In herd A, a progressive increase of IMI frequency was observed as lactation proceeded. Indeed, the overall frequency was 12% when DIM was <30 d and raised to 33% at the end of the follow-up period. At calving 3.4% of the cows had at least 1 quarter infected with *Staph.aureus* and at the end of the study, 66.7% of the cows were infected.

The geometric mean of SCC was below 32.000 cells/ml in negative quarters from herd B, while in herd A, negative quarters were in the range between 30.000 cells/ml and 100.000 cells/ml. Positive quarters showed a progressive increase of SCC in herd A, from 100.000 cells/ml up to 200.000 cells/ml at the end of lactation, while in herd B showed peaks up to 1.000.000 cells/ml at 0-30 DIM and at 180-210 DIM. A more detailed analysis of IMI and SCC data was recently reported (Zecconi *et al.*, 2003).

Teat thickness changes

Teat thickness in herd A had very small variations up to 240 DIM in both IMI positive and negative samples, then infected quarters showed larger variation in thickness in

Table 1. Distribution of IMI by aetiology and herds.

IMI aetiology	Herd A (n.)	Herd B (n.)	Total (n.)
Environmental Streptococci	28	25	53
Gram Negative	8	9	17
Others	18	5	23
Staphylococcus aureus	166	0	166
Coagulase Negative Staphylococci	89	62	151
Contaminated	11	16	27
Negative	1208	698	1906
Missing	0	1	1

comparison with negative quarters. These latter ones showed changes in thickness in the range 0, -5%, values that can be considered fully physiological. In herd B, IMI positive and negative quarters showed a different pattern from the beginning. Indeed, bacteriologically negative quarters showed a progressive decrease in thickness from -2% at the beginning of lactation to values in the range -3%, -9% at the end of the lactation, while infected quarters showed large and unpredictable variation in thickness along the whole lactation, as reported for conventional milking (Zecconi *et al.*, 1992; Hamann *et al.*, 1994).

Teat skin and apex score
The distribution of teat skin scores in the two herds considered showed to be different. Indeed, in herd A, scores 3 and 4, decreased their frequency as long as lactation proceeded, while scores 1 and 2 were the most frequent ones even if large variations could be observed. In herd B, scores 3 and 4 were also infrequent, while frequency of score 2 showed to be higher in the first 210 DIM in comparison with the last part of lactation, and score 1 frequency had the opposite pattern.

Teat apex scores patterns were also different in the two herds. Herd A showed an increasing frequency of score 2 as the lactation progressed up to 150 DIM when it reached a level of about 70% that was maintained for the remaining part of the lactation. Score 1 declined from a frequency higher than 50% in the first 30 DIM to 10% at the end of lactation. Score 3 frequency was in the range 5-10% during most of the lactation, to increase to values higher than 20% after 300 DIM.

In herd B the distribution of scores was more irregular with a higher frequency of score 2 in the first 180 DIM (range 40-90%), and frequencies in the range 30-60% in the remaining part of lactation. Score 1 frequency was very high at the beginning of lactation, but with few exceptions, it was between 20 and 30% during the lactation. Score 3 frequency was in the range 10-25% for the most part of lactation, while score 4 was practically observed only in cows with more than 300 DIM.

In table 2, the associations between each of the three different teat parameters (thickness, skin score and apex score) and IMI were assessed. Teat thickness and teat skin score didn't show any significant association with IMI occurrence, while teat apex score 3-4 showed a 3-times higher chance to have an IMI in comparison with teat with an apex score 1-2.

Relationship between milking parameters, biochemical parameters, cow and IMI
The second part of the study was focused on the relationships between milking parameters (duration, yield, flow) and some cow characteristics and IMI status. This part of the study was undertaken only in herd B and included 1605 milk samples and records of 6 days around sampling. The results of the ANOVA for repeated measurements show that IMI has a significant influence on milking duration and yield, while its interaction with DIM showed to be statistically significant for the previous parameters and for milk flow rate (Table 3).

The same approach was applied to assess the relationship between milk characteristics and the milking parameters. Only NAGase showed to be consistently influenced by milking parameters and IMI and their interactions. The status health, as expected, showed to influence SCC counts, while milk flow showed to significantly influence milk lysozyme concentration. The peak/average ratio (PAR) is the ratio between the recorded peak flow

Table 2. Association between teat tissues-related risk factors and IMI outcome.

Risk factor	Herd	OR	Lower confidence limits	Higher confidence limits	P=[1]
Teat Thickness	A	0.828	0.640	1.072	ns[2]
	B	0.864	0.581	1.284	ns
	Overall	0.838	0.675	1.041	ns
Skin Score	A	0.954	0.544	1.673	ns
	B	0.813	0.179	3.695	ns
	Overall	0.934	0.552	1.582	ns
Apex score	A	3.376	2.048	5.564	0.000
	B	2.905	1.327	6.362	0.006
	Overall	3.224	2.114	4.917	0.000

[1]Assessed by Pearson's χ2

[2] Not significant

Table 3. Relationship among milking parameters and cow and IMI characteristics. Summary of ANOVA for repeated measurements.

Source of variation	Milk yield (g)	Milking duration	Milk flow	Peak flow / Avg peak ratio
Days in milk (DIM)	*** (1)	***	***	
Calvings (AGE)				
IMI	***	***		
DIM*AGE		***	***	*
DIM*IMI	**	**	**	
AGE*IMI				

[1] * p<0.1; ** p<0.05; *** p<0.01

and the average flow. In so far, it could be considered as the result of the combination of other parameters such as yield, milking duration, DIM and age. This parameter showed to highly influence NAGase; indeed, the values observed in healthy animals were very similar independently of the age of the animal or the value of PAR. In infected animals, primiparous cows with a PAR>1.41 showed a NAGase activity nearly twice as high as the cows with PAR ≤1.41, while pluriparous cows showed an even more dramatic increase in NAGase activity as PAR increased (Table 4).

Discussion

The application of AMS in dairy herds induces changes in cow management, behaviour, nutrition and in the interaction between the milking machine and the udder. The individual

Table 4. NAGase activity by IMI status and Peak/Avg flow ratio.

Peak / avg flow ratio	Healthy		Infected	
	Primiparous	Pluriparous	Primiparous	Pluriparous
≤1.41	21.83	25.03	34.56	37.38
1.42-1.57	23.19	28.22	56.07	89.81
1.58-1.62	23.54	33.10	57.46	100.73
>1.62	17.22	29.50	44.49	64.89

quarter milking applied by AMS could decrease some of the possible negative effects of conventional milking such as overmilking. However, the increased milking frequency could have negative effects on teat tissues and skin. Overall the introduction of AMS in herd B, free from contagious pathogens, didn't influence the frequency of IMI and the SCC, as already shown in similar experiences (Svennersten-Sjaunja et al., 2000). In herd A, characterized by the presence of Staph.aureus IMI, the frequency of IMI showed to increase progressively, very likely as a consequence of the spread of infections during milking. The presence of a significant number of herds with Staph.aureus IMI in different countries emphasizes the importance to carefully evaluate the efficacy of preventive procedures in AMS milking.

Teat apex and skin conditions were overall maintained along the lactation. Teat skin conditions tend to decrease (increase in scores) while lactation proceeded, but at a level largely acceptable in both herds. Teat apex scores showed to increase (worse status) as days in milk increased with scores 3 and 4 more frequently observed when DIM are higher than 300. Teat thickness changes showed different patterns in the two herds, probably because of the different type of AMS. In general terms the decrease in thickness after milking was relatively small in herd A and in the range -2% and -9% in herd B, when bacteriological negative quarters were considered. These data suggest that, even when an individual quarter-milking process is applied, the interaction between milking machine and teat plays an important role. This was confirmed by the results of the second part of the study and particularly by the relationship between milking parameters and milk NAGase. Peak/average flow ratio showed to highly influence NAGase, particularly in infected animals and in pluriparous cows. These results suggest that milking of infected quarters could increase the level of udder inflammation probably by prolonged milking.

Conclusions

The results of this field trial confirm that AMS have no negative impact on IMI incidence, SCC and teat tissue conditions, when the initial cow health status and overall herd management are good. In presence of IMI, and when cows have more than 300 DIM the frequency of negative outcomes (increased SCC, increased NAGase activity, decrease in teat apex conditions) significantly increased and therefore improvement in AMS management and functionality are needed.

References

Casirani, G., Binda, E., Piccinini, R., Zecconi, A., 2002. Sviluppo ed applicazione di un sistema basato sulla valutazione dello stato del capezzolo. Obiettivi e Documenti Veterinari (10): 21-27.

Hamann, J., Burvenich, C., Bramley, A.J., Osteras, O., Woolford, M., Woyke, M., Haider, W., Mayntz, M., Ledu, J., 1994. Teat tissue reactions to machine milking and new infection risk. Bull. Int. Dairy Fed. 297: 1-43.

Hamann, J., Mein, G.A., 1990. Measurement of machine-induced changes in thickness of the bovine teat. J Dairy Res. 57: 495-505.

Kitchen, B.J., G. Middleton, and M. Salmon. 1978. Bovine milk N-acetyl- beta -D-glucosaminidase and its significance in the detection of abnormal udder secretions. Journal of Dairy Research. 45(1):15-20.

Metcalf, J.A., J.I. Gallin, W.M. Nauseef, and R.K. Root. 1986. Laboratory manual of neutrophil function. Raven Press, New York.

National Mastitis Council., 1999. Laboratory handbook on bovine mastitis. National Mastitis Council Inc., Madison WI, USA.

Neijenhuis, F., H.W. Barkema, H. Hogeveen, and J.P.T.M. Noordhuizen. 2000. Classification and longitudinal examination of callused teat ends in dairy cows. J. Dairy Sci. 83:2795-2804.

Sala, S., Zecconi, A., 2002. Development of an epidemiological software for intramammary infection diagnostic data recording and analysis. In: Proc. Ann. Meet. National Mastitis Council, Orlando Fl, USA, 41:156-157.

SPSS Statistical Package for Social Science. 2002. SPSS Advanced Statistics 11.5. SPSS Inc., Chicago, IL, USA.

Svennersten-Sjaunja, K., Berglund, I., Pettersson, G., 2000. The milking process in an automatic milking system, evaluation of milk yield, teat conditions and udder health. pp 277-288 in Proc. Int.Symp. on Robot Milking, Lelystad NL. Wageningen Pers, Wageningen, Netherlands.

Zecconi, A., Hamann, J., Bronzo, V., Ruffo, G., 1992. Machine-induced teat tissue reaction and infection risk in a dairy herd from contagious mastitis pathogens. J Dairy Res. 59: 265-271.

Zecconi, A., R. Piccinini, G. Casirani, E. Binda, and L. Migliorati. 2003. Effects of automatic milking system on teat tissues, intramammary infections and somatic cell counts. Italian Journal of Animal Science. 2(4):275-282.

BLIND QUARTERS IN AUTOMATIC MILKING SYSTEMS - CAUSES AND CONSEQUENCES

Torben Werner Bennedsgaard, Morten Dam Rasmussen & Martin Bjerring
Danish Institute of Agricultural Sciences, Foulum, Denmark

With conventional milking blind quarters may be a nuisance when attaching the milking clusters and may result in suboptimal placement of the cluster during milking. Furthermore blinding of quarters may reduce milk production and be a risk for infection if the quarter is leaking. With automatic milking each quarter is milked separately and the manual work of attaching the clusters are automated.

In a study of 30 Danish AMS herds the prevalence of cows with one or more blind quarters was found to be 12% compared to 8% in 120 herds with conventional milking (Figure 1). The higher prevalence of cows with blind quarters may both reflect difference in the risk of blinding a quarter or in the risk of culling a cow because of a blind quarter.

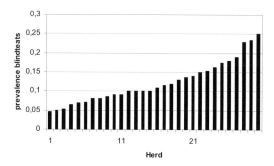

Figure 1. Prevalence of blind quarters in 30 Danish dairy herds with automatic milking systems.

Farmers in 22 herds were interviewed about the causes of blind quarters.

The main reason for having a blind quarter was mastitis during lactation (48%) where as events before introduction of the cows to the AMS accounted for 29% (Figure 2). Farmers mentioned that mastitis in some cases might have been related to problems with milking

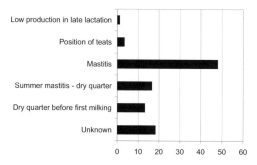

Figure 2. Causes of blind quarter in 30 Danish dairy herds with automatic milking systems.

the cows, and that the mastitis case with subsequent increased somatic cell count and reduced production was only one of the reasons for blinding a quarter.

In 40% of the herds a blind quarter could be the only reason for culling before introduction of AMS according to the farmers. After introduction of AMS no farmers considered a blind quarter as primary reason for culling.

CHANGES IN COW LOCOMOTION IN UK HERDS CONVERTING FROM CONVENTIONAL TO AUTOMATED MILKING

J. Dearing & J.E. Hillerton
Institute for Animal Health, Compton, United Kingdom

As part of the EU project on Automatic Milking a study was carried out in The Netherlands, United Kingdom and Denmark on "Health of dairy cows milked by an automatic milking system". In each country 15 farms were recruited that were changing from conventional to automatic milking. These farms were studied intensively. General aspects of herd health were studied on these farms from a few months before until one year after installation of the AMS. Attention was being paid to udder health, teat condition, lameness, body condition and fertility.

This poster examines the changes in locomotion score in UK farms changing from conventional to automatic milking and the implications of these changes.

Some 15 farms planning to install robots for automatic milking were recruited. They were monitored for 6 months prior to installation and up to 12 months post installation of the AMS. Protocols for locomotion scoring were discussed and practised with all members of the research team to ensure a good level of quality control throughout the project and therefore make the variations in health between farms and between countries comparable.

On each UK farm, 50% of cows were scored for locomotion, on average every 6-8 weeks, so allowing for seasonal effects and stage of lactation. A score of 1-4 was allocated:- from one indicating virtually no impediment to gait or feet to four indicating severe leg or foot lameness.

On farms where the data collection is complete locomotion has not been seen to be affected overall in the transition. Risk factors including a zero-grazing, foot baths and foot trimming and changing feeding are being considered.

A deteriorating fertility has been found on all the farms but this does not seem to be related to the locomotory activity of the cows.

A STUDY OF THE EFFECT OF LAMENESS ON VOLUNTARY MILKING FREQUENCY OF DAIRY COWS MILKED AUTOMATICALLY

Simon Grove, Torben Werner Bennedsgaard, Morten Dam Rasmussen & Carsten Enevoldsen
[1]Helle dyrlæger, Helle, Denmark
[2]Danish Institute of Agricultural Sciences, Foulum, Denmark

A high voluntary milking frequency is essential for the full benefit of an automatic milking system. In this study the effect of lameness on the voluntary milking frequency was estimated.

Eight dairy farms using automatic milking were visited twice. Cows with a daily milk yield of more than 15 kg were divided into three groups. Lazy cows were the 10 cows on the farms with the longest mean time between milking. Fetched cows were cows fetched for milking in 5-day period prior to the visit that were not included in the group of lazy cows. Cows that were not among the 15 cows with the longest mean time between milking and that had not been fetched for milking in the 5-day period prior to the visit was considered normal.

The effect of lameness on the mean time between milking and the frequency of milking interval longer than 12 hours were studied for the cows considered normal. The mean time between milking increased when the locomotion score increased (high locomotion score = abnormal locomotion). The frequency of milking interval longer than 12 hours increased as well when the locomotion score increased. It was also found that increase in daily milk yield decreased the mean time between milking and the frequency of milking interval longer than 12 hours, and that lower parity cows had shorter mean time between milking and fewer milking interval longer than 12 hours than older cows.

Frequency of lameness and the odds ratio of being lame between normal and lazy cows were estimated. The frequency of lameness was 11% for normal cows and 19% for lazy cows and the estimated odds ratio was 1.81. These results were close to statistical significance (p=0.0637).

Odds ratio of being lame between normal and fetched, but not lazy, cows were estimated to 1.74 but the difference was not statistically significant.

The results indicate that control of lameness is essential to obtain a high milking frequency.

INFLUENCE OF AN AUTOMATIC MILKING SYSTEM (VMS®) ON MILK SECRETORY ACTIVITY OF HEALTHY AND DISEASED UDDER QUARTERS

J. Hamann, H. Halm & R. Redetzky
Department of Hygiene and Technology of Milk, School of Veterinary Medicine Hannover, Hannover, Germany

Goal of the study

Milk secretion is, in addition to factors such as genetics, feeding and milking regime, mainly determined by the udder health status. This statement is broadly accepted for conventional milking. The present study compares the secretory activity of healthy and diseased quarters under automatic milking conditions to estimate the importance of udder health.

Material and methods

A group of 40 high-yielding German Holstein Frisian cows at different lactation stages and numbers were milked with an automatic milking system (Voluntary milking system VMS®; DeLaval, 42 kPa vacuum, 60 cycles/min pulsation rate, 65% pulsation ratio). Over 400 days, survey was carried out every 20 days and each session included 24 hours of continuous sampling. The mean secretory activity of all quarters (n = 6194 samples) was determined in relation to the udder health categories (IDF) Normal Secretion NS (n = 3212), Latent Infection LI (n = 1592), Unspecific Mastitis UM (n = 703) and Mastitis M (n = 687). A detailed analysis of two SCC related subgroups - 44 continuously healthy quarters (SCC < 100,000 /ml, n = 1179 samples) versus 113 continuously diseased quarters (SCC > 200,000 /ml, n = 942 samples) - was performed.

Results

The mean secretion rate of NS-quarters (317 g/h) differed significantly from that of M-quarters (248 g/h) by 22% (p < 0.05). The analysis of the influence of irregular milking intervals on both SCC related subgroups is summarized in table 1.

An additional analysis of variance confirmed the significance of the different milking intervals for both healthy and diseased quarters (p < 0.0001). Both quarter groups showed secretory reductions of approx. 40% between interval < 6 h and > 12 h. Comparisons between healthy and diseased quarters within each milking interval showed significant differences (p < 0.05), with the exception of the interval < 6 h.

Table 1. Milk secretion rate [g/h] in relation to milking interval and quarter health.

Milking interval	< 6 h	6 - 8 h	8 - 10 h	10 - 12 h	> 12 h
Healthy quarters	385 ± 148	318 ± 134	276 ± 116	274 ± 120	238 ± 102
Differences*	A	B	C	C	D
Diseased quarters	352 ± 188	265 ± 122	252 ± 133	213 ± 116	200 ± 103
Differences*	A	B	B	C	C
Healthy vs. diseased**	n.s.	< 0.001	< 0.05	< 0.001	< 0.001

*Ryan-Einot-Gabriel-Welsch multiple range test (groups with different letters differ significantly, p < 0.05)
**Student t-test (p values)

Implication

The mean secretion difference between continuously healthy and diseased quarters of approx. 15% was not compensated by increased milking frequencies (2.86 milkings/cow/day).

RISK FACTORS FOR TEAT END CALLOSITY IN DAIRY FARMS OPERATING WITH AUTOMATIC MILKING SYSTEMS (AMS)

Ilka C. Klaas, M. Bjerring & C. Enevoldsen
Danish Institute of Agricultural Sciences, Foulum, Denmark

Milking duration is reported to influence the degree of teat end callosity in dairy cows. Increased frequency of milking in an AMS may increase machine-on time and thus increase the degree of teat end callosity.

The aim of our study was to identify risk factors for teat end callosity in cows milked by automatic milking systems.

The study was carried out in 8 Danish farms operating with AMS. The first author visited the farms 4 times and assessed teat end callosity of 40-50 randomly chosen cows. The teat of a cow with the highest degree of callosity was scored on a 5-point scale ranging from 'no callosity' until 'rough and thick callosity ring'.

The following data from the AMS was collected at each farm visit: number of milkings, connection fails, milk yield and milk duration. The average number of milkings in the eight herds ranged from 2.9 to 3.4 milkings per day. A milking was defined as milking where milk yield was registered and as well included 3.5% of milkings with connection fail of at least one teat cup. Even though farms restricted number of milkings to a maximum of five per cow and day, cows with partial connection fail experienced a higher milking frequency with a maximum of 9 milkings per dag. 1020 cows were scored for teat end callosity. 69.2% showed teat end callosity; in 5.4% a thick callosity ring was observed. Only two cows had rough teat end callosity.

A logistic regression model was fitted with rough and thick callosity ring as outcome variable. Risk factors tested in the model were parity, days in milk (DIM), day of scoring, milk yield, milking duration, connection fail, and number of milkings. Preliminary results show that the risk for thick and rough teat end callosity increased with parity, DIM, number of milkings per day and milking duration. Further, cows with partial failure of cluster attachment were at higher risk for thick and rough teat end callosity.

Results show that the overall degree of teat end callosity in the studied herds is low, indicating a good teat end condition. Because higher milking frequency increases the risk for thick or rough callosity rings, special attention should be paid to cows with partial failure of cluster attachment.

INTRODUCTION OF AMS IN ITALIAN HERDS: UDDER SYMMETRY, TEAT THICKNESS AND HEALTH IN AN AUTOMATIC MILKING SYSTEM

Luciano Migliorati[1], Federico Calza[1], Giacomo Pirlo[1], Giuseppe Casirani[2] & Alfonso Zecconi[2]
[1]Dairy Cattle Section, Animal Production Research Institute, Cremona, Italy
[2]Dep. Animal Pathology, Infectious Diseases Laboratory, University Milan, Milan, Italy

Automatic milking system (AMS) raise some concerns regarding milk quality, cow welfare and health, and rate of culling cows whose mammary conformation is not suitable for this kind of technology.

This paper reports the preliminary results of a study focusing on the interaction between udder symmetry and teat tissues conditions.

The herd includes 50 cows selected among the original herd of 80 cows, previously milked with a herringbone milking parlour. The AMS machine is a single-stall robot (VMS [TM], DeLaval). The teat conditions and the symmetry of the udder were considered after a 20 days adaptation period, samples of the teat were taken twice in the first month and then monthly up to the 9th month of AMS. Milking time and teat cup attachment time were calculated for each cow for all the period.

Teat thickness was assessed by a cutimeter. Teat conditions were assessed by digital picture scoring by a panel of 2 people using reference images.

Symmetry of the udder was classified according to a scale of values, from 1 to 5, for each trait that has to be evaluated. This method allows to put in evidence the differences among animals. This method makes it possible to express a judgement of desirability or undesirability. The best value is 3. The greater the difference from the value of 3 the less the desirability of the udder morphology, (1-2= ± inclination of floor toward the fore part; 3=parallel to ground; 4-5± inclination of floor toward the rear part). The measurement of teat thickness showed that for the whole period, the changes of teat thickness, measured before and after milking, were within the limits considered as acceptable (±10%) in terms of risk for infection with teats of the udder=3 and front teats of the udder =4. On the rear teats of the udder=4, the changes thickness were sometime out of the acceptable limit of ±10%. The AMS induced an overall reduction of thickness, probably as a consequence of the application of mechanical forces that reduce fluids in teat tissues. Then each quarter thickness data were analysed by udder symmetry. When udder symmetry was =3, thickness pattern showed to be different from the one observed in the symmetry group=4. This difference was particularly evident on rear teats. Indeed, the group =4 showed markedly higher variations both for milking time and thickness on rear teats, than group =3. Rear teats of group =4 showed about 60 sec higher milking time than group =3, although similar yield of milk. Moreover, group =4 showed higher recognizing time of the rear teat, in particular on the left. Probably, with udder score =4, position of teats are less easily recognized by robot because the distance between the end of the lowest teat and the ground is shorter (25/35 cm), with more shadow. Besides, the attachment of teat cups to this kind of rear teats that are naturally inclined, determines a certain obstruction of teat orifice. This causes a less emission of milk per unit of time. In this way, animals with udder symmetry =3 have a shorter milking time than animals with symmetry =4.

In conclusion, the results of the first part of the research confirm that AMS has a better impact when the udder and teats are suitable and a worse impact with irregular ones. The reduction of teat thickness changes, attachment time and milking time raised on group =3, suggest a lower stress and a better health status of teats and udder.

Automatic milking – A better understanding

EFFECTS OF MILKING INTERVAL ON TEAT CONDITION AND MILKING PERFORMANCE WITH WHOLE-UDDER TAKE OFF

F. Neijenhuis[1], J.E. Hillerton[2], K. Bos[1], O. Sampimon[3], J. Poelarends[1], C. Fossing[4] & J. Dearing[2]

[1]Applied Research, Animal Sciences Group, Wageningen UR, Lelystad, The Netherlands
[2]Institute for Animal Health, Compton, United Kingdom
[3]Animal Health Service, Deventer, The Netherlands
[4]Institute for Animal Sciences, Foulum, Denmark

Introduction

As part of the EU project on Automatic Milking (AM) a study was carried out on "Health of dairy cows milked by an automatic milking system". A separate experiment was conducted to determine the effects of milking intervals on teat condition in the short-term in a conventional milking parlour. Milking with shorter intervals may lead to incomplete recovery of teats. This could lead to an accumulation of teat trauma. So a measurable risk may occur when using an AM system, if milking intervals are short.

Materials and methods

A Latin-square designed experiment in a conventional milking parlour used 12 cows which were allocated randomly for a period of 4 days, being milked with whole-udder cluster take off every 4, 8, or 12 hours. After two days habituation, milk flow profiles and ultrasound images of teat dimension just before and just after were gathered.

Results and discussion

Milk yield increased 16% per day when milking frequency was three times a day or more. The overall increase in milk yield per day when cows were milked more than twice a day was only found for first parity cows. Milking four times a day, with whole- udder take off, almost doubled the machine-on time and not increased milk yield compared to milking three times a day. The efficiency of milk harvesting decreased with increasing milking frequency.

Shorter milking intervals resulted in less increase in teat wall thickness, less reduction in teat cistern width and less shortening in teat canal length in response to milking. However, before milking the teats milked following short intervals had thicker teat walls, narrower teat cisterns and longer teat canals. This indicates either that short milking intervals are insufficient to allow recovery of teat tissue between milkings or are a result of less filling of the teat cistern with milk.

In this experiment whole-udder take off was practised, however most of the AM systems use quarter take off. This may reduce the effects on teat condition because of a shorter low flow period for individual quarters at the end of milking. In this experiment front teats had a larger increase in overall teat width than rear teats. With whole-udder take off the low flow period in front quarters will last longer compared to rear quarters.

Conclusions

The changes in teat condition found in this experiment after short milking intervals points to recovery times for teats being too short. Care should be taken to avoid too short milking intervals, to adjust stimulation to udder filling, and to milk only those cows with sufficient udder filling. This experiment reports a worst-case scenario because milking cows with an AM system includes longer pre-treatment and/or lag time before milking and in general quarter detachment is applied. However, too short intervals (4h) are hazardous to teat condition and moreover do not contribute to an increased system capacity in kg per day. A general advise might be to give a cow access to the AMS when milk interval is at least 6 to 7 hours.

MILK LEAKAGE IN AUTOMATIC AND CONVENTIONAL MILKING SYSTEMS

K. Persson Waller[1,2], T. Westermark[1], T. Ekman[2] & K. Svennersten-Sjaunja [3]

[1]Department of Ruminant and Porcine Diseases, National Veterinary Institute, Uppsala, Sweden

[2]Department of Obstetrics and Gynaecology, Swedish University of Agricultural Sciences, Uppsala, Sweden

[3]Department of Animal Nutrition and Management, Swedish University of Agricultural Sciences, Uppsala, Sweden

Introduction and objectives

Milk leakage between milkings can increase the risk for udder infections and mastitis in dairy cows. Increasingly more automatic milking systems (AMS) are introduced in Swedish dairy farms. However, studies indicate that the udder health of such farms can deteriorate. The reasons for this are not clear, but preliminary observations indicate that milk leakage may occur more often in such systems than in conventional milking systems (CMS). However, comparative data on the incidence of milk leakage in the various systems are not available. Therefore, the aim of this 2-year study was to compare the occurrence of milk leakage in one AMS in comparison to two other groups of cows managed using a CMS. In addition, we wanted to identify cow and management factors associated with higher risks of milk leakage.

Research procedures

The cows in one AMS with cows housed in a freestall barn, one freestall barn (FSB) and one tie-stall barn (TSB) at the university farm were used in this study. In the two latter barns the cows were milked at regular intervals in a herringbone milking parlor. In each of 2-years, all cows (n=230 total; 46 present both years) were observed at 2 h intervals during six 24 h periods. If milk leakage was observed, the activity and position of the cow was recorded. From the database of the herd, individual cow data was collected.

Results and discussion

A significantly higher proportion of milk leakage was observed in the AMS compared with the CMS. Milk leakage was recorded at least once in 39.0% (AMS), 13.2% (FSB) and 9.7% (TSB) of individual cows and in 16.2% (AMS), 3.8% (FSB) and 2.2% (TSB) of 24-h cow days studied. Overall, milk leakage was not related to milk production, parity, stage of lactation, or registration of heat. However, in AMS, 62% of primiparous and 28% of multiparous cows leaked milk at least once. Milk leakage occurred more often in rear udder quarters and was most often observed when cows were lying down. The interval from previous milking at the time of observation of milk leakage varied, especially in AMS, and approximately 20% of the observations occurred less than 4 h after milking. A large proportion of those were associated with disturbances at the previous milking. Milk flow rate was higher in udder quarters leaking milk than in other quarters. The cow somatic cell count (SCC) and udder disease score (based on the geometric mean of the last two or three monthly SCC; the higher the score, the higher probability for mastitis) were numerically, but not significantly, higher in cows with milk leakage. More studies are needed on the etiology of ML and on the occurrence and possible effects on udder health of ML in commercial farms.

IMPACT OF TEAT DIP COMPOSITION ON TEAT CONDITION AND UDDER HEALTH

M.D. Rasmussen[1], L. Foldager[1] & T.C. Hemling[2]
[1]Danish Institute of Agricultural Sciences, Foulum, Denmark
[2]DeLaval, Kansas City, Missouri, USA

The teat skin of dairy cows is serving as a protection against environmental attacks from weather, bedding, chemicals, and milking. Poor teat condition may lead to poor milking performance and new infections with mastitis. Consequently, great effort has been put into improvement of the teat skin and a reduction of pathogens on the teat skin. The objective of the experiment was to evaluate the influence of milking frequency and teat skin conditioner on teat skin condition of cows milked and post sprayed automatically.

Three groups of 40-45 cows were automatically milked with DeLaval VMS™ at the Danish Cattle Research Centre. In all 140 cow participated in the experiment. The teats of the cows were sprayed automatically post milking with either A: DeLaval Proactive™ (0.15% iodine, 2% emollient) or B: DeLaval Proactive™ Plus (0.15% iodine, 8% emollient. Treatments were switched between VMS 1 and VMS 2 and 3 in periods of 4 weeks. The cows on VMS 3 were left unsprayed for period 3. All teats of all cows were scored for teat skin and teat end condition before the experiment started. All cows visiting the robots between 6 to 16 were scored once per week throughout the study. Teat skin condition and teat end condition were scored immediately before milking. Scoring of teat skin condition has to be done on dry skin and therefore before milking. It would have been better to score teat end condition after milking but that interfered with the automatic teat spraying system and could not be carried out. Visual appearance and CMT-score of foremilk of all quarters were done weekly. Milk yield, milk flow, and time of milking were automatically recorded by the DeLaval VMS.

The use of teat spray B with 0.15% iodine and 8% emollient improved teat skin condition compared with application of teat spray A with only 2% emollient of cows milked automatically. The difference in teat skin condition was noted within few week of use. Interval between milkings had no influence on skin condition. There was no difference in teat end condition between the two treatments. Milking frequency or interval between milkings did not affect teat end condition scored immediately before milking. Long intervals between milkings lowered CMT-score on foremilk.

ANIMAL HEALTH OF HERDS CONVERTING TO AUTOMATIC MILKING SYSTEM

F. Skjøth[1], M.D. Rasmussen[2], L.A.H. Nielsen[1] & K. Krogh[1]
[1]Danish Cattle, Aarhus N, Denmark
[2]Danish Institute of Agricultural Sciences, Foulum, Denmark

The first automatic milking systems (AMS) in Denmark were installed in 1998 and since then a large number has been installed each year. Currently, there are 259 farms milking automatically in Denmark. World-wide the number of AMS farms has passed 2000. The installation of automatic milking may be accompanied by a re-building of the barn or may be installed in a totally new barn. Anyhow, the installation of AMS is synchronised with a change in management system and probably in housing system as well. The objective of the present study was to analyse the effects of the changes in housing and milking systems with focus on animal health.

The data material consists of information on 404 Danish cattle farms, which have had a change in production facilities at some known time between 1st August 1996 and 1st January 2002. The changes made can be introduction of new (or rebuild) free-stall barn (234 cases) or the introduction of AMS (105 cases) or the combination of both (65 cases). Data from further 404 Danish cattle farms, selected by random from the rest of the population of Danish cattle farms with milk yield recording, were included to form a control group. Data collected in the Danish National Cattle Database on these 808 farms were extracted and compiled. The data includes information concerning herd and cow level milk quality, yield, and composition, as well as calving and replacement information. Cow level milk data originates from monthly milk yield recordings. Data are extracted to cover the period 9 to 3 month prior and 3 to 9 months after the production change. In this manner the two periods' covers the same calendar time, separated with a period of 6 months. In this period it is assumed that management routines are changed to prepare the change in production system, as well as cows and manager are getting used to the new system.

The poster will present a variety of results on several groups of disease categories based on veterinary diagnosis reports, as well as indicators of acute and chronic mastitis based on cow-level somatic cell count data. Possible hypotheses are discussed how the observed effects relate to the type of change in production facility, as the design allows to investigate the separate effects of change in housing system and change in milking system.

QUARTER MILK FLOW PATTERNS IN DAIRY COWS: FACTORS INVOLVED AND REPEATABILITY

V. Tančin[1,2], A.H. Ipema[2], D. Peškovičová[1], P.H. Hogewerf[2], Š. Mihina[1] & J. Mačuhová[3]
[1]Research Institute of Animal Production, Nitra, Slovak Republic
[2]Agrotechnology and Food Innovations BV, Wageningen, The Netherlands
[3]Institute of Physiology, FML, Tech. Univ. Munchen, Freising, Germany

The objectives of our study were to describe the variation of quarter milk flow parameters and to determine the factors (parity, stage of lactation, peak flow rate, somatic cell count, time of milking, day of experiment, quarter position) that affect these parameters the most. Additionally, repeatabilities (r^2) of the quarter milk flow traits (duration of milk flow, time to reach peak flow, peak flow rate, and duration of milk flow in single phases - increase, plateau, decline and blind phase) were calculated. Repeatability of total milk yield and milk yield in single phases was calculated, too. The data from 39 Holstein cows, in their first to third lactation and free of clinical mastitis, were used for statistical analysis. A total of 1656 curves of quarter milk flows were recorded during six consecutive days. At the last evening and morning milking samples of milk from each quarter were collected for determination of somatic cell count (SCC).

The effect of parity, stage of lactation and day of experiment influenced only some of the studied parameters - milk yield, duration of the milk flow, duration and yield of plateau phase. Peak flow rate, quarter position, time of milking (morning and evening) and SCC significantly affected most of the measured parameters. From the most important differences within the factor peak flow rate we can mention the shorter duration of milk flow, higher milk yield on one side and longer duration of decline and blind phase for the quarters with high peak flow rate when compared with the low ones. Within the factor SCC, there was a significantly longer duration of increase, decline, and blind phase and lower milk yield in quarters with high SCC. However, there were no differences in peak flow rate of two SCC groups of quarters.

The highest r^2 were for total milk yield and yield of plateau phase 0.53 and 0.50, resp. The r^2 for the duration of decline and blind phases were 0.40 and 0.48 resp. The lowest r^2 were calculated for the duration of increase phase, and milk yield of the increase and blind phase 0.26, 0.12 and 0.21, resp. Peak flow rate, SCC, time of milking and front-rear position influenced the values of r^2 of traits to various extent.

Conclusion. In this experiment the quarter milk flow and yield parameters were not always influenced by the studied factors. Peak flow rate significantly influenced all measured traits only. Peak flow rate, SCC, time of milking and front-rear position influenced the values of r^2 of traits to various extent. Possible relationships between the quarter milk flow parameters and health of quarter could be considered. Our data contribute to the knowledge concerning the further development of machine milking equipment based more on the acceptance of biology of the quarters to milk cows faster and to keep the udder healthy.

STUDY ON THE INCIDENCE OF KETOSIS IN DAIRY COWS IN AN AUTOMATIC MILKING SYSTEM VERSUS A CONVENTIONAL MILKING SYSTEM

Christoph Wenzel & Alexander Nitzschke
Clinic for Ruminants and Pigs (Internal Medicine and Surgery), Justus-Liebig-University, Gießen, Germany

Introduction

The prevalence of ketosis in cows during the first 2 months of lactation ranges from 7 to 32 per cent without differentiation between primary and secondary ketosis. This major metabolic disorder leads to decreased feed intake, lower milk yield and an increased risk of periparturient disease. Ketosis is known to be influenced by the husbandry system. Therefore, a study is currently underway to determine whether the incidence of ketosis is influenced when cows are kept in automated husbandry.

Materials and methods

The study is being conducted over a time period of one year, from May 2003 until May 2004. From a herd of cows milked by an automated system (n = 60), samples are taken every three weeks from all cows in early lactation (≤ 70th day of lactation). These cows make up the experimental group. From 5 randomly selected cows at the same stage of lactation in a conventionally milked herd (n = 60) samples are taken every three weeks, too. These cows serve as controls. All cows are either Red Holstein or Black Holstein breed. Urine and blood samples are taken from all cows. Using test strips, the urine is tested for ketone bodies. Blood serum is evaluated for glucose, beta-hydroxybutyrate (BHBA), urea and aspartate aminotransferase (AST). If urine ketone bodies are detected (≥ 4 mmol/l), the cow is clinically examined.

All cows are kept in the same stable (loose-housing), with the compartments for the experimental and control groups being separated by the feeding table. Both herds are given the same feed (a part-mixed ratio consisting of grass silage, maize silage, hay, brewers' grains). The experimental group receives concentrate in the milking stall (Lely Astronaut®), and the control cows have access to two concentrate feeders.

Mean and standard deviation are calculated for all samples of each group. Non-parametric tests are used to investigate differences between both milking systems.

Results and discussion

Preliminary results from the first six months are presented.

2% of all cases in the automatic group show a concentration of ketone bodies ≥ 4 mmol/l in the urine, while 11% in the controls show the same level. Clinical investigations revealed no primary cause. No significant differences between the two groups were found in serum concentration of glucose, BHBA, urea and AST, however. This preliminary results didn't support the hypothesis that cows which have all basic requirements freely accessible to them and are therefore left alone have a higher risk of ketosis than those with a conventional daily routine. Until now, it appear as if all cows have been able to take in enough concentrate to cover metabolic energy needs.

ABNORMAL MILK

DETECTION AND SEPARATION OF ABNORMAL MILK IN AUTOMATIC MILKING SYSTEMS

Morten Dam Rasmussen
Danish Institute of Agricultural Sciences, Foulum, Denmark

Abstract

This paper gives an overview of the work done in WP3 including definitions of normal and abnormal milk, suggestions for reference methods, and tests of AMS models according to these requirements. The general conditions of hygiene in milk production in the EU are defined by the Commission Directive 89/362/EEC (1989) but not all elements apply to automatic milking. The following text is proposed to be included in the coming EU Hygiene Directive:

Milking must be carried out hygienically ensuring in particular: - that milk from an animal is checked for abnormalities by the milker or a method achieving similar results and that only normal milk is used for human consumption and that abnormal, milk with a withholding period, and undesirable milk are excluded.

The definition of abnormal milk caused by clinical mastitis is proposed to be based on the homogeneity of the milk and not on the colour since the colour of the milk changes with breed, stage of lactation, feedstuffs etc. The reference method is suggested as filtration of the milk through a filter with a pore size of 0.1 mm, and milk where clots are clearly visible in such a filter is then defined as being abnormal. Incidences of watery and yellowish milk may or may not be detected by this method. The current AMS models have systems to produce alarm lists of cows that should be checked for abnormalities in their milk and at present the systems are not intended for automatic diversion of milk. Five different AMS-models were tested in six herds. The sensitivities for detection of abnormal milk in the six herds varied from 13 to 50% when calculated for the actual milking, from 22 to 100% for the test days, and from 43 to 100% when calculated for the previous week. Specificities for the same time periods were found to be 87-100%, 85-100%, and 35-100%, respectively. The sensitivities and specificities are generally too low for automatic diversion of abnormal milk and it seems as if most of the models could benefit from application of more sophisticated algorithms or measurements more directly related to the definition of abnormal milk.

Introduction

The general conditions of hygiene in the milk production in the EU are defined by the Commission Directive 89/362/EEC (1989) and Chapter III-4 reads:

Before the milking of the individual cow the milker must inspect the appearance of the milk. If any physical abnormality is detected, milk from the cow must be withheld from delivery.

Fulfilment of this directive is presently a problem with automatic milking systems because normally a human is not present and visual inspection of foremilk is not performed. Technical solutions may replace visual inspection for detection of abnormal milk, either before or during milking, and subsequent separation. However, unequivocal and generally accepted definitions of normal and abnormal milk are not available. In order for AMS companies to develop sensors to detect abnormal milk, a precise definition of abnormal or unacceptable milk is needed. The definition has to apply not only to automatic milking but to conventional milking as well. A workshop was held at the Danish Institute of Agricultural Sciences on November 27, 2002 in order to give input to the coming EU-hygiene directive concerning this matter (Rasmussen, 2002). Participants from outside of the EU were invited to this workshop to make the definition applicable worldwide and give a broad input to the present paper.

Workshop on definition of abnormal milk at time of milking

The main purpose of the workshop was to present background material for a definition, discuss the intention and consequences of the definition, and finally outline agreements and disagreements. Those parts of the definition where there was consensus at the workshop are clearly stated below. Based on the discussion at the workshop, the author recommends:
1. There should be no double standards. The requirements for milk quality produced under conventional and automatic milking conditions should be the same.
2. Milk at time of milking can be classified in four categories:
- *Normal milk*: Milk suitable for human consumption.
- *Abnormal milk*: Milk which differs from normal milk in respect of colour or homogeneity.
- *Milk with a withholding period*: Milk which, prior to the milking of the animal, is known to be unfit for human consumption following treatment of the animal with antibiotics or other veterinary products requiring that the milk must be withheld from sale for such use.
- *Undesirable milk*: Milk which, prior to the milking of the animal, is known to be unsuitable for human consumption, e.g. colostrum, high somatic cell count.
3. If clots appear, the milk is abnormal. The reference method is proposed to be the appearance of clearly visible clots on a filter with a pore size of 0.1 mm.
4. Milk that has changed in colour because of the level of red blood cells is regarded as abnormal milk.
5. Milk from the first 72 hours after calving with at least two daily milkings (the colostrum period) is regarded as undesirable. Milk may be withheld for a longer period if it still appears discoloured.
6. The cell count of milk should not be included in the definition of abnormal milk at time of milking.

Re 1. There was consensus at the workshop that there should be no double standards. The requirements shall apply to all milking conditions and not be special for automatic milking. The author suggests that the following text for the coming EU Hygiene Directive may apply:

Milking must be carried out hygienically ensuring in particular: - that milk from an animal is checked for abnormalities by the milker or a method achieving similar results and that only normal milk is used for human consumption and that abnormal milk, milk with a withholding period, and undesirable milk are excluded.

This text implies that milk should be inspected and that abnormal, milk with a withholding period, and undesirable milk should not be delivered for human consumption. Milk with a withholding period and undesirable milk are conditions known prior to milking and may or may not be checked or monitored during milking. The sensors used for automatic milking systems to detect abnormal milk should be as good as the milker. The skills of the milker are not defined and the workshop proposed to use the level of an experienced milker to give the reference level for sensitivity and specificity. The author recommends that the specificity should be >99% to be well accepted by farmers. A sensitivity of about 80% has been reported for conventional milkers. It was noted that there are no educational requirements for becoming a milker, but this should not be used to alter the requirements for detection of abnormal milk. The dairy industry may have additional requirements of bulk milk.

Re 2. A classification and definition of milk was drafted at an ISO meeting for development of international standards for automatic milking (ISO/TC 23/WG 1 Automatic milking installations) in the week before the workshop. The text was further discussed and revised at the workshop. The benefit of the definition is to help to distinguish between conditions where milk is known to be unfit for human consumption either prior to the milking of the individual cow (undesirable milk and milk with a withholding period) or at the milking (abnormal milk). The definition of abnormal milk is not a food safety issue. It is the responsibility of the farmer to deliver only normal milk but it can only be checked for abnormalities at the time of milking. Consequently, reliable sorting mechanisms are needed for unattended milking methods. The workshop discussed the rejection of abnormal milk at the quarter or cow level. The main opinion was that if milk from any quarter is abnormal, all milk from that cow should be considered abnormal. Discard of abnormal quarter milk only may be attractive but the thought of milk coming from cows carrying an infection may harm the image of the milk.

Re 3. It is important that the reference method for classification of the milk is based on science, is applicable, repeatable, and objective. The workshop was in favour of defining abnormal milk caused by clinical mastitis on the homogeneity of the milk and not on the colour since the colour of the milk changes with breed, stage of lactation, feedstuffs, etc. The author proposed a reference method to be based on filtration of the milk through a filter with a pore size of 0.1 mm. Milk where clots were clearly visible in such a filter was then defined as being abnormal. Incidences of watery and yellowish milk may or may not be detected by this method. The workshop questioned the reference method and additional information concerning the consequences of using the filter as a reference method is needed and will be provided.

The current standard is to inspect the appearance of foremilk. Clinical mastitis develops due to invasion of pathogens through the teat canal. It may happen that clinical signs are not seen in the foremilk but will appear later on and the workshop proposed to include all milk from the quarter in the definition, i. e. if clots are detected at any stage of the milking the milk is abnormal. It is not practical to base the daily judgement on milk fractions other than the foremilk. However, the frequency of cows with no clots in the foremilk but clots appearing late into the milking is expected to be very low and clots are likely to appear at the foremilking of subsequent milkings.

Re 4. There was consensus at the workshop that milk coloured by blood should be regarded as abnormal milk. The frequency of visible blood in the milk is low. The reference level for detection has not yet been defined. The reference being white milk, test panels of consumers and professionals can detect samples with about 0.1% of blood, but even 1% of blood does not show clearly in a black strip cup which is the conventional reference method at foremilking. The percentage of haemoglobin in blood, i.e. red blood cell count, has to be taken into account when determining the reference.

Re 5. There is an overlap between the physiological phases of colostrum and milk production. Colostrum is a normal secretion of a postpartum udder in early lactation, but it cannot be regarded as "normal milk" according to the above definitions. At the fourth day after calving, the secretion from most cows will appear as normal milk when milking twice daily. The workshop did not fully agree on the number of days to withhold colostral milk but 3 full days is regarded as a minimum withholding period. Local or national regulations may require a longer withholding period.

Re 6. There was consensus at the workshop that the cell count of milk should not be included in the definition of abnormal milk at time of milking. A high cell count is a clear indicator of inflammation in the udder but cannot be required to be measured at every milking for determination of abnormal milk. It is still recommended that cell count is part of the milk quality survey of bulk milk.

Selection of herds for detection of abnormal milk

The Danish distributors of AMS were contacted to appoint potential herds with at least 100 cows and a technically well functioning AMS. Technicians from the distributors visited the herds before the days of testing and were present during the test hours. Herds with bulk milk SCCs of 300,000 cells/ml or more were selected since such herds can be expected to have cows with subclinical as well as clinical mastitis. For herds with more than 100 cows we expected to be able to find at least 10 cow milkings with abnormal milk from at least five different cows. One herd was selected for each of AMS models 1-4 and two herds for no. 5. The models are kept anonymous in the tables. Only one model of AMS was equipped with a colour sensor to automatically divert milk with blood but the test for diversion based on colour has not been carried out at present.

Data were collected from the five AMS models present in Denmark, i.e. DeLaval VMS, Fullwood Merlin, Gascoigne Melotte, Insentec Galaxy, and Lely Astronaut. The six selected herds were sampled for a various number of hours from 13 to 48. At least 50 cows with normal milk was sampled twice in the herds we sampled for only a short time. Cows were foremilked in the milking box just before the automatic milking. Normally, the interval from cow identification to start of movement of the robot arm is very short but for some models it was possible to add a time lag in order to have enough sampling time. For other models one person handled cows of one box each so that sampling could start as soon as the cow entered the box. Foremilking was done into a four-chambered strip cup with 0.1 mm filters mounted at the outlet. A CMT-scoring plate was collecting the foremilk from each quarter. Visual scoring was done during foremilking. A small amount of water was run through the filters to remove foam before the visual inspection of the filters. CMT-scoring was also done immediately after foremilking. Two consecutive milkings without clots on the filter, no visual abnormality, and low CMT-score were needed to classify cows and quarters as normal whereas any milking with clots on the filter and a CMT-score >3 was rated as abnormal. The remaining unclassified cows and quarters were either omitted (first milking but otherwise normal) or discarded (CMT-score >3 or visually changed in colour but no clots on the filter).

Detection of abnormal milk in tested herds

Sampling was carried out in the six herds for a time period of 13 to 48 hours and resulted in collection of foremilk scorings of 169 to 623 cow milkings (Table 1). A large percentage of the samples were omitted, especially in the herds with a short sampling time. About 5 to 15% of the cow milkings were discarded because the CMT-score was 4 or 5 or they were visually changed in colour but with no clots on the filter. The number of cow milkings with normal milk was from 47 and up and the number of cow milkings with abnormal milk was from 4 to 18.

All herds had an alarm list based on conductivity and the results are presented in table 2. The number of discarded samples that appeared on the alarm lists varied. One to five cow milkings with abnormal milk matched directly the alarm based on conductivity and 2 to 13 did not.

Table 3 presents the sensitivities and specificities calculated for the actual milkings, for the test days, and for the previous week (including the test day). Sensitivities were generally low for the actual milking and increased when looking at a longer time span. Specificities were generally high at the actual milking and dropped when looking at a full

Table 1. Number of cows and milkings in the tested herds.

Model	Herd	AMU	Cows	Hours of sampling	No. of cow milkings					Quarters	
					Total	Drop	Discard	Normal	Abn.	Normal	Abn.
1	1	2	79	48	350	89	46	206	9	936	9
2	2	3	145	13	222	113	35	56	18	256	22
3	3	4	116	13	178	90	21	61	6	243	6
4	4	2	100	14	192	104	26	47	15	227	15
5	5	2	105	16	169	102	7	54	6	223	6
5	6	3	184	36	623	184	69	366	4	1526	7

Table 2. Number of cow milkings during the test day(s) of each herd and the number of discarded and abnormal milkings and divided into being on the alarm list or not at the actual milking. Alarm systems were based on conductivity.

	No alarm				Alarm list			
Herd	Total	Discard	Normal	Abnorm.	Total	Discard	Normal	Abnorm.
1	250	41	203	6	11	5	3	3
2	102	33	56	13	7	2	0	5
3	74	14	55	5	10	7	2	1
4	73	19	41	13	15	7	6	2
5	54	4	47	3	13	3	7	3
6	406	43	361	2	33	26	5	2

week. Herd 5 had the lowest specificity for the actual milking but it turned out that sensors were not calibrated sufficiently. Consequently, the specificity was very low when looking at the alarm list for a week. It could be speculated that the relatively high numbers of abnormal cow milkings found in Herd 2 could be a result of the relatively low sensitivity found in this herd. However, five of the 14 cows with abnormal milk were separated manually. For Herd 4, the sum of sensitivity and specificity was 100% indicating that correct appointment of abnormal quarters was purely random. From plots of Herd 5, it seemed as if sensors of this herd were drifting, which may explain the poor specificity compared with Herd 6.

Table 3. Sensitivity (SE) and specificity (SP) for appointing abnormal and normal milk from cows during an actual milking, the day(s) of testing or the previous week. Alarm systems were based on conductivity.

	Abnormal		Actual milking		Day(s) of test		Previous week	
Herd	Cows	Milkings	SE	SP	SE	SP	SE	SP
1	5	9	33	99	100	89	100	85
2	14	18	28	100	36	100	43*	100*
3	6	6	17	96	33	92	100	55
4	9	15	13	87	22	85	67	62
5	5	6	50	87	60	87	100	35
6	2	4	50	99	100	99	100	83

* Data only available for three days.

Discussion

Automatic diversion or alarm list

Sensitivities were generally low when calculated for actual milkings and certainly too low for automatic diversion of abnormal milk. Herd 1 had a sensitivity of 33% for the actual milking, but the sensitivity increased to 100% when calculated for the two sampling days, which is fine for an alarm list (attention or alert list). Herds 2 and 3 were still low in sensitivity when looking at the test day (24 hours). Herd 3 improved its sensitivity to 100% when looking at a one-week window. However, it could be suspected that this high sensitivity was due to a very poor specificity (55%) where about half of the cows in the herd were alerted within a week. Such a low specificity will hide the true abnormal cows and probably make the farmer "immune" to the alerts. The model used for Herd 2 seemed very restricted in its calculations of alerts and consequently kept a very high specificity even when looking at a three-day window (100%). The numbers calculated for Herds 1 and 6 give a fair balance between alerting all abnormal cows and keeping a reasonable specificity when used for an alarm list. Herd 2 will miss too many cows with abnormal milk but have less work checking cows that are truly normal (none abnormal but two discarded). The low specificity of Herd 5 indicates a need for an alarm system for malfunctioning of sensors.

A drop in milk yield or flow may indicate that the cow is sick. However, the daily variations are large in these parameters and uncritical use of these as indicators of udder problems may of course flag these but at the cost of a very low specificity. For Herd number 3 such an alarm system is useless looking at the whole week and experiencing a specificity of 12% (data not shown). Herd 1 was better in the same time span (specificity 65%) but still too low for practical use if all udders of flagged cows should be checked. The alarm lists for udder health should focus on this matter and a high sensitivity should not be achieved at the cost of poor specificity.

Classification of cows and quarters

There are several ways of classifying quarters and cows into having clinical mastitis, subclinical mastitis or being healthy. The definition of abnormal milk in this context is milk that differs in homogeneity and colour from that of normal milk. This means that only quarters and cows with clinical mastitis are rated as abnormal and that normal milk may originate from cows with subclinical mastitis. Changes in homogeneity may not show at every milking and some quarters have teat canal infections that cause clots in the foremilk but do not cause an inflammatory response of the quarter. Consequently, classification of quarters and cows into normal and abnormal milk can be difficult if based on single milkings. We could choose to base the classification on multiple milkings or use inflammatory parameters to support the decision. During the test, appointment of quarters with abnormal milk was supported by CMT-score of the foremilk. Likewise, the prerequisites for appointing quarters with normal milk were no visual appearance of any clinical sign and a low CMT-score at two consecutive milkings, which increases the probability of having a true normal status (Rasmussen *et al.*, 2002). This procedure, however, left a group of cows and quarters with high CMT-scores, watery or yellowish milk. This was a relatively large group compared with the number of cows with abnormal milk. Many of the discarded cows were flagged on the alarm list but not included in the calculation of sensitivity and specificity. Consequently, the sensitivities and specificities probably look better than they actually are. Automatic sorting of milk is not the primary objective for an alarm list based on conductivity where flagging of cows with potential mastitis problems is more important.

Confidence intervals for sensitivity and specificity

Some of the herds only had very few cases of abnormal milk. The confidence interval will be relatively large for small numbers and in the herd with six cow milkings with abnormal milk only two should be pointed out in order not to discard the hypothesis of a sensitivity of 80% with a 95% confidence. However, only one cow was alerted at the actual milking in Herd 3 and for this herd the sensitivity was <80%. Likewise, 10 cows should have been pointed out in Herd 2 in order not to reject the hypothesis of a sensitivity of 80% where only 5 cow milkings were flagged at the actual milking. Finding 12 out of 20 cow milkings with abnormal milk will give a sensitivity of 60%, which then for a 95% confidence interval will include a sensitivity of 80%. If the 80% sensitivity is a minimum requirement, then at least 19 out of 20 cow milkings with abnormal milk should be diverted. If this is the case then it will be required that the sensors measure the property of abnormal milk rather directly. This does not leave much room for development of sensors measuring correlated properties of abnormal milk. I propose that at least 16 out of 20 cow milkings with abnormal milk should be detected, which ensures that the sensitivity is 62.5% or better for a 95% confidence interval. For an 80% confidence interval, the minimum guaranteed sensitivity will then be 68.5%.

Another way to increase sensitivity is to decrease specificity. However, a low specificity for automatic sorting will directly influence the economy of milk production since normal saleable milk will be discarded. Farmers will not accept low specificities for automatic sorting and to a certain extent not for an alarm list either since the proportion of normal to abnormal cow milkings then increases. It is proposed that the specificity for automatic sorting should be >99%. To give that guarantee with a 95% probability, all 50 out of 50 normal cow milkings should test normal. Testing 99 as normal out of 100 truly normal cow milkings will only assure that the specificity is >97%, and testing 199 normal out of 200 truly normal cow milkings will give a minimum specificity of 98.5%. Calculation of specificity on the quarter level improves the statistical power. Appointing 995 quarters as normal out of 1000 truly normal quarter milkings will, with a 95% confidence interval, guarantee that the specificity is >99%. I propose that more than 200 cow milkings should test normal in order to calculate the specificity safely at the cow level, which will also improve the precision of specificity calculations on a quarter basis. Calculations at the quarter level should not be confused with the fact that diversion of abnormal milk still is at the cow level.

Recommendations and conclusions

It is quite clear from the relatively low sensitivities that the current systems cannot be used for automatic sorting of milk. Alarm systems should focus more on the definition of abnormal milk if a higher sensitivity shall be achieved without lowering the specificity. Automatic sorting based on a low specificity will discard a lot of milk and relatively much for herds with a low prevalence of clinical mastitis (Rasmussen, 2003). The farmers will not accept low specificities (Ouweltjes, 2004). Fuzzy logics can be used to improve sensitivity and specificity of systems using conductivity as the main source of information for mastitis detection (de Mol and Woldt, 2001; de Mol et al., 2001). Some of the systems may obviously benefit from adopting and implementing such calculation models.

References

Mol, R.M. de, W. Ouweltjes, G.H. Kroeze and M.M.W.B. Hendriks. 2001. Detection of estrus and mastitis: field performance of a model. Applied Engineering in Agriculture. 17:399-407.

Mol, R.M. de and W.E. Woldt. 2001. Application of fuzzy logic in automated cow status monitoring. J. Dairy Sci. 84:400-410.

Ouweltjes, W. 2004. Demands and opportunities for operational management support. Operational management on farms with automatic milking systems. EU project "Implications of the introduction of automatic milking on dairy farms (QLK5-2000-31006) as part of the EU-program "Quality of Life and Management of Living Resources", Deliverable D28, 36 pp.

Rasmussen, M.D. (ed.), 2002. Definition of normal and abnormal milk at time of milking. Internal report no. 169 for Workshop of the EU-project (QLK5-2000-31006): Implications of the introduction of automatic milking on dairy farms. November 27, 2002, 102 pp.

Rasmussen, M.D. M. Bjerring and F. Skjøth. 2002. Visual inspection of foremilk. In Rasmussen, M.D. (ed.) Internal report no. 169 for Workshop of the EU-project (QLK5-2000-31006): Implications of the introduction of automatic milking on dairy farms. November 27, 2002, 47-62.

Rasmussen, M.D., 2003. Consequences of definitions of acceptable milk quality for the practical use of automatic milking systems. EU project "Implications of the introduction of automatic milking on dairy farms (QLK5-2000-31006) as part of the EU-program "Quality of Life and Management of Living Resources", Deliverable D6, 25 pp.

IMPACT OF AUTOMATIC MILKING ON EXCRETION OF ANTIBIOTIC RESIDUES

Karin Knappstein, Gertraud Suhren & Hans-Georg Walte
Federal Dairy Research Institute, Institute for Hygiene and Food Safety, Kiel, Germany

Abstract

Prevention of antibiotic residues in milk is an important aspect in milk production to ensure health protection for the consumer and high quality of milk for dairy processing. Limited information is available on the potential impact of milking frequencies associated with Automatic Milking (AM) on excretion of veterinary drugs in milk.

Under experimental conditions the excretion of antibiotics in milk was studied in healthy cows (somatic cell count in composite milk below 100 000/ml) after intramammary treatment with 4 different commercially available drugs at milking frequencies of 3, 2 and 1.5 times per day. The concentrations of antibiotic residues in cow composite milk sampled at every milking time were determined by HPLC methods and compared to the Maximum Residue Limits (MRL) for each compound.

For drugs containing cefquinome or penicillin or a combination of penicillin, nafcillin and dihydrostreptomycin significantly ($p<0.05$) shorter excretion periods were observed in cows milked 3 times per day compared to cows milked 1.5 times per day. For one drug containing ampicillin and colistin differences were not significant. Higher concentrations of residues in milk were determined with shorter intervals between treatment and next milking.

After intramammary treatment of clinical mastitis with cefquinome no significant difference of excretion times was detected between cows milked 2 times (n=15) or 1.5 times (n=4) per day. In these cows only milk yield had a significant influence on excretion time with shorter excretion times in high yielding cows ($p<0.05$).

The indicated withholding periods for milk were sufficient for all drugs. Nevertheless it is recommended to milk treated cows at least twice per day, because prolonged excretion was determined in healthy cows milked less frequently.

Introduction

Prevention of antibiotic residues in milk is necessary to ensure health protection of consumers and to avoid failures during milk processing. In practice residues of antibiotics in milk are most often determined by microbial inhibitor tests. Treatment of mastitis with antibiotics is one of the most important causes associated with positive results of inhibitor tests on bulk tank milk (Fabre et al., 1995). In conventional milking management factors like accidental milking of treated cows, not attending the withholding time and failures during milking are the main reasons for antibiotic residues in bulk milk together with insufficient cleaning of milking equipment (Schällibaum, 1990). Rasmussen et al. (2003) reported a higher risk for inhibitor positive tests to be associated with Automatic Milking

Automatic milking – A better understanding

(AM). Like in conventional milking management failures were the main reasons for residues in milk.

Little is known about the impact of milking intervals on the excretion of antibiotic residues in milk. Previous investigations were mainly focused on the effect of stripping on concentrations of antibiotics in milk after treatment of mastitis. Schluep and Heim (1980) found no difference for cefacetrile excretion in quarters stripped 4 times within 10 hours between treatment and next regular milking compared to longer intervals in quarters of the same cow. In contrast, Henschelchen and Walser (1983) determined significantly shorter excretion periods for procain-penicillin G and oxytetracycline when cows were stripped in 2-hour intervals after treatment. Both groups made their investigations with healthy cows.

The distribution of antibiotics in the udder is dependent on the vehicle used in drugs for intramammary infusions as well as on the physico-chemical properties of the antibiotic. High molecular weight, low lipid solubility and high degree of ionization at pH of milk prevent antibiotics from permeation of the blood milk barrier after intramammary treatment (Ziv, 1975). The distribution of antibiotics changes due to increased pH of milk in quarters with acute clinical mastitis.

The aim of the study reported here was to determine the excretion of antibiotics in milk in dependence on milking frequency. Four drugs were studied in healthy cows milked with three different milking frequencies. One drug was selected for further analysis in cows with clinical mastitis. The concentration of residues in milk at the end of the withholding period was compared to Maximum Residue Limits (MRL) set by Council Regulation 2377/90/EEC.

Materials and methods

Excretion trials

Excretion trials were performed under experimental conditions simulating milking frequencies as observed in AM systems.

Dairy cows

Trials in healthy cows German Holstein, groups of 5 (4 to 6) cows, lactation number 1 to 6, days after calving 41-299, somatic cell count (SCC) in composite milk <100 000/ml at three samplings in weekly intervals before beginning of treatment trial (in the following referred to as healthy cows), average milk yield per cow 21.4-37.8 kg per day, comparable milk yield between three groups tested with one drug, body weight 534-800 kg.

Trials in cows with clinical mastitis German Holstein, groups of 12 and 4 cows, lactation number 1 to 5, days after calving 1-422, SCC in composite milk between 3.2×10^5/ml and 7.0×10^6/ml, average milk yield per day between 10.3 and 37.9 kg, body weight 610-851 kg. The number of clinical cases of mastitis per cow was: one case (7 cows), 2 cases (5 cows), 3 cases (1 cow); in 2 cows 2 quarters were affected at the same time. Infections with the following pathogens were determined: *Staphylococcus aureus* (2 quarters), coagulase-negative staphylococci (3 quarters), *Streptococcus uberis* (5 quarters), *Escherichia coli*/coliform bacteria (5 quarters), enterococci (2 quarters), coryneform bacteria (2 quarters) and mixed infections (2 quarters), no pathogen detected in 3 quarters.

Treatment

Drugs The following drugs were used for intramammary treatment of healthy cows:

- Cobactan® LC (Hoechst Roussel Vet, now Intervet Int., Unterschleissheim, DE), 75 mg cefquinome (CEF) per injector (as CEF-sulfate), withholding time for milk: 5 days, MRL in milk: 20 µg/kg
- Procain-Penicillin G 3 Mio. (WDT, Garbsen, DE), 1898 mg Penicillin G (PEN) per injector (as procain-benzyl-PEN), withholding time for milk: 5 days, MRL in milk: 4 µg/kg
- Nafpenzal® MC (Intervet Int, Boxmeer, NL), 180 mg penicillin, 100 mg nafcillin and 100 mg dihydrostreptomycin (DHS) per injector (as PEN-sodium, NAF-sodium and DHS-sulfate), withholding time for milk: 5 days, MRL in milk: PEN - 4 µg/kg, NAF - 30 µg/kg, DHS - 200 µg/kg
- Omnygram® (Virbac S.A., Carros, F), 866 mg ampicillin (AMP) and 82.5 mg colistin (COL) per injector (as AMP-trihydrate and COL-sulfate), withholding time for milk: 6 days, MRL in milk: AMP - 4 µg/kg, COL - 50 µg/kg

Only Cobactan® LC was used for the excretion studies in cows with clinical mastitis.

Treatment Healthy cows: 4 udder quarters per cow (worst case), one injector per quarter, 3 treatments within 24 hours (Cobactan® LC and Omnygram® - milking 2x/day) or 48 hours (Procain-Penicillin G 3 Mio., Nafpenzal® MC and Omnygram® - milking 3x and 1.5x/day), respectively. Milking times during the treatment period were adjusted to allow application of drugs in similar intervals for the 3 groups of cows tested per drug.

Cows with clinical mastitis: Treatment of quarters with clinical symptoms of disease, one injector Cobactan® LC per quarter, 3 treatments within 24 hours; additional treatment of one cow with 26 ml Cobactan® 2.5% i.m. equiv. to 650 mg cefquinome

Milking frequency Healthy cows: Per drug 3 groups of 5 cows tested with milking frequencies: 2 times per day (interval 14/10 hours, reference), 3 times per day (interval 8 h), 1.5 times per day (interval 16 h)

Cows with clinical mastitis: 2 times per day (17 cases), 1.5 times per day (5 cases)

Sampling The experimental period included anamnesis (3 milkings or one milking for healthy and mastitic cows resp.), treatment period, indicated withholding period for milk plus 2 days. Sampling was extended if residues were detected more than 2 days after the end of the withholding period.

- quarter milk samples: for determination of somatic cell count (SCC) - at every milking; for determination of mastitis pathogens - healthy cows: one milking before start of treatment, cows with clinical mastitis: at every milking
- composite milk: for determination of SCC and residues of antibiotics, taken at every milking during the experimental period

Storage of samples Max. 60 h at 6 °C; composite milk - max. 3 weeks at -20 °C for further investigations; lyophilisation and storage at 6 °C for later re-examinations if necessary

Laboratory analysis
Udder health
SCC was determined according to IDF Standard 148A:1995. Mastitis pathogens were identified according to the guidelines of the German Veterinary Association (DVG, 2000).

Detection of residues of antibiotics
Qualitative detection Screening tests (microbial inhibitor tests, receptor assay, ELISAs) with sufficient sensitivities to detect antibiotic residues in milk at MRL concentrations were applied (Report D11, Knappstein *et al.*, 2003).

Quantitative detection For quantitative detection of residues HPLC methods were applied (Suhren and Knappstein, 1998, 2003, Suhren and Walte, 2003).

Determination of withdrawal time
The withdrawal time was determined by the Time-to-Safe-Concentration (TTSC) method according to the guidance of the European Agency for the Evaluation of Medicinal Products for the determination of withdrawal periods for milk (EMEA, 1998).
Cows with clinical mastitis who received treatment on 2 quarters or additional parenteral treatment were excluded from calculations.

Analysis of variance
In order to determine which factors have a systematic influence on the withdrawal time an analysis of variance was carried out by use of the GLM (General Linear Model) procedure of SAS, release 8.01. The linear model had the following equation:

$$Y_{ijkl} = \mu + mf_i + dac_j + ln_k + b_1(X_{ijkl}) + b_2(X_{ijkl}) + e_{ijkl}$$

where Y_{ijkl} is the dependent variable (first time in hours when the antibiotic content fell below the MRL), μ the overall mean, mf_i the effect of the ith milking frequency (3, 2, 1.5), dac_j the effect of jth days after calving (healthy cows: $\leq$100 d, >100 d; cows with clinical mastitis: $\leq$60 d, >60 d), ln_k the effect of the kth number of lactation (1, >1). Milk yield and SCC as continuous variables were used as covariate with b_1 the slope for milk yield, b_2 the slope for SCC; e_{ijkl} the random residual error. In addition, body weight (kg) was included as influencing factor for Nafpenzal® MC and Omnygram® experiments as well as for Cobactan® LC trials in cows with clinical mastitis.

Results and discussion

Excretion of residues in healthy cows
The excreted amount of residues in milk in percent of the total amount applied is summarized for all 4 drugs in dependence on milking frequency in figure 1.
Highest amounts of all drugs were excreted via milk in cows milked 3 times per day. No general tendency between milking frequency and excreted amount could be derived. In accordance with the pharmaco-kinetic properties of AMP the amount excreted via milk was very low. In contrast, high amounts of DHS (up to 80%) and COL (65%) were excreted via

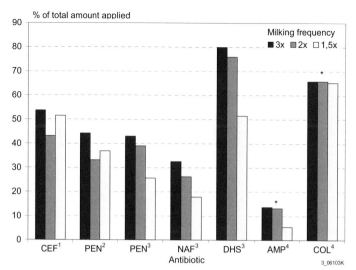

% of total amount applied

*Figure 1. Excreted amount of antibiotic residues in milk in % of total amount applied in dependence on milking frequency; [1] Cobactan® LC, [2] Procain-Penicillin G 3 Mio, [3] Nafpenzal® MC, [4] Omnygram®, * deviating treatment scheme.*

milk, which is probably due to the low lipid solubility of both drugs. For COL almost no difference was observed between the 3 groups of cows.

In figure 2 examples for the excretion of residues in milk of individual cows due to different milking frequencies are presented.

Although the concentrations of PEN excreted in milk were similar for all cows at the beginning of treatment, more variation was seen after several milkings. Whereas the concentration was below the MRL after 72 hours in all cows milked 3 times per day, this was only the case after 112 h in cows milked with intervals of 16 hours. No relation was observed between milk yield and excretion time.

The influence of interval between treatment and next milking on the concentration in milk is shown for selected antibiotics in figure 3.

Higher concentrations in milk were observed when cows were milked in shorter intervals after application of the drug. This has to be regarded when cows are milked more frequently

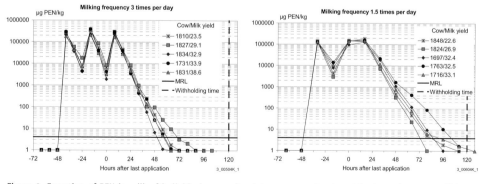

Figure 2. Excretion of PEN in milk of individual cows after intramammary treatment in dependence on milking frequency.

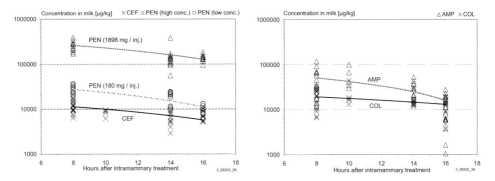

Figure 3. Concentrations of different antibiotics in milk in dependence on interval between treatment and milking.

in AM systems. The comparison of two drugs containing PEN in different concentrations showed that concentrations in milk also increased with higher dosage of treatment. Deviations from recommended treatment schemes (off-label use) may therefore influence concentrations of antibiotics in milk as well as duration of excretion in milk.

The calculated withholding times according to the TTSC method are summarized for all four drugs in figure 4.

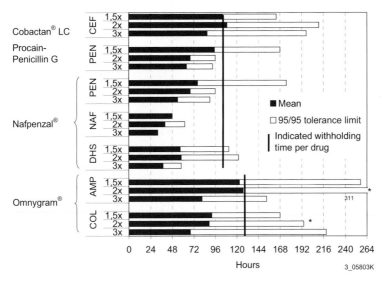

Figure 4. Withholding times for antibiotics in 4 commercial drugs (intramammary application) calculated by TTSC method in dependence on milking frequency; * deviating treatment scheme.

Withholding periods tended to be shorter with increasing milking frequency. The mean withholding periods were in accordance with the indicated withholding periods set for the respective drugs. If the 95/95 tolerance limit was calculated, in several cases the indicated withholding period was not sufficient, due to the large variation between individual cows and high safety factors. The reason for the high safety factor is the low number of cows (4 to 6 cows per group). To reduce the safety factor the use of a minimum number of 19 cows is recommended by EMEA (1998) for determination of withholding times. For PEN-treated cows milked 2 and 3 times per day and for NAF- and DHS-treated cows milked 1.5 times and 3 times the withholding times were sufficient even with the 95/95 tolerance limit.

Shorter excretion periods in cows milked more frequently as determined by TTSC method were confirmed by the analysis of variance (Table 1).

For all antibiotics the excretion times in cows milked 1.5 times versus 3 times were significantly different except for Omnygram®. On average also longer excretion periods of AMP and COL were observed in cows milked less frequently. Probably due to the large variation between excretion times in individual cows the differences were not significant for these two antibiotics.

From the different behaviour of antibiotic compounds in the two drugs containing more than one antibiotic it is confirmed, that not only the vehicle of the drug is important for excretion times in milk, but also the properties of the antibiotic itself.

Other factors of significant influence on excretion time were days after calving for CEF (shorter excretion times in cows <100 d after calving) as well as number of lactation for PEN and DHS in Nafpenzal® LC experiments (shorter excretion times for cows in first lactation). The latter was no longer significant when body weight was included into the analysis.

Table 1. Withdrawal time (in hours) for the different antibiotic drugs in healthy cows (Least Square Means (LSQ$_M$) and standard error).

	Milking frequency per day[1]		
	3x	2x	1.5x
Cobactan® LC			
CEF	71.8 ± 6.5 [a]	111.6 ± 4.8 [b]	99.9 ± 5.6 [b]
Procain-Penicillin G 3 Mio.			
PEN[2]	64.7 ± 4.6 [a]	65.7 ± 4.5 [ab]	96.3 ± 4.5 [b]
Nafpenzal® MC			
PEN[3]	51.1 ± 4.5 [a]	61.4 ± 4.7 [ab]	76.7 ± 5.3 [b]
NAF	31.8 ± 1.6 [a]	38.8 ± 1.7 [b]	48.0 ± 1.9 [c]
DHS	34.8 ± 3.9 [a]	48.9 ± 4.0 [b]	56.7 ± 4.6 [b]
Omnygram®			
AMP	82.9 ± 27.8 [a]	n.a.	123.5 ± 18.5 [a]
COL	62.3 ± 20.3 [a]	n.a.	69.5 ± 13.5 [a]

[1] Different letters within the same row show significant differences between milking frequencies (p<0.05) [2] 1898 mg PEN per injector, [3] 180 mg PEN per injector in a drug combination, n.a. = not applied

Excretion of residues in cows with clinical mastitis

Only a limited number of excretion studies was possible in cows with clinical mastitis. The drug Cobactan® LC was selected for these studies, because in healthy cows large variability of excretion times as well as concentrations in milk exceeding the MRL close to the end of the withholding period were determined.

The excretion of CEF in milk in percent of the total amount applied varied between 4.4% and 46.7% for cows milked two times per day and between 11.1% and 28.9% for cows milked 1.5 times per day. The average excretion rate in milk was 26.7% and 18.0%, respectively and much lower than those observed in healthy cows (43.1% for cows milked 2 times, 51.5% for cows milked 1.5 times per day). Probably changes in permeability of the blood milk barrier or in pH of milk followed by increased absorption of antibiotics from the udder are the reasons for these differences. The extremely low excretion rate of 4.4% was observed in one cow which had severe mastitis and fever. This may be explained by severe damage of the blood udder barrier thus allowing passage of a higher percentage of the antibiotic to blood serum.

The concentrations of CEF in milk after treatment are presented for individual cows milked 2 times per day in figure 5.

Large variation was found for excretion times in milk of individual cows. The time until concentrations fell below the MRL was between 38 and 72 hours. In none of the cows the MRL was exceeded at the end of the withholding time.

Also for cows milked 1.5 times per day large variation was observed. Times until concentrations in milk fell below the MRL varied between 48 and 64 h (data not shown).

Different excretion times were determined in individual cows treated repeatedly for different cases of mastitis with at least two weeks between different cases (Table 2).

The withdrawal times calculated by TTSC method and by analysis of variance are summarised for the two groups milked with different milking frequencies in table 3.

The withdrawal times in cows with mastitis were much shorter than in healthy cows milked with the same milking frequencies after treatment with Cobactan® LC. One reason is probably the lower total amount of drug used for treatment of only one quarter compared

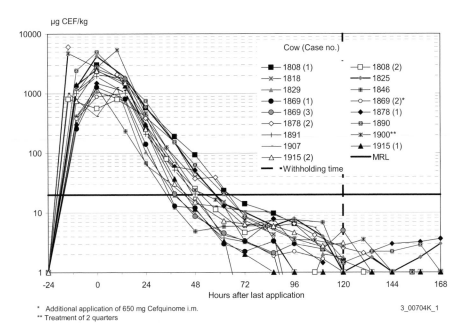

*   Additional application of 650 mg Cefquinome i.m.
** Treatment of 2 quarters

3_00704K_1

Figure 5. Excretion of CEF in milk of cows with clinical mastitis, milking frequency two times per day.

Table 2. Variation in excretion times of CEF in milk in individual cows with more than one case of clinical mastitis.

	First time after last application (hours) when concentration in milk was below MRL		
Cow[1]	1st case	2nd case	3rd case
1808	72	58	-
1868	48	48	-
1869	38	48[2]	48
1878	62	72	-
1907	48	64	-
1915	48	48	-

[1] milking frequency two times per day, except 1907 (1st case - 2 x, 2nd case 1.5 x per day) and 1868 (both cases 1.5 x per day), [2] cow received additional treatment with 650 mg CEF i.m.

Table 3. Withdrawal times for CEF (hours) in cows with clinical mastitis in dependence on milking frequency per day.

Milking frequency per day	TTSC method		Analysis of variance	
	Mean	95/95[1]	$LSQ_M \pm se$[2]	sign.[3]
2 times (n=15)	54	66	57.7 ± 3.0	a
1.5 times (n=4)	52	108	45.5 ± 6.4	a

[1] 95/95 tolerance limit, [2] LSQ_M=Least Square Mean, se=standard error, [3] different letters within column show significant differences (p<0.05)

to the worst case studies with treatment of 4 udder quarters in healthy cows. In addition, a lower percentage of the total amount of CEF was excreted via milk in cows with mastitis. Therefore it can be concluded that larger amounts of CEF were absorbed into blood serum and excreted via other ways. This was especially the case in one cow with fever where less than 5% of the antibiotic were excreted via milk.

The means calculated by the TTSC method were very similar for the two groups milked with different intervals. When the 95/95 tolerance limits are included it has to be considered that for the group milked 1.5 times per day a large safety factor had to be included due to the low number of cows in this trial (n=4) versus 15 cows in the group milked 2 times per day. Between the two groups no significant differences of withdrawal times were determined by analysis of variance. The only factor of significant influence on excretion time was the average daily milk yield per cow, with shorter excretion times in high yielding cows (p<0.05). This factor was no longer significant when body weight was included into the analysis.

Conclusions

The excretion time of antibiotic residues in milk of healthy cows varied with different milking frequencies after intramammary treatment. For CEF, PEN, NAF and DHS significantly shorter excretion times were observed in cows milked 3 times per day compared to cows milked 1.5 times per day. Longest excretion times were observed for all antibiotics except for CEF in cows milked less frequently with 1.5 milkings per day.

Treatment trials in cows with clinical mastitis showed large variation in excretion times of individual cows and much shorter excretion times than in worst case studies in healthy cows. Although no significant differences were determined between groups milked 2 times or 1.5 times per day it should be considered that only one drug was tested in a low number of cows. A more pronounced influence of milking frequency may be observed for other drugs. Although results from studies in healthy cows can not necessarily be transferred to cows with clinical mastitis controlled milking of cows at least 2 times per day is recommended after treatment with antibiotics.

If cows in AM systems are milked in shorter intervals after treatment higher concentrations of residues in milk have to be expected. This could increase the risk for carry over of antibiotics into milk of the next cow milked at the same place, especially if failures in cleaning procedures occur.

According to the results presented prolonged excretion of residues in milk of individual cows in connection with milking frequencies deviating from regular milking times twice per day seem to be of minor importance for positive results of inhibitor tests in bulk tank milk when recommendations for treatment are followed.

Acknowledgements

The work was funded by the European Commission within the EU project "Implications of the Introduction of Automatic Milking on Dairy Farms (QLK5-2000-31006)" as part of the EU-programme "Quality of Life and Management of Living Resources".

The content of this publication is the sole responsibility of its publisher, and does not necessarily represent the views of the European Commission nor any of the other partners of the project. Neither the European Commission nor any person acting on behalf of the Commission is responsible for the use, which might be made of the information presented above.

References

Council Regulation (EEC) No. 2377/90 of 26 June 1990 laying down a Community procedure for the establishment of maximum residue limits of veterinary medicinal products in foodstuffs of animal origin. Official J. European Communities No. L 224/1, 18.8.90

DVG Sachverständigenausschuss "Subklinische Mastitis", 2000. Leitlinien zur Entnahme von Milchproben unter antiseptischen Bedingungen und Leitlinien zur Isolierung und Identifizierung von Mastitiserregern. Giessen, März 2000.

European Agency for the Evaluation of Medicinal Products (EMEA) - Committee for Veterinary Medicinal Products, 1998. Note for Guidance for the Determination of Withdrawal Periods for Milk. EMEA/CVMP/473/98-final.

Henschelchen, O. and K. Walser, 1986. Effect of milking interval on antibiotic residues in milk after intramammary treatment. Tierärztliche Umschau 838: 83-85.

International Dairy Federation (IDF), 1995. Milk, Enumeration of somatic cells, Method C, Fluoro-opto-electronic method. IDF Standard 148A:1995.

Knappstein, K., G. Suhren, H.-G. Walte, 2003. Prevention of antibiotic residues. Influences of milking intervals and frequencies in automatic milking systems on excretion characteristics of different antibiotics in milk. Report D11, EU project Implications of the introduction of automatic milking on dairy farms. http://www.automaticmilking.nl

Rasmussen, M.D. and P. Justesen, 2003. The frequency of antibiotics in milk from herds with automatic milking systems. Proceedings of Satellite Symposium Effects of the robotic milking system on milk quality with reference to typical products, August 30, 2003, Rome, Italy, p. 318.

Schällibaum, M., 1990. Antibiotikatherapie und Rückstände in der Anlieferungsmilch. Swiss Vet 7: 7-9.

Schluep, J. and H. Heim, 1980. Ausscheidung in der Milch von Cefacetril und Penicillin G nach intramammärer Applikation zu verschiedenen Zeitpunkten. Schweizer Archiv für Tierheilkunde 122: 39-43.

Suhren, G. and K. Knappstein, 1998: Detection of incurred dihydrostreptomycin residues in milk by liquid chromatography and preliminary confirmation methods. Analyst 23: 2797-2801.

Suhren, G. and K. Knappstein, 2003. Detection of cefquinome in milk by liquid chromatograph and screening methods. Analytica Chimica Acta 483: 363-372.

Suhren, G. and H.-G. Walte, 2003. Experiences with the application of method combinations for the detection of resiudes of antimicrobial drugs in milk from collecting tankers. Milchwissenschaft 58: 536-540.

Ziv, G., 1975. Pharmacokinetic concepts for systemic and intramammary antibiotic treatment in lactating and dry cows. Proceedings IDF Seminar on Mastitis Control, April 7-11, 1975, Reading, UK, pp. 314-340.

SELECTION OF COWS FOR TREATMENT OF UDDER INFECTIONS IN AMS HERDS

Torben Werner Bennedsgaard, Susanne Elvstrøm & Morten Dam Rasmussen
Danish Institute of Agricultural Sciences, Foulum, Denmark

Abstract

The selection of cows for veterinary treatments of udder infections in 20 Danish Dairy herds using automatic milking systems (AMS) was described by interviews and data analysis. The methods used to identify cows with mastitis were related to the type of AMS installed in the herd. With one type of AMS the farmers primarily relied on the alarm lists on conductivity from the AMS, whereas the farmers with another type of AMS primarily used their own observations to detect cows with mastitis. 5 out of the 12 farmers with the second type of AMS found that mastitis was detected at a later time after introduction of AMS. Despite the limited use of the information from the AMS to detect cows with mastitis the number of cows treated for mastitis increased from 0.44 treatments per cow year to 0.61 treatments per cow year during the first year after introduction of AMS. The time of treatment in relation to stage of lactation was unchanged. The udder health accessed by analysis of the individual cow somatic cell counts (SCC) only showed small and temporary changes after introduction of AMS.

Introduction

The objective of this study was to evaluate the identification of cows with abnormal milk and the decision process leading to veterinary treatment in Danish dairy herds with AMS.

In herds with AMS, the identification of cows for separation of milk or veterinary treatment of udder infections is no longer primarily based on the visual inspection of the milk. The farmer must select cows for clinical examination based on the notifications lists from the AMS or observation of the behaviour of the animals. Furthermore, the introduction of a quality control program may include special procedures like the California Mastitis Tests to be routinely used in groups of animals (Rasmussen et al., 2001)

The changes in identification principles may lead to new principles for veterinary treatment of udder infections. Acute infections may be possible to detect at an earlier phase whereas chronic infections may be overlooked. (De Mol and Ouweltjes, 2001; Milner et al., 1997).The treatment success may be influenced by the changes in time of infection detection and may affect the farmers' threshold for treatment.

A change in udder health and treatment policy was proposed based on a study of 808 Danish dairy herds (Rasmussen et al., 2003). It was found that the frequency of cows with acutely elevated somatic cell counts (SCC) increased whereas the frequency of cows with chronically elevated SCC decreased after introduction of AMS in both new and existing barns. In herds with new barns with conventional milking, the opposite development was seen.

The frequency of filed mastitis treatments was unchanged in the herds introducing AMS, but decreased in herds with new barns using conventional milking.

Material and methods

Twenty herds introducing AMS between June 1999 and March 2003 were included in the study. The herds were selected based on geographical location in the western part of Denmark and willingness to participate in an interview.

Questionnaire

The farmers were interviewed to evaluate the identification of cows with abnormal milk and the decision process leading to veterinary treatment. During the interview, a questionnaire was filled out. The questionnaire included both general questions about herd health management and use of the AMS and a discussion of the five latest cases of mastitis treatment and culling.

Data on disease treatment and production

Data on veterinary treatments, reproduction and milk production were available from the Danish Cattle Database. Some key figures for the herds are presented in table 1.

Table 1. Herd characteristics of 20 Danish AMS herds in 2003.

	Mean	p10-p90
Herd size	110	55-143
Milk production (kg ECM/cow year)	9,020	6,500-10,500
Avg. calving number	2.1	1.8-2.5
Udder treatments per 100 cow years	56	21-93
Calculated bulk tank SCC	280	190-370
Culling rate	42	22-54

The results of the interviews were combined with an analysis of data of veterinary treatments and monthly test milkings to describe the principles leading to selection of cows for veterinary treatment.

Results

Questionnaire

The farmers were asked to what extent they used the different alarm lists from the AMS. Significant differences were seen between the two most common types of AMS. In the herds with type 2 AMS, only half the farmers used the lists for conductivity and fluctuations in milk production whereas all farmers with type 1 AMS used the list for conductivity to identify cows with mastitis (Tables 2, 3, 4). When asked how they identified cows with mastitis, the farmers with type 2 AMS mentioned their own observations in the herds (Table 4). These farmers also thought that mastitis was detected at a later stage after introduction of AMS

(Table 5). Thirteen of the farmers had expected a reduction in the use of veterinary medicine after introduction of AMS. However, only six farmers felt there had been a reduction, whereas five had experienced an increase in the use of veterinary medicine (Table 6). The average mastitis treatment incidence was 0.44 per cow year during the last year before introduction of AMS in the herds and 0.61 per cow year in the first year after. The estimated bulk tank somatic cell count (SCC) based on the individual cow SCC increased from 270,000 in the last year before AMS to 290,000 in the 6 month after whereupon it decreased to 275,000 (Figure1).

Table 2. Use of alarm lists for conductivity from 3 types of AMS.

	Used often	Sometimes	Limited use	Not used
Type 1	6	1		
Type 2	4	2		6
Type 3		1		

Table 3. Use of alarm lists for fluctuations in milk production from 3 types of AMS.

	Used often	Sometimes	Limited use	Not used
Type 1	2		4	
Type 2	3	2	7	
Type 3			1	

Table 4. Importance of different methods to diagnose mastitis.

	Most important	Second most important
Type 1	Robot-list (7/7)	Own observations (4/7) CMT-test (2/7)
Type 2	Own observations (7/12) Robot-list (4/12)	CMT-test (4/12) Robot-list (4/12)
Type 3	Own observations	Robot-lists

Table 5. Farmer thinks that mastitis is detected at least as early as before AMS.

	Yes	No
Type 1	7	0
Type 2	7	5
Type 3	0	1

Table 6. Expectations and farmers' opinion of the use of antibiotic treatment in connection with AMS.

AMS	Expected			Realized		
	Increase	Reduction	Unchanged	Increase	Reduction	Unchanged
Type 1	0	4	2	3	2	2
Type 2	0	9	3	2	4	6
Type 3	0	0	1	0	0	1

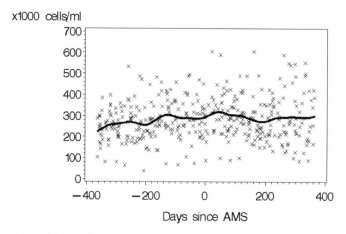

Figure 1. Estimated bulk tank somatic cell count calculated from individual cow SCC from one year before to one year after introduction of AMS.

Data on disease treatment and production

The time of mastitis treatment in relation to stage of lactation showed no changes after introduction of AMS (Figure 2). Also the stage of lactation for increase in somatic cell counts (acute score, Rasmussen *et al.*, 2001) was unchanged after introduction of AMS. At each monthly test milking in early lactation, about 35 percent of the cows with previously low SCC had an increase in somatic cell count indicating a new infection compared with about 20 percent in late lactation (Figure 3).

Discussion

The methods for detection of cows with mastitis were closely related to the type of AMS installed in the herds. Rasmussen & Bjerring (2003) have evaluated the sensitivity and specificity of the different types of AMS for detection of abnormal milk. AMS type 1 was found to have a high specificity compared with type 2, the sensitivity being lowest for type 1, however. The result of the present study shows the farmers' reaction to the predictive value of the lists from the AMS. The farmers probably react to the low specificity by combining the information from the alarm lists with their own observations of the cow's

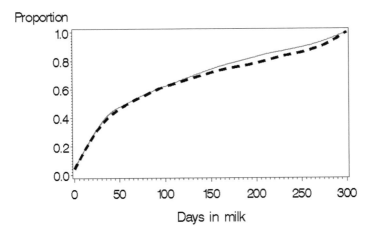

Figure 2. Days in milk at treatment for mastitis before (broken line) and after (full line) introduction of AMS.

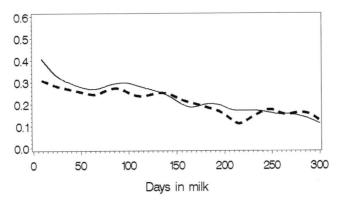

Figure 3. Cows with first case of acutely elevated somatic cell counts as proportion of cows at risk before AMS (broken line) and after AMS (full line).

behaviour and appearance, or simply by relying only on their own observations if the predictive value of the alarm lists is too low.

According to the Danish quality control system, CMT-tests as control measure are used as a standard procedure for detection of infected quarters immediately before installation of AMS and at least two times within the first three months with AMS. The result of the interviews shows that CMT is still considered as important tool for identifying infected cows in at least 6 of the 20 herds.

Despite a temporary increase in the estimated bulk tank SCC in the first months after introduction of AMS, the small changes in the proportion of cows with acutely elevated SCC and the constant estimated bulk tank indicate that udder health is only moderately affected in the herds after introduction of AMS (Figure 3). However, the use of veterinary treatments for mastitis increased by about thirty percent in the first year with AMS.

Despite the limited use of the information from the AMS in detection of cows with mastitis, especially for AMS type 2, the farmers were able to maintain the udder health in the herds. However, since many of the herds changed to a new barn at the same time as introducing the AMS an improvement of the udder health would have been expected (Rasmussen *et al.*, 2003).

References

De Mol, R.M. and W. Ouweltjes. 2001. Detection Model for Mastitis in Cows Milked in an Automatic Milking System. Preventive Veterinary Medicine 49[1-2]: 71-82. 2001.

Milner, P., K.L. Page, and J.E. Hillerton. 1997. The effects of early antibiotic treatment following diagnosis of mastitis detected by a change in the electrical conductivity of milk. Journal of Dairy Science 80[5]: 859-863.

Rasmussen, M.D., J.Y. Blom, L.A.H. Nielsen, and P. Justesen. 2001, Udder health of cows milked automatically. Livestock Production Science 72[1-2]: 147-156.

Rasmussen, M.D., F. Skjøth, L.A.H. Nielsen and P. Justesen. 2003, Animal health of herds converting to automatic milking. EAAP 2003 Rome, Session CM6.5, Cattle, management and Health.

Rasmussen, M.D. and M. Bjerring. 2003. Definition of normal and abnormal milk at time of milking - Possibilities of automatic milking systems to detect and separate milk based on quality. Deliverable D7.

Automatic milking – A better understanding

INFLUENCE OF VARYING MILKING INTERVALS ON MILK COMPOSITION - A PHYSIOLOGICAL APPROACH ON SECRETION OF "NORMAL MILK"

J. Hamann & H. Halm

Department of Hygiene and Technology of Milk, School of Veterinary Medicine Hannover, Hannover, Germany

Abstract

A group of approx. 40 high-yielding German Holstein Frisian cows were milked robotically (VMS®, DeLaval) throughout 400 days. During 20 session days (20-day intervals), samples (quarter foremilk [QFM], quarter composite milk [QCM] and cow composite milk [CCM]) were drawn continuously for 24 hours. Cell count (SCC) determination (Fossomatic) was done in all milk fractions, culturing in QFM. The milk constituents NAGase activity, lactate, fat, protein, and lactose were analysed in QCM.

Data analyses (SAS, PROC GLM) were directed to evaluate milk secretion, SCC and milk composition in relation to the milking interval (MI). The overall SCC in QCM was 4.75 lg (= 56,234 cells/ml).

The mean milking frequency (MF) over 24 h was 2.72 with an average MI of 9:01 ± 2:32 hours and a mean milk yield of 26.87 ± 10.59 kg/cow/day. Distribution of all 1,802 milkings was as follows: MF1: = 0.93%; MF2: 38.04%; MF3: 49.44%; MF4: 11.04% and MF5: 0.47%.

The lactation stage (days in milk, DIM) had a marked influence on the udder secretion rate, being independent from MI (4 - 6 h, 6 - 8 h; 8 - 10 h; 10 - 12 h, 12 -14 h). Milk secretion fell to approx. 50% between DIM < 100 and DIM > 299. The quarter secretion rate (MI: 6 h = 356 g/h; MI: > 12 h = 232 g/h) showed highly significant differences between all MI and a non-linear pattern in relation to the MI applied.

All milk constituents but lactate showed non-linear, MI-dependent secretion patterns. Values for NAGase and lactose for MI > 10 hours suggested inflammation, but mean SCC amounted to 4.75 lg. Due to its insensitivity to MI, lactate could be useful for automatic mastitis detection, also when the cell count level is < 100,000 cells/ml.

Introduction

The application of automated milking systems enables milking events at a voluntary basis with irregular milking intervals (MI). The magnitude and extent of effects on milk yield and milk composition by varying MI is not well established. The majority of published data focus on CCM instead of QCM. Furthermore, either they use extended sampling intervals (up to several months) or, when sampling interval was reduced, trials lasted little (Weiss and Bruckmaier 2002, Poelarends et al., 2002, Wangler et al., 2002). Milk yield and composition is mainly influenced by udder health at quarter level (Hamann and Reinecke 2002). Therefore, only QCM data should be used for comparing effects between different milking systems. Moreover, the definition for normal milk with normal milk composition under conventional conditions has been defined with a SCC threshold of 100,000 cells/ml and healthy udder quarters (Doggweiler and Hess 1983, Tolle *et al.*, 1971, Hamann 2003).

This study was performed to evaluate the interaction between irregular MI and milk secretion (udder/quarter level) and the corresponding levels of SCC and some milk constituents under robotic milking conditions (VMS®).

Material and methods:

A group of 40 high-yielding German Holstein Frisian cows at different lactation stages and numbers were milked robotically (VMS®: voluntary milking system, DeLaval; vacuum 42 kPa, pulsation rate 60 c/min, pulsation ratio 65%). Over 400 days, survey was carried out every 20 days, and each session included 24 hours of continuous sampling. In addition to milk yield SCC, NAGase, lactate, fat, protein and lactose were determined in 6194 quarter composite milk samples (QCM). Means were calculated for five different MIs in a range between < 6 h to > 12 h using a general linear model (PROC GLM, SAS 1996) for repeated measurements. Signficances were calculated using the Student-Newman-Keuls-test and the Student-t-test for impaired samples.

Results

Physiological references during the lactation
Table 1 summarises some physiological cow data and mean values of MI, MF and SCC throughout the study.

Between two and four milkings a day, lactation number and stage tend to decrease while yield increases.

As figure 1 shows, the udder milk secretion rate [g/h] is reduced, regardless of MI, by approx. 50% between DIM < 100 and DIM > 299.

Milking interval and secretion rate
Significant (at least $p \leq 0.05$) differences occurred between the quarter secretion rates at all MI and also between the front and hind quarters (Table 2).

Table 1. *Lactation-related physiological data (milking frequency [MF] in relation to yield, lactation no., days in milk, milking interval [MI] and somatic cell count [SCC]) throughout the study (400 days; 20 session days).*

MF [times/day]	1	2	3	4	5
cow n = (%)	6 (0.90)	254 (38.31)	327 (49.32)	73 (11.01)	3 (0.45)
Lactation No.	3.17 ± 1,17	2.81 ± 1.40	2.40 ± 1.42	2.23 ± 1.22	2.33 ± 1.53
Days in milk	148 ± 155	214 ± 106	182 ± 98	139 ± 91	18 ± 91
Yield/session [kg]	11.5 ± 6.7	20.8 ± 8.1	29.2 ± 9.4	36.9 ± 11.3	35.1 ± 10.0
Milkings n =	6	508	981	292	15
MI [h:min]	12:24 ± 2:21	10:48 ± 2:33	8:45 ± 2:04	6:52 ± 1:35	5:28 ± 0:50
SCC/ml CCM [lg]	5.13 ± 0.72	4.98 ± 0.52	4.79 ± 0.50	4.90 ± 0.56	5.37 ± 0.18

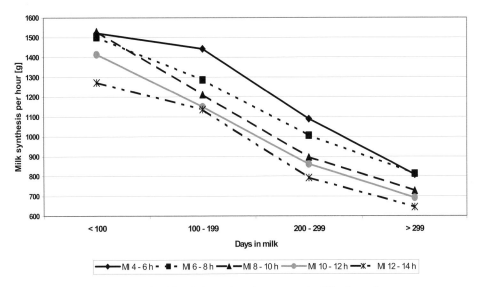

Figure 1. Milk secretion rate at cow level in relation to lactation stage and milking interval.

Table 2. Milk secretion [g/h] at quarter level (all, front, rear); n = 7,044 milkings at 20 session days.

Milking interval [h]	< 6	6 - 8	8 - 10	10 - 12	> 12
All, n =	638	2172	2090	1183	961
All, xa ± sd*	356 ± 195[a]	310 ± 131[b]	286 ± 129[c]	260 ± 124[d]	232 ± 108[e]
Front, n =	314	1085	1048	596	488
Front, xa ± sd	321 ± 214[a]	279 ± 105[b]	256 ± 108[c]	233 ± 104[d]	201 ± 88[e]
Rear, n =	324	1087	1042	587	473
Rear, xa ± sd	389 ± 168[a]	342 ± 145[b]	316 ± 141[c]	288 ± 136[d]	263 ± 117[e]
Rear vs. front**	‡‡‡	‡‡‡	‡‡‡	‡‡‡	‡‡‡

* = different letters within lines differ significantly
** = Student t-test for impaired samples (‡‡‡ = p < 0,001, ‡‡ = p < 0,01, ‡ = p < 0,05, n.s. = not significant)

The comparison of the measured quarter secretion rates per hour with the expected linear secretion rate is shown in figure 2.

It was seen that the application of irregular milking intervals resulted in a non-linear secretion rate displaying higher levels for shorter and lower levels for longer intervals.

Milking interval and related SCC and milk composition

Table 3 (QCM) and table 4 (CCM) detail the interactions between MI and milk composition.

All milk constituents except lactate showed a non-linear, MI-dependent secretion pattern.

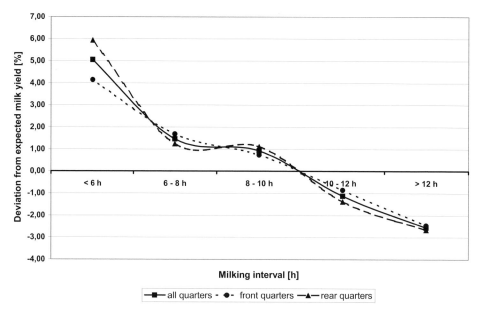

Figure 2. Percentage deviation of milk yield from linearity of milk secretion; n = 7,044 quarter milkings at 20 session days.

Table 3. Mean values of QCM components in relation to different milking intervals.

Milking Interval	< 6 h	6 - 8 h	8 - 10 h	10 - 12 h	> 12 h
n = milkings	490	1880	1859	1081	884
SCC [lg$_{cells/ml}$]	4.77 ± 0.52[a]*	4.72 ± 0.54[b]	4.73 ± 0.55[b]	4.78 ± 0.56[a]	4.79 ± 0.55[a]
NAGase [U]	0.19 ± 0.27[e]	0.22 ± 0.27[d]	0.26 ± 0.31[c]	0.33 ± 0.30[b]	0.37 ± 0.32[a]
Lactate [lg$_{\mu mol/l}$]	1.43 ± 0.42[a]	1.43 ± 0.45[a]	1.43 ± 0.51[a]	1.42 ± 0.52[a]	1.48 ± 0.49[a]
Fat [%]	4.41 ± 1.14[a]	4.39 ± 1.03[a]	4.28 ± 1.09[b]	4.05 ± 1.05[c]	4.00 ± 1.16[c]
Protein [%]	3.26 ± 0.30[d]	3.35 ± 0.32[b]	3.35 ± 0.36[b]	3.32 ± 0.31[c]	3.39 ± 0.37[a]
Lactose [%]	4.83 ± 0.22[a]	4.81 ± 0.24[b]	4.79 ± 0.25[b]	4.74 ± 0.32[c]	4.73 ± 0.28[c]

* = different letters within lines indicate significant (at least p < 0.05) differences

Principally, the tendencies observed in these values is comparable to that of QCM values (Table 3). Yet, also here, values of NAGase activity and lactose exceeded the physiological range despite that the SCC mean ranged < 100,000 cells/ml.

Discussion

Robotic milking is intended to apply higher daily milking frequencies than conventional milkings. However, several studies have shown that this did not necessarily translate into increases in milk yield (Kremer and Ordolff 1992, Reinecke 2002, Wirtz *et al.*, 2002). It is to be expected that the secretory activity may also contribute to milk composition. The

Table 4. Mean values of CCM components in relation to different milking intervals.

Milking Interval	< 6 h	6 - 8 h	8 - 10 h	10 - 12 h	> 12 h
n = milkings	150	519	498	284	231
SCC [lg$_{cells/ml}$]	4.92 ± 0.51^a	4.82 ± 0.52^b	4.85 ± 0.53^{ab}	4.88 ± 0.51^{ab}	4.92 ± 0.49^a
NAGase [U]	0.22 ± 0.28^c	0.23 ± 0.25^c	0.28 ± 0.30^b	0.35 ± 0.26^a	0.39 ± 0.31^a
Lactate [lg$_{\mu mol/l}$]	1.52 ± 0.39^a	1.46 ± 0.45^a	1.48 ± 0.47^a	1.46 ± 0.49^a	1.52 ± 0.49^a
Fat [%]	4.41 ± 1.12^a	4.42 ± 0.93^{ab}	4.36 ± 0.94^{ab}	4.18 ± 0.97^b	4.21 ± 1.11^b
Protein [%]	3.33 ± 0.37^b	3.37 ± 0.33^{ab}	3.37 ± 0.39^{ab}	3.33 ± 0.32^b	3.40 ± 0.39^a
Lactose [%]	4.84 ± 0.19^a	4.82 ± 0.19^a	4.80 ± 0.21^a	4.75 ± 0.25^a	4.72 ± 0.25^c

* = different letters within lines indicate significant (at least $p < 0.05$) differences

data in table 1 and figure 1 indicate the lactation stage (DIM) as the most substantial influence on milk secretion. Although a linear secretory activity was expected under robotic milking conditions due to the relatively short MI, the results however confirmed that the secretion rate under robotic milking conditions is non-linear (Table 2 and figure 2). The MI-related values of milk constituents in QCM and CCM indicated comparable tendencies in the changes of the parameters. Strikingly, the values obtained for NAGase activity and lactate at MI > 10 h would be classified as pathological in both QCM and CCM when obtained under conventional circumstances. Yet, SCC means remained < 100,000 cells/ml, showing that well-established so-called physiological ranges for typical milk components also depend on the MI (Halm 2003).

This arouses problems when it comes down to appropriate online mastitis diagnosis. If a MI-dependent parameter is to be used, it must be evaluated strictly in relation to the corresponding MI. This implies the need to establish new physiological ranges. The other way to solve the problem would be to choose a MI-independent parameter in the first place, e.g. lactate.

References

Doggweiler, R. and E. Hess 1983: Zellgehalt in der Milch ungeschädigter Euter. Milchwissenschaft 38: 5 - 8, 1983

Halm, H., 2003. Zum Einfluss eines automatischen Melkverfahrens auf Milchmengenleistung und Milchinhaltsstoffe hochleistender DH-Kühe unter Berücksichtigung von Laktationsstadium und Eutergesundheit. School for Veterinary Medicine, Hannover (thesis)

Hamann, J. 2003: Definition of the physiological cell count threshold based on changes in milk composition. IDF-Mastitis Newsletter, 25, 9 - 12, 2003

Hamann, J. and F. Reinecke, 2002. Machine milking effects on udder health - comparison of a conventional with a robotic milking system. In: Wageningen Pers, First North American Conference on Robotic Milking, Proc., 17-27

Kremer, J.H. and D. Ordloff, 1992. Experiences with continuous robot milking with regard to milk yield, milk composition and behaviour of cows. In: Ipema, A.H., A.C. Lippus and W. Rossing (editors), Prospects for automatic milking, Wageningen Pers, EAAP publication 65

Poelarends, J., D. J. Willemsen and K. de Koning 2002: Relationship between MI and somatic cell count. In: Wageningen Pers, First North American Conference on Robotic Milking, Proc., IV82-IV85

Reinecke, F., 2002. Untersuchungen zu Zellgehalt und N-acetyl-beta-D-glucosaminidase-Aktivität (NAGase) in Viertelanfangsgemelken sowie zur Leistungsentwicklung von Kühen bei Anwendung eines konventionellen oder eines automatischen Melkverfahrens. School for Veterinary Medicine, Hannover (thesis)

SAS Institute Inc., 1996. SAS/Stat Software, Changes and Enhancement through release 6.12.

Tolle, A., Heeschen, W., Reichmuth, J. and Zeidler, H. 1971: Counting of somatic cells in milk and possibilities of automation. Review article., No. 166, Dairy Scienece Abstracts 33: 875 -979, 1971

Wangler, A., P. Sanftleben and O. Weiher 2002: Milk yields and constituents under conditions of different milking frequencies in automatic milking systems. In: Wageningen Pers, First North American Conference on Robotic Milking, Proc., V76-V79

Weiss, D. and R. M. Bruckmaier, 2002: Influence of the milking interval on milk composition in automatic milking systems. In: Wageningen Pers, First North American Conference on Robotic Milking, Proc., IV78-IV81

Wirtz, N., E. Oechtering, E. Tholen and W. Trappmann, 2002. Comparison of an automatic milking system to a conventional milking parlour. In: Wageningen Pers, First North American Conference on Robotic Milking, Proc., III50-III55

COST OF DISCARDING MILK WITH AUTOMATIC SEPARATION OF ABNORMAL MILK

Diederik Pietersma[1] & Henk Hogeveen[2,3]
[1]Department of Animal Science, McGill University, Montreal, Québec, Canada
[2]Business Economics Group, Wageningen University, Wageningen, The Netherlands
[3]Department of Farm Animal Health, Faculty of Veterinary Medicine, Utrecht University, The Netherlands

Abstract

The goal of this research was to investigate theoretically how the costs associated with discarding abnormal milk and the number of undetected abnormal milkings vary with the detection system's performance. Factors evaluated were: mastitis incidence (20 and 60 cases per 100 cows per year), sensitivity (0 through 100%), specificity (99, 99.5 and 100%), separation of abnormal milk at the quarter and cow milkings level, and farmer intervention to detect mastitis after the first or third abnormal milking. The calculations were performed assuming quarter level detection models for abnormal milk. Reducing the specificity increased the cost of discarded milk due to increased false positives. This increase in cost was higher for cow level separation than for quarter level separation. A high mastitis incidence level resulted in higher cost due to the increase in milk discarded during the withholding period. Increasing the sensitivity rapidly reduced the number of undetected abnormal milkings, especially given farmer intervention after three abnormal milkings and a high incidence of mastitis. This research suggests that the specificity of an automated system to discard abnormal milk must be very high to prevent economic losses due to incorrectly discarding normal milk and the subsequent reluctance of farmers to use such a system.

Introduction

Detection of abnormal milk is an important part of automatic milking (AM). Abnormal milk has been defined as "milk which differs from normal milk in respect of colour or homogeneity" (Rasmussen, 2003). The cause of abnormal milk is usually mastitis, which may result in clots, flakes, or a watery appearance. Currently used AM systems do not automatically detect and exclude abnormal milk from delivery to the bulk tank. Instead, the system generates a warning that a cow may have mastitis based on the sensor signals of the most recent milking. For a confirmed case of mastitis, the dairy farmer programs the AM system to withhold the milk for the mastitis treatment period. However, in the near future, AM systems will likely be required by legislation to automatically keep abnormal milk separate from the normal milk (Rasmussen, 2003; Jepsen et al., 2003). Investigation of the effects of implementing fully automated separation of abnormal milk with AM systems is therefore needed.

An automatic system for the separation of abnormal milk should be able to correctly discard a reasonable proportion of abnormal milkings, but should have very few normal milkings mistakenly detected as abnormal (i.e. false positives) to limit the amount of normal milk discarded unnecessarily (Hogeveen and Lankveld, 2002; Rasmussen, 2003). This type of performance is quite different from the performance of mastitis detection systems. The latter act as a screening tool to focus the farmer's attention to the cows most likely to have mastitis and a moderate number of false positives can be considered as acceptable as no automated actions are taken. Hillerton (2000) estimated that a human milker in conventional milking systems might be able to detect 80% of the quarters with abnormal milk (sensitivity), while classifying 100% of quarters with normal milk as normal (specificity). Rasmussen (2000) investigated the detection of visually abnormal milk with AM using electrical conductivity sensors. At 70% sensitivity, the tested systems had many false positives and the performance was considered inadequate. In proposed legislation, a sensitivity of 80% has been suggested (Jepsen *et al.*, 2003). Due to the very low prevalence of mastitis at the quarter milkings level, the specificity must be very high to limit the number of false positives. Rasmussen (2003) proposed a specificity larger than 99% to make the system acceptable to dairy farmers.

Initial calculations of the cost of discarding milk associated with automatic separation of abnormal milk with AM were done by Hogeveen and Lankveld (2002). Their results suggested that substantial costs can be expected with detection algorithms that have a specificity of 99.3% or less. The goal of this study was to further investigate theoretically the application of automatic separation of abnormal milk. Specific objectives were 1) to evaluate the effects of mastitis prevalence, sensitivity and specificity of detection models, and separation of milk at the quarter and cow milkings level on the cost of discarding milk; and 2) to explore the effects of prevalence and sensitivity on the number of undetected abnormal milkings.

Materials and methods

A spreadsheet was used to calculate the expected costs of discarded milk and number of undetected abnormal milkings for AM systems that automatically separate abnormal milk.

Classification performance

The detection of abnormal milk represents a classification problem with two classes. The class of interest, i.e. abnormal milk, is usually referred to as the positive class, while the other class is referred to as negative. A 2x2 contingency table or confusion matrix is often used to distinguish among the different outcomes when testing two-class classification models using cases labelled with a so-called "gold standard" (Pietersma *et al.*, 2003; Swets, 1988). Table 1 shows such a matrix in which true positives (TP) and true negatives (TN)

Table 1. Confusion matrix for two-class classification.

	Actual class is positive	Actual class is negative	Total (predicted)
Predicted as positive	TP	FP	TP + FP
Predicted as negative	FN	TN	FN + TN
Total (actual)	TP + FN	FP + TN	

are correct classifications, a false positive (FP) is an actual negative case, incorrectly predicted as positive, and a false negative (FN) is an actual positive case, predicted as negative. With these possible outcomes, the sensitivity is defined as TP / (TP + FN) and the specificity is defined as TN / (FP + TN). The confusion matrix was used to calculate the expected number of discarded milkings (TP + FP) and undetected abnormal milkings (FN).

General assumptions

Most abnormal milkings, perhaps as much as 99%, are attributable to either mastitis or colostrum. The latter is programmed to be separated for the first few days after calving. On rare occasions, the abnormality is caused by the presence of blood giving the milk a slightly red appearance (Rasmussen, 2003). This research was limited to the separation of abnormal milk due to clinical mastitis. The calculations were performed for a herd of 100 dairy cows, producing 8500 kg milk per cow per calendar year. These cows were assumed to be 320 days in milk (excl. colostrum) per calendar year with 2 milkings per day. Thus a total of 256000 quarter milkings with on average 3.3 kg milk were involved. The cost per kg discarded milk was assumed to be 0.15 €, which was regarded as the variable costs of producing 1 kg of milk under Dutch circumstances. Each mastitis case was assumed to include a six day withholding period due to antibiotic treatment. These programmed abnormal milkings do not need to be classified by a detection model, but the withheld milk was included in the total amount of discarded milk. The calculations were performed assuming quarter level detection models for abnormal milk, with sensitivity and specificity also pertaining to quarter milkings.

Each mastitis case at the cow level was assumed to be detected eventually, perhaps only after several FN milkings. If an abnormal milking has not been detected by the automatic separation system, the mastitis warning system may have put the cow on the attention list followed by detection, treatment, and programmed discarding of milk by the farmer. For the calculations of the cost of discarding milk, it was assumed that, on average, the farmer will have intercepted undetected mastitis cases after the second abnormal milking. For example, given a detection model for abnormal milk with 80% sensitivity, 20% of the mastitis cases will not be detected at the first abnormal milking. At the second milking, after signs of mastitis appeared, these 20% of the mastitis cases are again submitted to the detection model, with 0.8*0.2 or 16% being detected. Following the second milking, the farmer is assumed to detect the remaining 4% of mastitis cases. In reality, some mild cases of mastitis may never be detected and instead be dealt with by the cow's immunity system. Although these cases were excluded from the calculations in this study, they increase the total number of undetected abnormal milkings.

Factors affecting amount of discarded milk

In this study, two levels of mastitis prevalence were considered: 20 and 60 cases per 100 cows per year. With the amount of discarded milk calculations, two levels of sensitivity were considered: 80%, which is equivalent to human milkers (Hillerton, 2000) and 40%, which is expected to be much easier to attain for detection systems fixed at a very high specificity. Specificity was evaluated at three levels: 99, 99.5 and 100%. Separation of abnormal milk at the quarter level was compared with discarding the milk of all four quarters (cow level separation). Each abnormal cow milking was assumed to include, on average, 1.33 quarters predicted as abnormal. The amount of milk discarded using cow level separation was thus expected to be three times the amount calculated at quarter level separation.

Factors affecting number of undetected abnormal milkings

The number of abnormal milkings that are not detected and end up in the bulk tank (FN) were calculated for two mastitis prevalence levels: 20 and 60 cases per 100 cows per year. With the assessment of undetected abnormal milkings, sensitivity was evaluated for the entire range from 0 to 100%, with 0% representing AM systems without automatic separation of abnormal milk. Regarding farmer intervention for undetected abnormal cases, two possibilities were evaluated. The first assumed that the mastitis warning system has 100% sensitivity, i.e., the farmer is able to detect mastitis cases that were missed by the automatic separation system right after the first abnormal milking. The second possibility assumed that, on average, farmer intervention takes place after three undetected abnormal milkings.

Results and Discussion

Cost of discarded milk

In table 2, the calculation results are shown for five different scenarios. The first and second show the effect of increasing the specificity from 99 to 99.5%. Scenarios two and

Table 2. Calculation of amounts and cost of discarded milk per 100 cows per year for five scenarios at quarter level separation of abnormal milk.

Scenario	1	2	3	4	5
Assumptions					
Mastitis cases /100 cows /year	20	20	20	60	60
Quarter milkings withheld	960	960	960	2880	2880
Quarter milkings to classify	255040	255040	255040	253120	253120
Sensitivity detection system (%)	80	80	40	80	40
Specificity detection system (%)	99	99.5	99.5	99.5	99.5
Results					
True Positives (#)	20	20	20	60	60
False Negatives (#)	5	5	19	14	58
True Negatives(#)	252465	253740	253726	251780	251737
False Positives (#)	2550	1275	1275	1265	1265
Undetected abnormal milk (kg)	16	16	64	48	191
Correctly discarded milk (kg)	66	66	66	199	199
Incorrectly discarded milk (kg)	8467	4234	4233	4201	4200
Discarded antibiotics milk (kg)	3188	3188	3188	9563	9563
Total discarded milk (kg)	11721	7488	7487	13963	13962
Cost (€/year)	1758	1123	1123	2094	2094

tree show the effect of changing the sensitivity from 80 to 40%, while four and five explore the same change in sensitivity for a higher mastitis incidence level.

Figure 1 shows the result for all evaluated scenarios. At 100% specificity, the costs were similar for quarter and cow level separation. These costs represented milk discarded for the withholding period and the detected milking. At a high prevalence, costs were higher due to the increase in milk discarded during the withholding period. Reducing the specificity increased the cost of discarded milk due to increased false positives.

Cow level separation was approximately 1300 and 2600 € per 100 cows per year more expensive than quarter level separation at 99.5 and 99% specificity, respectively. However, while quarter level separation may be the more economical alternative, cow level separation may be preferable to assure that all milk from an infected cow is withheld from the food processing chain (Rasmussen, 2003). Calculations at 40% sensitivity resulted in an insignificant decrease in cost of discarded milk (Table 2). To prevent economic losses due to incorrectly discarding normal milk, the specificity of an automated system to discard abnormal milk must be very high, perhaps as high as 99.5%. Too much milk discarded unnecessarily will make farmers reluctant to using automated separation systems for abnormal milk.

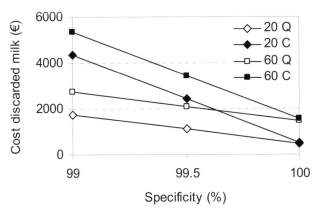

Figure 1. Total cost (€ per 100 cows per year) due to discarded milk against specificity, given yearly mastitis incidence of 20 or 60 cows, separation at the quarter (Q) and cow (C) level, and 80% sensitivity.

Undetected abnormal milkings

Figure 2 shows the calculated number of FN and the FN relative to all abnormal milkings, both plotted against the sensitivity of the detection system. At 0% sensitivity (no automated milk separation) and farmer intervention after the first or third abnormal milking, the number of FN was equivalent to one and three times the incidence of mastitis at the cow level. Increasing the sensitivity of an automatic system for abnormal milk separation results in a linear reduction of the number of undetected abnormal milkings for farmer intervention after the first milking and a larger but declining reduction of FN for farmer intervention after three milkings. At 0% sensitivity, 50 and 75% of the abnormal milkings (FN + TP) are undetected, for farmer intervention after the first and third abnormal milking, respectively (Figure 2). Increasing the sensitivity of the detection system decreases this proportion at an increasing rate.

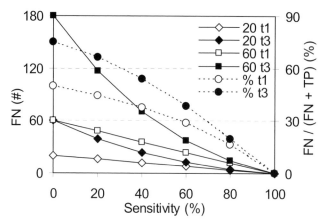

Figure 2. Total number of undetected abnormal milkings per 100 cows per year (FN) and proportion of all abnormal milkings (FN / FN + TP) against sensitivity, given yearly mastitis incidence 20 or 60 cows and farmer intervention after one (t1) or three (t3) abnormal milkings.

Trade-off between sensitivity and specificity

To achieve acceptable performance in terms of cost of discarded milk, the specificity might have to be at least 99.5%. However, currently available detection systems may not be able to achieve 80% sensitivity at this very high level of specificity. From the perspective of reducing the amount of abnormal milk in the bulk tank, implementing in the short term an automatic separation system for abnormal milk with 99.5% specificity and a low sensitivity of, e.g., 40% would already reduce the number of undetected abnormal milkings substantially, especially in situations where farmer intervention is expected to take place only after several abnormal milkings and with a high incidence of mastitis (Figure 2). This study did not distinguish between severe and mild abnormalities regarding the classification performance. However, severely abnormal milkings are likely the least difficult to detect for an automatic separation system. Thus, most of the severe abnormalities are expected to be detected at even overall low sensitivities, while most of the FN are expected to involve less obvious abnormalities that are more difficult to distinguish from normal milk.

Further research will be required to develop abnormal milk detection systems with very high values for both sensitivity and specificity. Improved sensors that measure aspects closely correlated with signs of abnormal milk may be needed. In addition, detection performance is expected to be improved through the use of detection models that integrate multiple sensor signals and other on-farm data. Developing such models requires sensor data from a very large number of milkings to include a sufficient number abnormal cases and may benefit from the application of data mining techniques to extract meaningful classification rules from these data and from the use of Bayesian networks to incorporate existing domain knowledge regarding mastitis detection.

Conclusions

The results of this research suggest that the specificity of detection systems for automatic separation of abnormal milk needs to be very high, perhaps as high as 99.5%, to avoid substantial economic losses due to incorrectly discarding normal milk. A high mastitis incidence level leads to higher costs of milk discarded during the withholding period and many more undetected abnormal milkings than a low incidence of mastitis. Quarter level separation of abnormal milk can lead to substantially lower cost of discarding milk than cow level separation. Implementation of an automatic separation system with a low sensitivity would already lower the number of undetected abnormal milkings substantially.

References

Hillerton, J.E. 2000. Detecting mastitis cow-side. In: Proc. Nat. Mastitis Council 39th Ann. Meeting. National Mastitis Council, Madison, WI, USA: 48-53.

Hogeveen, H. and J.M.G. Lankveld. 2002. Economics of milk quality - some starting points for discussion. In: Proc. Workshop Definition of normal and abnormal milk at time of milking, Foulum, Denmark: 81-89.

Jepsen, L., B. Everitt, C. Cook, R.H. Oost, H. Hogeveen and J.B. Rasmussen. 2003. Quality Management at Farm Level - Code of Good Hygienic Practices for Milking with Automatic Milking Systems. Bulletin of IDF No. 386. International Dairy Federation, Brussels, Belgium.

Pietersma, D., R. Lacroix, D. Lefebvre and K.M. Wade. 2003. Induction and evaluation of decision trees for lactation curve analysis. Computers and Electronics in Agriculture 38: 21-36.

Rasmussen, M.D. 2000. Evaluation of methods for detection of abnormal milk during automatic milking. In: Hogeveen, H. and A. Meijering (editors), Proc. Int. Symp. On Robotic Milking, Wageningen Pers, Wageningen, The Netherlands: 125.

Rasmussen, M.D. 2003. Definition of normal and abnormal milk at time of milking; consequences of definitions of acceptable milk quality for the practical use of automatic milking systems. Deliverable D6 of EU project 'Implications of the introduction of automatic milking on dairy farms' (QLK5-2000-31006).

Swets, J.A., 1988. Measuring the accuracy of diagnostic systems. Science 240: 1285-1293.

THE USE OF SPECTRAL PHOTOMETRY FOR DETECTION OF MASTITIS MILK

Martin Wiedemann[1] & Georg Wendl[2]
[1]Landtechnischer Verein in Bayern e.V., Freising, Germany
[2]Bavarian State Research Center for Agriculture, Institute of Agricultural Engineering, Farm Buildings and Environmental Technology, Germany

Abstract

When using automatic milking system reliable sensors are necessary for detecting altered milk resp. ascertaining the udder health. The objective of this research was to prove the possibility to detect mastitis milk by using the spectral photometry in combination with the established measurement of electrical conductivity (EC). Therefore two experiments were carried out. A preceding inspection of about 250 quarter composite milk sample resulted a high correlation (r=0.62) between milk colour and milk fat content in the red wave length of visible light (band 620-700 nm). The results of the first experiment revealed a significant difference (P<0.01) between the spectral reflectance (SR) of quarter foremilk (first 500 ml) with low somatic cell count (SCC) (<100,000 som. cells/ml) and increased SCC (>200,000 som. cells/ml) in the blue wave length (band 400-520 nm). The possibility to differ milk with low SCC from those with increased SCC disappeared when using the SR of the samples after start of milk ejection and the samples of quarter composite milk. In the second experiment a number of more than 1,100 quarter foremilk samples (16 cows, 9 days, 280 milkings) were analysed to ascertain the state of udder health by using the EC in combination with SR. A diagnostic test showed sensitivities of 55% (threshold: 200,000 som. cells/ml) resp. 71% (threshold: 500,000 som. cells/ml) at a specificity of 95%. The sensitivity could be increased about 10% if a "or-" combination of EC and SR was used.

The results of this research provided a better detection rate than the present sensor systems in practice. In both cases (EC or SR measurement) it was essential to do the measuring in the first fractions of the quarter milk.

Introduction

Today the consumer demands for a high level of food production. Comply with legal requirements an inspection of milk at any milking time has to be done to ensure that milk for consumption is produced by healthy cows. However it is not possible to check the foremilk with regard to abnormalities by a milker in automatic milking systems. Therefore sensor systems which are able to detect any variations in milk are needed. In future those systems should not only give information about the milk condition, in fact they have to decide whether the milk is marketable or not. With respect to these requirements it was investigated if the spectral photometry as a simple to use technique can support these necessary decisions in addition to the EC measurement.

Former investigations have shown that changes in milk like blood or colostrum can be detected by using the SR (Espada & Vijverberg, 2002; Wiedemann & Wendl, 2003).

Furthermore a few authors found that bovine mastitis can cause changes in milk colour (Espada & Vijverberg, 2002; Ouweltjes & Hogeveen, 2001; Ordolff, 2001). However Bergann & Schick (1998) pointed out a high correlation between milk colour - based on the CIE-L*a*b* system - and the milk fat content (r>0.9).

On account of this fact it was very important to find out the most fat affected wave lengths in the visible light. The main intention of this research was to descry a possibility for detecting altered (mastitis) milk with a well experienced method like spectral photometry and to compare and combine the milk colour with the well known electrical conductivity.

Method and material

The study was carried out on the research farm "Hirschau" (TU Munich-Weihenstephan) using a VMS® that milked 45 cows (Red Holstein x Simmental). The research was split up into two experiments.

In order to get the quarter milk fractions and the foremilk samples the construcion of a sampler which does not influence the normal automatic milking routine, was needed. By using this sampler it was possible to branch off about 10% of each quarter milk flow ("test portion"). Because of the well known effects of teat cleaning on stimulation and milk ejection (Bruckmaier & Hilger, 2001) no mechanical udder cleaning was done and the attachment of the teat cups was realized as fast as possible to get still the cisternal milk; the four teat cups were attached within 45 s.

In part A the quarter specific milk fractions of 40 animals (with and without secretion disorders) were collected for two days. Altogether 48 milkings were sampled and analysed. The main interest of this part was to scan the SR and EC of the fraction-based milk samples according to the udder health (expressed on SCC of the first milk fraction). The quarter milk yield was divided in three fractions: fraction 1 (first 500 ml), fraction 2 (middle part of milking) and fraction 3 (last 300-500 ml). A sample of the quarter composite milk was taken as well.

In part B the quarter foremilk samples of 16 cows (with and without secretion disorders) were collected and analysed over a period of nine days (280 milkings). The udder health was classified on base of the quarter specific SCC of the test portion of the first 500 ml of milk. The values of SR and EC were recorded out of the same test portions. The results about detecting conspicuous quarters were presented in a diagnostic test (sensitivity, specificity).

SR was measured by using the instrument *spectro-color* (DrLange, Germany) in the reflectance mode under laboratory conditions. Before warming all milk samples to a uniform temperature of 37°C to do the SR measurement, aliquots of 2 ml were frozen for further analyses. To get one reflectance value the arithmetic averages of SR in the different bands (400-520 nm, 530-610 nm, 620-700 nm) were calculated.

EC was measured at 25°C using the LDM electrode from WTW (LDM 130, 82362 Weilheim, Germany). The samples were also examined for fat, protein, lactose and SCC in a standard test in an official laboratory (MPR Bayern, Wolnzach).

For statistical analyses the milk samples were grouped according to SCC as follows:

SCC I: <100,000 som. cells/ml   *SCC II*: 100,000-200,000 som. cells/ml
SCC III: 200,000-500,000 som. cells/ml   *SCC IV*: >500,000 som. cells/ml

These classes were created to comply with conclusions of several authors: Hamann & Zecconi (1998) categorised a SCC with more than 100,000 som. cells/ml in quarter foremilk as disturbed milk secretion whereas Smith (1995) defined the threshold at 200,000 som. cells/ml.

Data are presented as means ± SEM. Results were analysed using the repeated measures analysis of the MIXED procedure and tested for significance ($P<0.01$) using Least Significant-Difference Test (LSD) of the SAS program package (version 8.02).

Results

The preceding investigation of more than 250 composite milk samples (from part A and B) showed the following correlations between fat content ($\log_e$) and the SR in the specific bands:

$r_{(400\text{-}520\ nm)}=0.25$, $r_{(530\text{-}610\ nm)}=0.54$, $r_{(620\text{-}700\ nm)}=0.62$.

These results prompted to take only the blue band (400-520 nm) for detection of mastitis milk into account. The findings are similar to former results (Wiedemann & Wendl, 2003).

The trend of SR values during milking, depending on the time of taken the sample and the status of udder health, was investigated in part A. Figure 1 shows the pattern of the first fraction of the four quarter individual milkings (LF-1, RF-1, LR-1, RR-1) and additionally of the following three fractions of the diseased right rear quarter (RR-2 to RR-4).

The first milk fractions of the non diseased quarters did not show any noticeable deviation, while the SR of the diseased quarter presented a negative deviation of about 4% from the healthy quarters in the band between 400 to 520 nm. The SR of the later milk fractions of the diseased quarter (milk ejection has already been taken place) adapted to the pattern of the milk fractions of the healthy quarters (LF, RF, LR) in band 1. Therefore it was no longer possible to distinguish the diseased from the healthy quarters. The large

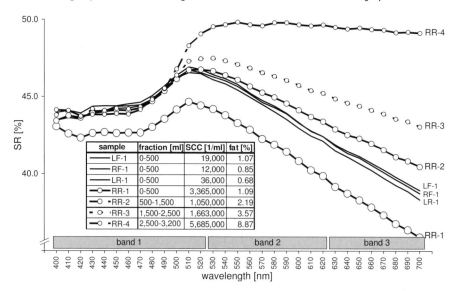

Figure 1. Spectral reflectance (SR) of quarter individual milking fractions of the udder diseased cow 497 (March 4th 2003; 17.2 kg of milk).

deviation in SR of RR-1 to RR-4 in the bands higher than 520 nm are mainly affected by the fat content (1.1 to 8.9%).

On the other milkings (n=48) the same occurrence could be observed (Figure 2). Only in the first 500 ml of milk a significant difference between healthy and diseased quarters was found (*P<0.01*). The average SR in the band 400-520 nm of the following milk fractions (2 and 3) and the quarter composite milk samples did not allow a clear differentiation based on the spectral photometry.

For the same samples the EC was measured as well. In figure 3 the means ± SEM of the milk fractions in the different SCC-classes are shown. The EC of SCC IV quarters differed significantly in the fraction 1 and 2 and in the quarter composite milk sample from SCC I (*P<0.01*). However with ongoing milking time the distance between the EC values of healthy and diseased quarters is getting smaller. In some cases the significant difference between the SCC-classes disappeared totally.

The measured results of both parameters (SR and EC) pointed out that a reliably expressiveness can only be reached if a recording is possible in the first individual quarter milk fraction (<500 ml). Recording the values of the quarter composite milk samples could not generate reliable values.

Based on the previous results of part A, 280 milkings were examined in part B to determine whether milk from diseased quarters is detected with the measuring of EC and of SR. Table 1 shows the distribution of the 1,120 quarter milkings and the means of SR and EC in the different SCC-classes.

A specificity of 95% could be reached by determining the thresholds for each parameter (absolute and relative (inter quarter comparison)). Hence the absolute limits for

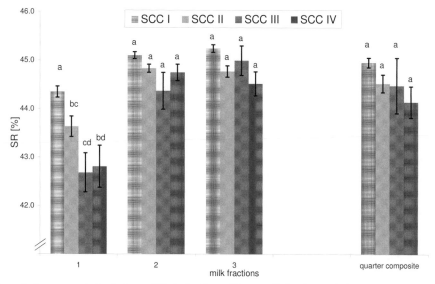

Figure 2. Average spectral reflectance (SR) in the three different milk fractions and the quarter composite milking (400-520 nm); data are clustered according to the SCC in the first milk fraction (SCC I <100,000 cells/ml (n=152); SCC II 100,000 - 200,000 cells/ml (n=17); SCC III 200,000 - 500,000 cells/ml (n=7); SCC IV >500,000 cells/ml (n=16)).

a, b, c, d means with different letters indicate differences between SCC-classes (P<0.01).

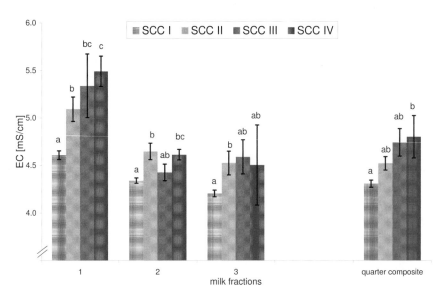

Figure 3. Average electrical conductivity (EC) in the three different milk fractions and the quarter composite milking; data are clustered according to the SCC in the first milk fraction (SCC I <100,000 cells/ml (n=152); SCC II 100,000 - 200,000 cells/ml (n=17); SCC III 200,000 - 500,000 cells/ml (n=7); SCC IV >500,000 cells/ml (n=16)). a, b, c means with different letters indicate differences between SCC-classes (P<0.01).

Table 1. Number of observations, means of spectral reflectance (SR) and electrical conductivity (EC) in the different SCC-classes. (a,b,c,d) means with different letters indicate differences between SCC-classes (P<0.01).

SCC-class	SCC range (1/ml)	Number of observations	SR (%)	EC (mS/cm)
I	<100,000	973	44.52[a]	4.71[a]
II	100,000-200,000	56	43.35[b]	4.92[b]
III	200,000-500,000	46	42.45[bc]	5.26[c]
IV	>500.000	45	41.84[cd]	5.98[d]

inconspicuous quarters (<200,000 or <500,000 som. cells/ml) were given for EC at 5.4 to 5.5 mS/cm and for SR (400-520 nm) at 41%. With the calculation of the inter quarter ratio a diseased quarter was also detected with a difference of EC of more than 10% and SR of more than 3.5% (Table 2). The use of the absolute values of SR are not very meaningful. With the inter quarter difference the sensitivity could be increased from 29 to 42% (200,000 som. cells/ml) resp. from 41 to 53% (500,000 som. cells/ml). The combination of both parameters allowed softer thresholds.

With the additional measurement of the SR it was possible to detect 55% (only EC: 44) out of the 91 quarters with more than 200,000 som. cells/ml resp. 71% (only EC: 61%) out of the 45 quarters with more than 500,000 som. cells/ml at a specificity of 95%.

Table 2. Results of diagnostic tests including the absolute and relative values of electrical conductivity (EC) and spectral reflectance (SR). The limit of SCC for conspicuous quarters was variably defined. The threshold for the parameters were reached when the specificity was up to 95%.

Conspicious quarters		>200 000 som. cells/ml (n=91)		>500 000 som. cells/ml (n=45)	
		Sensitivity (%)	Threshold	Sensitivity (%)	Threshold
Absolute	EC	45	5.4 mS/cm	60	5.5 mS/cm
	SR	29	41%	41	41%
	combination	(45)	5.5 mS/cm or 40.5%	(64)	5.5 mS/cm or 40.5%
Relative	EC	44	10%	61	11%
(inter quarter	SR	42	3.5%	53	4%
comparison)	combination	(55)	11% or 3%	(71)	13% or 4.5%

Conclusion

The experiments have shown that the SR is an other useful and simple to use tool for detecting mastitis milk when it is used in combination with EC. The described method of SR enables to disregard the strong influence of fat content on milk colour. The most meaningful results could be achieved if

- the spectral photometry measurement was done in the band of 400 to 520 nm and
- the milk samples were collected before milk ejection started; a recording of the parameters in the quarter composite milk was not helpful (for both: SR and EC).

The described method - adapted to an automatic milking system - gave much better results than those systems, which are in use at the moment (Köhler & Kaufmann, 2002).

The detection of milk from conspicuous quarters can be even more improved by using intelligent software, e.g. neuronal networks or fuzzy systems, and/or including more parameters like the ion concentration of milk (Köhler & Kaufmann, 2002; Wiedemann *et al.*, 2003).

Acknowledgments

We thank the farm Hirschau, Technical University Munich, for providing the animals. The study was supported by Bavarian Ministry of Agriculture and Forestry and by a grant from the Hanns-Seidel-Foundation.

References

Bergann, T. and M. Schick, 1998. Signifikante Beziehungen - Farbmessung beschreibt den Milchfettgehalt. Lebensmitteltechnik 6: 52-53.

Bruckmaier, R.M. and M. Hilger, 2001. Milk ejection in dairy cows at different degrees of udder filling. Journal of Dairy Research 68: 369-376.

Espada, E. and H. Vijverberg, 2002. Milk Colour Analysis as a Tool for the Detection of Abnormal Milk. In: Proceedings to The First North American Conference on Robotic Milking (Toronto), Wageningen: IV29 - IV38.

Hamann, J. and A. Zecconi, 1998. Evaluation of the electrical conductivity of milk as a mastitis indicator. In: Bulletin of the International Dairy Federation 334: 5-22.

Köhler, S. D. und O. Kaufmann, 2002. Statistical model for the identification of udder diseases in AMS. Landtechnik 6: 57-58.

Ordolff, D., 2001. Einsatz von Farbmessung zur Bewertung von Vorgemelken. In: Institut für Agrartechnik, Universität Hohenheim (editor), Proceedings to 5th International Conference: Construction, Engineering and Environment in Livestock Farming, Stuttgart: 218-223.

Ouweltjes, W. and H. Hogeveen, 2001. Detecting abnormal milk through colour measuring. Natl. Mastitis Council Ann. Meet. 40: 217-219.

Smith, K. L., 1995. Standards for somatic cells in milk: physiological and regulatory. In: IDF Mastitis Newsletter 144/21: 7-9

Wiedemann, M. and G. Wendl, 2003. Untersuchungen zur Erkennung von Veränderungen der Milch mit Hilfe der Farbmessung. In: Kuratorium für Technik und Bauwesen in der Landwirtschaft e.V. (editor), Proceedings to 6th International Conference: Construction, Engineering and Environment in Livestock Farming, Vechta: 124-129.

Wiedemann, M., D. Weiss, G. Wendl and R. M. Bruckmaier, 2003: The importance of sampling time for online mastitis detection by using the electrical conductivity or measuring the Na+ and Cl- content in milk. In: S. Cox (editor), Proceedings to 1st European Conference on Precision Livestock Farming (Berlin), Wageningen: 173-178.

AN ON-LINE SOMATIC CELL COUNT SENSOR

D.S. Whyte, R.G. Orchard, P.S. Cross, T. Frietsch, R.W. Claycomb & G.A. Mein
Sensortec Ltd., Hamilton, New Zealand.

Abstract

Measurement of the somatic cell concentration of raw milk is widely accepted as an indicator of mastitis and milk quality. However there are no commercially available automated on-line sensors to measure Somatic Cell Count (SCC) for on-farm use. This paper discusses the development of an automated SCC sensor based on measuring the DNA concentration and describes its performance.

Introduction

A common off-line test for SCC is the California Mastitis Test (CMT). The CMT was developed in 1957 by Schalm and Noorlander who modified the Whiteside test (Whiteside, 1939). The test involves the addition of an anionic surfactant to the milk. The reagent interacts with the DNA and proteins in somatic cells to form a gel.

The nature of the gel has not been extensively studied. It is known that the viscosity of the gel is correlated to the DNA content of the milk (Milne and DeLangen, 1977) and that DNase (Singh and Marshall, 1966) and proteases (Dounce and Monty 1969) stop gel formation. What has not been widely reported is the chemical mechanism leading to gel formation and the rheological properties of the resulting gel.

Christ (1962) proposed the following theory of gel formation. Negatively charged detergent heads bind to positively charged proteins. The reagent's carbon tail is then exposed in water. Since this carbon tail is strongly hydrophobic, it binds to the tails of other reagent molecules. The negatively charged heads of these molecules are then bound indirectly to other proteins via a water molecule and another pair of reagent molecules, thus forming the basis of the gel structure.

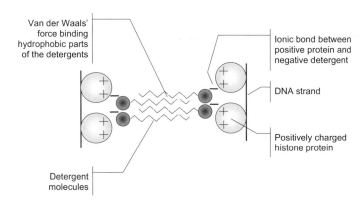

Figure 1. Schematic diagram of proposed theory of gel formation.

We have proposed a modification to this theory (outlined in figure 1). It is now commonly known that DNA is wrapped around proteins called histones before being packed into the chromosome complex. Our theory is that the reagent's negatively charged end forms an ionic bond with the positively charged histones. The reagent's hydrophobic carbon tail is then exposed in water and binds to the tails of other reagent molecules. The negatively charged heads of these molecules are bound directly to other histone molecules, forming the basis of the gel structure.

It has been reported that this gel is non-homogeneous, non-Newtonian, thixotropic (Whittlestone *et al.*, 1970) and rheodestructive (Milne, 1977). However no data have been published illustrating these claims. Microscopic examination of DNA-stained gel revealed its lack of homogeneity, with visible gaps between stained strands. Figure 2 shows the apparent viscosity of the gel obtained using a rotational viscometer at 12 rpm (DV-II PRO Digital Brookfield with LV spindle 1, Middleboro, MA, USA). There is a clear rheopectic phase where the gel forms and viscosity increases; a period where viscosity peaks; and a thixotropic phase where the gel breaks up. Since the thixotropic breakdown is non-reversible, it may also be called a rheodestructive breakdown. A final property of the gel that was observed is visco-elasticity, which can be demonstrated by the "Weissenberg effect," where a fluid 'climbs' a rotating rod (Tiu and Boger, 1983).

Attempts to automate or standardize the viscosity measurement of the DNA-detergent gel test have included:

- Rotary viscometers (Nichols and Phillips, 1972; Nageswararao and Calbert, 1969).
- Falling ball viscometers (Carre, 1970; Kiermeier and Keis, 1964).
- Rolling ball viscometers (Whittlestone and Allen, 1966; Whittlestone *et al.*, 1972).
- Flow cells used either as the Wisconsin Mastitis Test (Thompson and Postle, 1964) or Brabant Mastitis Reaction (Jaartsveld, 1962; Bottazzi, 1963).

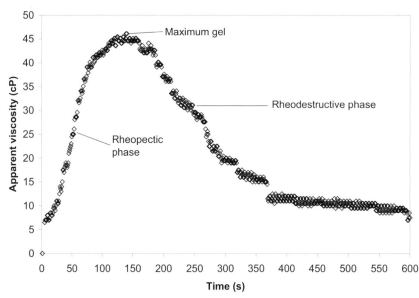

Figure 2. *Illustration of the time dependency of detergent-DNA gel formation and degradation, showing growth of gel (rheopectic phase), maximum gel formation and the rheodestructive phase.*

None of these systems were developed for on-line, on-farm use, probably due to the complexities of the gel and the limited markets for on-farm technology at the time.

Sensortec has undertaken extensive farmer consultations with groups in 7 countries and 11 markets, which illuminated the two likely major uses for an on-line SCC sensor. The first is early mastitis detection, when SCCs are beginning to rise, which could prompt an immediate action to maintain optimal animal health. The other major use is bulk tank SCC management, where farmers can identify, remove or treat problem cows. Different markets were then interested in several measurement levels depending on the market's testing thresholds for bulk tank SCC. From this, we determined a need for flexible band reporting, rather than strictly quantitative results.

The SCC thresholds for this trial were selected based on the market results. The system that best fits the most markets is a five-band scale: < 200; 200 - 500; 500 - 1500; 1500 - 5000; > 5000 kcells/ml (kcells = 1000's cells), however it is possible to fine-tune these thresholds to the requirements of each specific market.

Sensor description

Figure 3 shows a block diagram of the SCC sensor. Currently the sensor operates using the following procedure:

a. A 'start-of-milking' signal from the automatic milking system (AMS) initiates collection of a 4 ml milk sample or, in stand-alone mode, the 4 ml milk sample can be inserted manually. Accuracy of the sample volume is important as it affects the accuracy of the result.
b. The sample is precisely mixed with reagent in the mixing chamber.
c. The mixture is transferred to the measurement chamber.
d. The viscosity is determined by the time taken for the gel to flow out of the measurement chamber.
e. If the outflow passage is blocked, vacuum is applied to clear the blockage.
f. At the end of the measurement cycle, the mixing and measurement chambers are flushed to minimise any cross-contamination between samples.

Materials and methods

The sensor was tested in the laboratory to obtain a sensor calibration and was then tested on-line in the field to verify the calibration. All milk samples were submitted to an independent dairy testing laboratory (Testlink North, Hamilton, New Zealand) for SCC determination (CombiFoss 5000, FOSS Electric, Hillerød, Denmark).

For the lab testing, high and low SCC milk was obtained from a research dairy herd and combined in various proportions to obtain at least 30 measurements for each SCC band. A total of 13 original milk samples were used. To obtain 30 measurements below 200 kcells/ml, 16 additional milk samples were collected directly from individual cows in the research herd. The milk samples were tested in singlicate or duplicate by pipetting 4 ml of sample into the mixing chamber, with both duplicate measurements being used in the performance analysis. The remainder of the process (addition of reagent, transfer to measurement chamber, measurement and post-measurement cleaning) was completed automatically. The

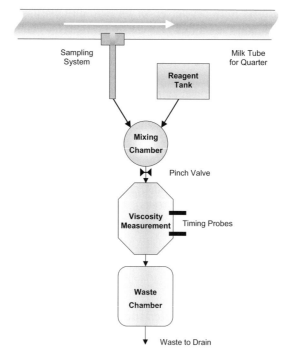

Figure 3. Block diagram of SCC sensor.

data obtained was used to calibrate time-to-drain thresholds which maximised the percentage of correct classifications.

For on-line testing, the sensor was placed on an AMS (Merlin, Fullwood, Ellesmere, UK) and 75 samples were measured. The sensor received a 'start-of-milking' signal and commenced automatic sampling 30 or 90 seconds thereafter. The entire process (sampling (4 ml), mixing, measuring, cleaning) was completed automatically. A reference sample was taken concurrently with the sensor obtaining its sample.

Results and discussion

Table 1 and figure 4 show the data leading to the thresholds chosen and the sensor performance in each band.

During lab testing the sensor correctly classified 95, 85, 76, 72, and 95% of samples in each of the five bands, from lowest to highest respectively. Of the 238 samples, only three were erroneous by more than one band, which we believe to be a key performance indicator of the sensor.

The high proportion of correct classifications in the highest band may be misleading because a large proportion of the samples in this band were far above 5000 kcells/ml. It should also be noted that the SCC distribution (for the lab testing) was not representative of a real population, where almost all samples would lie in the two lowest bands.

The SCC distribution in the on-line testing phase is a result of pseudo-random sampling of cows. The order in which cows were milked determined which cows were sampled.

Table 1. Summary of SCC sensor prototype performance: time-to-drain thresholds, the total number of milk samples (n), and the number of correct ($n_{correct}$) and proportion of correct (p) measurements from the calibration and on-line testing.

SCC band (kcells/ml)	Time-to-drain thresholds (s)	n (lab)	$n_{correct}$ (p) (lab)	n (on-line)	$n_{correct}$ (p) (on-line)
< 200	< 1.1	38	36 (95%)	64	62 (97%)
200 - 500	1.1 - 1.7	39	33 (85%)	2	2 (100%)
500 - 1500	1.7 - 6.2	59	45 (76%)	2	1 (50%)
1500 - 5000	6.2 - 59	60	43 (72%)	0	0 N/A
> 5000	> 59[1]	42	40 (95%)	1	1 (100%)
Overall	N/A	238	197 (83%)	69	66 (96%)

[1]The sensor time-out was set at 60 seconds.

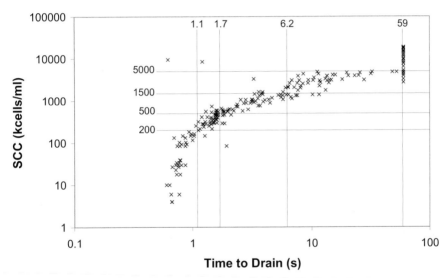

Figure 4. Calibration curve for SCC sensor prototype (lab testing only), showing the location of SCC band boundaries.

Therefore during on-line testing, unlike the lab testing phase where the distribution was artificially controlled, we observed a distribution representative of the herd. As a result we have obtained only five samples above 200 kcells/ml. It is almost certain that this estimate of sensor performance (96% overall) is unrealistically high - a performance similar to the lab calibration is probably more reasonable.

The SCC sensor prototype described here has a level of performance adequate for use as a mastitis control tool. We believe there is clear potential for the device to be commercialised for on-farm use.

References

Bottazzi, V., 1963. Contributo al metodi di riconoscimento del latte mastitico. Il Latte 37:755-759.

Carre, X., 1970. Effect of leucocytes on the viscosity of milk in the sodium lauryl sulphate test for mastitis. These 22, Ecole Nationale Veterinaire, De Lyon.

Christ Von W., 1962. Mit 3 Abbildungen über das Reaktionsvermögen von Proteinen mit synthetischen Netzmitteln als Grundlage zum Verständnis des California mastitis Tests. Deutsche tierärztliche Wochenschrift 4:108-110

Dounce and Monty, 1955. Factors affecting the gel-ability of isolated nuclei. Journal of Biophysical and Biochemical Cytology 1:155.

Jaartsveld, F.H.J., 1962. Contribution to diagnosis of mastitis in cattle in connection with the mastitis control. Netherlands Milk Dairy Journal 16:260-264.

Kiermeier, F., and K. Keis, 1964. Semi-quantitative modification of the Schalm test. Milchwissenschaft 19:65-69.

Milne, J.R., 1977. Observations of the Californian Mastitis Test (CMT) reaction. II. Photomicrographic studies of somatic cells and their reaction with surface active agents. New Zealand Journal Dairy Science. Technolnology 12:48-50.

Milne J.R. and De Langen H., 1977. Observations of the California mastitis test (CMT) reaction . I. The roles of Deoxyribonucleic acid (DNA) and milk protein in the reaction. New Zealand Journal of Dairy Science and Technology 12:44-47.

Nageswararao, G., and H.E. Calbert, 1969. A comparison of screening test to detect abnormal milk. Journal of milk food technology 32:365-368.

Nichols, G.M., and D.S.M. Phillips, 1972. A rotary viscometer for leukocyte count determinations in milk. Australian Journal of Dairy Technology 27:134-136.

Schalm, O.W., and D.O. Noorlander, 1957. Experiments and observations leading to development of the California Mastitis Test. Journal of the American Veterinary Medical Association 130:199-204.

Singh, B., and R.T. Marshall, 1966. Bacterial deoxyribonuclease production and its possible influence on mastitis detection. Journal of Dairy Science 49:822-825.

Thompson, D.I., and D.S. Postle, 1964. The Wisconsin Mastitis Test - an indirect estimation of leucocytes in milk. Journal of Milk Food Technology 27:271-275.

Tiu, C. and Boger, D., 1983. Rheology and Non-Newtonian Fluid Mechanics, June pg 24.

Whittlestone, W.G., and D.J. Allen, 1966. An automatic viscometer for the measurement of the California Mastitis Reaction. Australian Journal of Dairy Technology 21:136-139.

Whittlestone, W.G., L.R. Fell and H. de Langen, 1970. A viscometric method for the estimation of milk cell count. Journal of Milk Food Technology 33:351-354.

Whittlestone, W.G., H. de Langen and L. Cate, 1972. A simple semi-automatic viscometer for the estimation for somatic cells in milk. Milchwissenschaft 27:84-86.

Whiteside, W.H., 1939. Observation on a new test for the presence of mastitis in milk Canadian Public Health Journal 30:44.

SEEING RED: AUTOMATED DETECTION OF BLOOD IN MILK

D.S. Whyte, R.G. Orchard, P. Cross, A. Wilson, R.W. Claycomb & G.A. Mein
Sensortec Ltd, Hamilton, New Zealand

Milk that has changed in colour because of the presence of red blood cells is regarded as abnormal milk. The incidence of visible blood in raw milk is rare but, nevertheless, it should be monitored routinely in any AMS to monitor animal health and to maintain consumer confidence in milk.

Our search for a reliable reference method, or "gold standard", included colour scanning, atomic flame spectroscopy, particle counting, direct microscopic counts of Red Blood Cells (RBC), commercial dipsticks, and direct assay of haemoglobin (Hb) concentration. A standard based on a fixed percentage of blood would be unreliable due to variations in RBC content of blood and Hb concentration in the RBCs. A concentration of 0.1% raw bovine blood in milk may contain 5-10 million RBC/mL.

The output from a practical optical in-line sensor could be reported in terms of either Hb concentration or the RBC/mL of milk. Hb concentration provides the most accurate and precise output because Hb is directly responsible for the red colour typically associated with blood in milk. However, RBC count is likely to be more acceptable and more easily understood because consumers and farmers are already 'in tune' with visualizing cell counts. As an indication of the appropriate range, bovine milk containing 2-4 million RBC/mL has a visible reddish tinge. Milk with 10 million RBC/mL is clearly red. To correct for variations in Hb content of the RBCs, sensors could be programmed with an average Hb density value obtained for different breeds of cows or specific regions or countries if desired.

A sensor was developed to optically measure in-line concentrations of blood in milk from each quarter of 28 cows milked in NZ's first AMS. Most of these cows were freshly calved or in early lactation. Each day for a period of 16 days, either composite udder milk samples were collected for comparison by means of an auto sampler (Shuttle, Lely Industries NV) or separate quarter samples were taken from the individual milk tubes during the first minute of milking. All samples were tested within 24 hours to determine blood concentration of the milk. All of 6 samples with >10 M RBC/mL were identified (a Sensitivity of 100%) while 473 of 475 samples with < 10 M RBC/mL were correctly identified as negative (a Specificity of 99.6%).

Precision of the sensor for a range of variables is shown in the table on the next page. Temperature effects are estimates over the range expected in the field (0-40°C ambient, 20-40°C for milk)

RBC count (Mcells/mL)	Precision (1 S.D, Mcells/mL)[1]	Variation with sensor temperature (Mcells/mL/°C)	Variation with milk temperature (Mcells/mL/°C)
0	1.16	-0.07 ± 0.04 [2]	0.03 ± 0.01 [3]
10	1.18		
25	3.80		
100	22.11		

[1] includes variation between sensors, between samples and for Hb density.
[2] 1 standard deviation of sensor temperature gradient between sensors.
[3] 1 standard deviation of milk temperature gradient between samples.

DETECTION OF TISSUE DAMAGES CAUSED BY MILKING MACHINES USING CONDUCTIVITY MEASUREMENT

Kerstin Barth

Institute of Organic Farming, Federal Agricultural Research Centre (FAL), Westerau, Germany

Although electrical conductivity (EC) is only indirectly related to the somatic cell count in milk, conductivity measurement is one of the main components for detection of mastitis in automatic milking systems. Nevertheless, EC measurement reveals changes in tissue permeability, which can not necessarily attributed to mastitis. It seems probable that all mechanical strains might lead to an increase of EC. The daily application of the milking machine causes the main mechanical stress on the udder tissue. If the machine performs in an equal intensity on each teat, the effect of machine milking on EC is compensated as long as the difference method is used for data evaluation. Thus, an unbalanced strain of the quarters would lead to significant EC differences on single quarter positions. A study carried out in a herd milked in a rotory parlour three times per day showed significant higher conductivity readings on the right front position caused by the twisted milking cluster. Therefore, the aim of this study was to investigate the influence of a twisted cluster on conductivity readings during lactation.

In a herd of 120 lactating cows EC was monthly measured on foremilk and the position of the cow in the parlour was registered. Cows were milked two times per day in a 2 x 5 Tandem parlour. The forces applied on the teats by the twisted cluster were measured with a special device developed by the Deutsche Landwirtschaftsgesellschaft (DLG). In the middle and at the end of lactation quarter strippings were gained from cows (n = 19), which were often standing at the same parlour side. Teat positions of these cows were measured in the middle of lactation. The distance between the two front and two rear quarters was 20 and 10 cm, respectively. The front and rear teat of each udder half were 14 cm apart and the distance between the surface and the lowest teat end was 51 cm. Teat length was 55 and 45 mm for the front and rear teats, respectively. Analyses of variance revealed the interactions of parlour side (attachment of the cluster from the left or right side of the cow) and quarter position on EC readings (Figure 1). Stripping yield per quarter differed

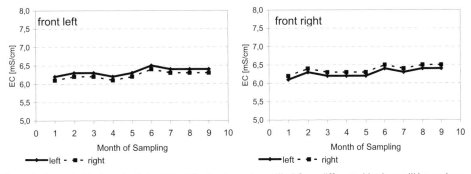

Figure 1. Mean electrical conductivity (EC) of the front quarters milked from different sides in a milking parlour.

only at the front left position (p < 0,091). Nevertheless, the results suggest that the continuous measurement of quarter electrical conductivity in milking robots might be used for detection of tissue strain caused by the milking machine as well as for monitoring of mastitis.

INTRODUCTION OF AMS IN ITALIAN DAIRY HERDS: THE DETECTION OF CLINICAL AND SUBCLINICAL MASTITIS BY AMS SYSTEMS

E. Binda, G. Casirani, R. Piccinini & A. Zecconi
Dipartimento di Patologia Animale Igiene & Sanità Pubblica Veterinaria, Milano, Italy

The introduction of automatic milking systems (AMS) in Italian dairy herds increased the importance of an accurate monitoring of milk quality and udder heath, as a consequence of reduced interaction between cows and milkers. All the commercial AMS have different on-line devices to assess milk quality and to divert abnormal milk. However, many farmers are reluctant to rely on electronic devices to monitor cow health status. The aim of this study is to assess the accuracy of the approaches proposed by the different AMS software and to explore the opportunity to apply different methods to forecast the outcome of subclinical mastitis. The study was undertaken in 4 herds, two equipped with Lely Astronaut™ AMS and the others with VMS™, DeLaval AMS. Overall 2127 quarter milk samples were taken and analysed following NMC procedure for bacteriological assays and IDF for somatic cell counts. The quarter status was defined as follows: a bacteriologically positive sample with SCC ≥200.000 cells/ml was defined as *infected*; a bacteriologically positive sample with SCC <200.000 cells/ml as *latent IMI*; a bacteriologically negative sample was defined as *healthy* if SCC ≤100.000 cells/ml and as *inflamed* when SCC >100.000 cells/ml.

In farms with Lely Astronaut™ AMS data regarding milk yield, conductivity and mean number of milkings were collected 3 days before a 3 days after milk sampling by printing reports and manual recording of the data on a database.

In farms with VMS™, DeLaval AMS data automatically recorded by AMS software (milking duration, milk flow, conductivity and yield), 6-10 milkings before and 6-10 milkings after milk sampling, were retrieved. A further index was calculated: peak/average ratio (PAR) which is the ratio between the recorded peak flow and the average flow. Data were analysed by the means of ANOVA for repeated measurements and $\chi 2$ statistics on SPSS 11.5 (SPSS, 2002).

The results showed that:

- The detection systems applied in the different AMS are different and generally incomparable;
- The alarms proposed by the software have an insufficient accuracy with both low levels of sensitivity or specificity;
- The overall accuracy levels differ not only among AMS but also among herds even when the same AMS is used;
- The different parameters recorded differ in sensitivity and specificity among herds even when the same AMS is used;
- Milking frequency affect the accuracy of the different parameters and therefore this value should included in the algorithm for mastitis detection.

Three of the five cases of clinical mastitis and one of the three times when milk was visually changed were detected with AMS1 at sampling. All four cases of clinical mastitis were detected with AMS2 at the time of sampling or just before. Of the 14 quarters with colostrum (AMS2) four were detected at sampling and six during the 10 milkings before; all based on the analysis of milk colour. From the three quarters with visually changed milk two were detected by AMS2 on the basis of milk colour.

Sensitivity of the detection methods used in the AMS needs to be improved. Before the systems work adequately, conventional detection methods should be applied regularly. The amount of false attentions should be clearly reduced (AMS1), in order to keep up the motivation of the farmer to check the cows that are on the attention list.

PROTEOLYSIS OF MILK PROTEIN FROM UNINFECTED GLANDS AND GLANDS INFECTED WITH *STREPTOCOCCUS UBERIS*

Lotte B. Larsen[1], Morten D. Rasmussen[2], Martin Bjerring[2] & Jacob H. Nielsen[1]
[1]Department of Food Science, Danish Institute of Agricultural Sciences, Foulum, Denmark
[3]Department of Animal Health and Welfare, Danish Institute of Agricultural Sciences, Foulum, Denmark

Protein degradation and presence proteolytic enzymes were studied in milk from cows before and after experimental inoculation with *Streptococcus uberis* into one quarter. Milk from both infected and uninfected glands was analysed for three subsequent milkings after infection and compared with preinfection level. Level of proteolysis, determined as free amino terminals in skimmed milk quarter samples by fluorescamine assay, was found to increase with the somatic cell count, and furthermore was higher in milk from infected quarters compared with uninfected. Analysis for responsible proteases showed that both plasmin and plasminogen-derived activities and activity from the lysosomal protease cathepsin D were higher in milk from infected quarters than in milk from uninfected. Further analysis of casein by Tris-Tricine SDS-PAGE showed, however, that proteolysis of caseins occurred not only in the casein fraction isolated from milk from infected glands, but also in the milk from uninfected glands from 3rd milking after infection. This was accompanied with a relative increase in plasmin activity not only in infected glands compared with uninfected, but also in the uninfected neighbour gland when compared with preinfection level of the same gland. Therefore, the deterioration of milk protein was not restricted to the infected quarter, but also occurred in the uninfected quarter even though he response in SCC was found to be restricted to the infected gland. Therefore, this study indicates that separation at quarter level based on e.g. SCC of milk in automatic milking systems cannot be recommended, as the milk quality in the contralateral gland may be compromised.

COMPOSITION OF MILK CLOTS

Lotte B. Larsen[1], Morten D. Rasmussen[2], Martin Bjerring[2] & Jacob H. Nielsen[1]
[1]Department of Food Science, Danish Institute of Agricultural Sciences, Foulum, Denmark
[3]Department of Animal Health and Welfare, Danish Institute of Agricultural Sciences, Foulum, Denmark

Milk from infected glands may contain visible signs of blood in terms of altered colour or even blood clots, but in most instances the clots or flakes appear with a white or yellow colour. Based on observations on filtration experiments (100 µm pores) of mastitis milk it truly appears that in some mastitis milk samples red blood clots are present, but filtration of other samples results in presence of light coloured unidentified precipitated material in the retained fraction (Rasmussen *et al.*, 2002). There may be more than one explanation for the occurrence of light coloured flakes and clots in milk from infected glands, and probably the effects act together or are more or less pronounced in different types of mastitis milk (Rasmussen & Larsen, 2003). To approach the underlying mechanisms milk clots were isolated from mastitis milk, and subjected to protein composition analysis by SDS-PAGE. Analysis of clots from three different animals showed that the protein composition of milk clots from these animals was relatively similar. The analysis showed that casein was not the main component, and furthermore that the protein composition was significantly different form the protein composition of milk whey, even though some whey proteins were present. This indicates that clot formation in the mastitic milk analysed in this study was not a result of casein precipitation caused by the elevated proteolytic activity in mastitis milk.

References

Rasmussen, M. D., Madsen, & Bjerring, M. (2002). Available tests and reference methods. In "Definition of normal and abnormal milk at the time of milking". Danish Institute of Agricultural Sciences Internal report no. 169, 63-73.

Rasmussen, M. D. & Larsen, L. B. (2003). Milking hygiene: New issues and Opportunities from automatic milking. *Italian Journal of Animal Science, in press.*

A METHOD TO DETECT FLAKES AND CLOTS IN MILK IN AUTOMATIC MILKING SYSTEMS

B. Maassen-Francke[1], M. Wiethoff[1], O. Suhr[1], C. Clemens[2] & A. Knoll[2]
[1]WestfaliaSurge GmbH, Oelde, Germany
[2]Technische Universität München Institut für Informatik, Garching, Germany

The requirements of hygiene in milk production in the EU are defined by the Commission Directive 89/362/EEC (1989). But up to now, not all demands are met by automatic milking systems. The directive claims that "the milker must inspect the appearance of the milk." In reference to automatic milking systems WP 4 proposes that "milk from an animal is checked for abnormalities by the milker or a method achieving similar results". At a workshop in Denmark Foulum held on November, 27th 2002 under the leadership of the manager of workpackage 4, milk was defined as abnormal if clots appear. The workshop had consensus that there should not be a double standard for conventional and automatic milking. To make sure that abnormal milk is separated it is necessary to detect clots and flakes in the milk automatically. Now a sensor system has been developed as a functional prototype which can detect clots and flakes in milk.

The technology consists of a sensor as a hardware component and a software component. The basic procedure is that milk is passed over a sample carrier which is located in a measuring chamber. The fluid passes off the carrier while flakes and clots deposit on it. After a certain time the surface of the carrier is checked and depending on the results of the sensor the milk can be discarded or can be led to the collection unit.

The detection method is performed with an optical sensor composed of a digital camera with a resolution high enough to distinguish flakes of a size from 0,1mm and more. The camera is able to take one or more photos of the carrier surface. With the help of the adjacent image processing the abnormality of the milk can be identified.

Even with a satisfactory cleaning there can be other particles than flakes and clots in the milk which can be mistaken for flakes. A conglomerate of all particles can be found as well in the caught milk fraction on the carrier. Therefore it is necessary to distinguish between clots and other particles like straw particles, sawdust, grain of sand and nevertheless foam or white spots caused by reflection. Each particle has a geometrical skeletal structure and special colour of its own, so it can be classified by those parameters.

Studies were carried out screening several hundred milk samples of which 59 with 2114 detectable objectives were evaluated. Criteria for the algorithm of detection were 20 parameters of which the size, the intensity of the colour, the roundness of the shape, the compactness, the body structure of the objectives are the most important ones. Preliminary results are very promising as about 90% of all objectives were classified correctly. A detailed principle of the measurement will be demonstrated.

MAKING SENSE OF IN-LINE SENSING FOR MILK CONDUCTIVITY

G.A. Mein[1], R.A. Sherlock[2] & R.W. Claycomb[1]
[1]Sensortec Ltd, Hamilton, New Zealand
[2]SmartWork Systems Ltd, Christchurch, New Zealand

Attempts to measure the electrical conductivity (EC) of milk as a means of detecting mastitis date back to the first half of the last century. Those early efforts (in common with several conductivity devices on the market today) were unsuccessful, however, because of the normal biological variations in conductivity of milk that have nothing to do with mastitis. The value of EC as a mastitis detection tool has been disappointing, principally because the conductivity of milk is influenced by variations such as those listed below.

- From cow to cow within a herd, from herd to herd, from breed to breed
- Over the course of a lactation and in response to varying milking intervals
- Over the course of a milking in response to changing fat content of the milk
- In response to varying feed types and intake amounts
- In response to changes in milk temperature
- As a result of dirt, milkstone or milkfat buildup on the sensor electrodes.

A quarter conductivity sensing system has been developed in an attempt to eliminate or control all the non-mastitic variables that can affect milk conductivity in order to focus solely on those factors that cause an increase in conductivity as a consequence of tissue damage within the cow's udder.

Examples of recent results

1. The results of a recent study in New Zealand's only automatic milking herd for the period Aug 1, 2002 to March 1, 2003 showed that:
- All 8 quarters with clinical mastitis were correctly identified by quarter and date
- 12 of the 13 quarters infected with a major mastitis pathogen were correctly identified by quarter and date
- Only 7 of 26 quarters infected with minor pathogens appeared on the alarm list.

If the presence of a major pathogen is defined as a "True positive", the Sensitivity was 92% (12 of 13 quarters) and Specificity was 95% (391 of 411 quarters).

2. EC values were continuously collected via EC sensors installed in six AMIs on four Danish farms. These data were analysed and compared with clinical mastitis records derived from monthly herd health visits by milk inspectors from the Danish Dairy Board throughout a period of 7-8 months from Dec 1, 2002 to July 31, 2003. A "true positive" was defined as a cow with visible clots in the foremilk of one or more quarters at a monthly herd visit. "True negative" cows had no visible signs of clots in their milk and low CMT (scores of 1

or 2). Cows with CMT scores 3, 4 or 5 were discarded from the analyses according to the rules of the Danish approval test.

The overall results, using a single set of critical threshold settings for all four herds, showed that the EC sensor flagged 45 of 56 cows with a clinically affected quarter (Sensitivity 80%) but raised a false alarm for 7 out of every 100 cows that were truly negative (Specificity 93%). These results indicated the potential for further improvements by adjusting threshold values to suit individual herds and individual herd management objectives, if desired.

ELECTRICAL CONDUCTIVITY OF MILK AS AN INDICATOR TRAIT FOR MASTITIS

Elise Norberg & Inge Riis Korsgaard
Department of Animal Breeding and Genetics, Danish Institute of Agricultural Sciences, Foulum, Denmark

Introduction

Electrical conductivity (EC) of milk has been introduced as an indicator trait for mastitis over the last decades (Hamann and Zecconi, 1998). Detection models have mainly been based on the level of EC, but problems with false negative and false positive exists. Therefore, other EC traits may be considered for detection of mastitis. In this study, various EC traits were investigated for their association with udder health.

Material and methods

In total, 322 cows with 549 lactations were included in the study. Cows were, based on veterinary treatment and bacteriological samples, classified as healthy, clinically or subclinically infected and EC was measured on each quarter at 2 seconds interval during every milking throughout lactation. Four EC traits were defined; the inter-quarter ratio (IQR) between the highest and lowest quarter EC values (●), the maximum EC level for a cow (■), IQR between the highest and lowest quarter EC variation (▲), and the maximum EC variation for a cow (×). Values for the traits were calculated for every milking throughout the entire lactation. (Symbols in parenthesis refer to figure 1.)

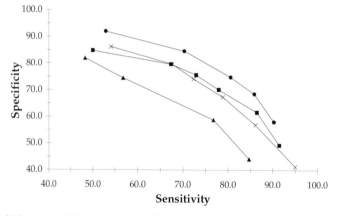

Figure 1. Sensitivity and specificity for detection of clinical mastitis.

Results

All EC traits increased significantly ($P < 0.001$) when cows were subclinically or clinically infected. A simple threshold test and discriminant function analysis was used to validate the ability of the EC traits to distinguish between cows in different health groups. Traits reflecting the level rather than variation of EC, and in particular the IQR, performed best to classify cows correctly. By using this trait, 80.6% of clinical and 45.0% of subclinical cases were classified correctly. Of the cows classified as healthy, 74.8% were classified correctly. However, some extra information about udder health status was obtained when a combination of EC traits was used. For more details see Norberg *et al.*, 2003.

Further possibilities

Detection of mastitis can be improved by considering better indicator traits or improving the statistical models and methods applied to the data. In Korsgaard *et al.* (2003), daily measurements of somatic cell score (SCS) were analysed with a time series model (local linear growth model with exceptions, see e.g. Smith and West (1983)). With this model they obtained on-line probabilities of mastitis status of a cow based on deviations from the normal pattern of SCS. When the probability exceeded a certain threshold, an alarm was triggered. This method is expected to improve the results when applied to EC.

References

Hamann and Zecconi. 1998. *Bulletin 334 Int. Dairy Fed.,* Brussels, Belgium.

Korsgaard *et al.*, 2003. *DJF Intern rapport* nr. 176, 19-32.

Norberg *et al.*, 2004. *J. Dairy Sci.* (accepted)

Smith and West. 1983. *Biometrics* 39: 867-878.

THE CHANGE OF MILK QUALITY DURING LACTATIONS

L. Pongrácz & E. Báder
University of West Hungary, Faculty of Agricultural and Food Sciences, Institute of Animal Breeding and Husbandry, Mosonmagyaróvár, Hungary

It is unacceptable to add abnorman milk to the bulk tank, so automatic separation of this milk must be realised because of higher demands on milk quality. The objective of this study was to gain current knowledge on the present situation of the milk quality during lactations.

Material and methods

Milk yield and SCC of red Holstein Friesian cows were used for the evaluation. 305 day of lactations were calculated from the database of the Livestock Performance Testing Ltd., Gödöll_ (Hungary) between April 1995 and January 2002 contained 4614 lactations. Herd-year-season-calving age at first calving effects were not taken into consideration. Biometric calculations were processed by Microsoft Excel 97 and Statistica softwares. Student's t-test was used for comparison of means. A P value of 0.1, 1, 5 and 10 (and NS=not significant) was considered as significant.

Results and discussions

Test day data were processed according to the somatic cell count. Taking into account the SCC lower and higher than 400,000 cells/ml during 8 test day procedures, at last 1175 and 117 observations (lactations) have left, respectively. The trend of decrease in both groups is worth to be studied just like results according to the number of lactations.

432 out of 1175 lactations (or cows) complete their first lactation produced good quality of milk that is 35.1% of the whole population. Among older animals (in lactation 9 and 10) there were no cow producing only high quality milk during the 8 test day procedures. However, cows producing only high somatic cell count milk during 8 test day procedures represent 53.0% of the whole population in lactation 3-5.

Conclusions

25% of the cows start their lactation with high somatic cell count. Till the second test day it drops to the half.

The ratio of healthy cows during the whole lactation is approximately 25%.

Losses in milk production associated with elevated SCC can be estimated, too. Reasons are lower yields and worst persistence. The differences were statistically significant.

References

Hogeveen, H. Van der Vorst, Y. Ouweltjes, W. and Slaghuis, B.A. (2001). Automatic milking and milk quality: A European perspective. Proc. 40th Annual Meeting of the National Mastitis Council. 152.

Pongrácz, L (2002). Some aspects of udder health related to milk quality and gravity of mastitis. PhD Theses, Mosonmagyaróvár. 120.p.

Rasmussen, M.D. (1999). Management, milking performance and udder health. In: Aagaard K. (Ed.) III. Future Milk Farming, Proc. FIL-IDF 25th Int. Dairy Congress. Aarhus. Denmark. 174-178.

IMPEDANCE BASED SENSOR FOR MASTITIS DETECTION IN MILK

Sonia Ramírez-García, Valérie Favry, Kim Lau, Gillian McMahon & Dermot Diamond
National Centre for Sensor Research (NCSR), Dublin City University, Dublin, Ireland

The electrical impedance of milk was used to determine mastitis in cows. New stainless steel electrodes were used as sensors. The electrodes were characterised and the optimum frequency and potential for the measurement of impedance in milk were determined, the final conditions being 100 MHz and 15 mV. It was found that the fouling of the electrodes was negligible over at least 5 days and the noise of the signal was approximately 0.8% of the measure at 18.3^0C. The temperature and pH of the milk was also monitored. The status of the milk was determined using impedance ratios, normalising against the quarter with highest impedance. The sensitivity and selectivity of the method using electrical impedance, pH, temperature and Cl^- concentration were both very good. The results presented in this paper demonstrate the efficacy of electrical impedance as a screening method for mastitis that can be implemented in automatic milking parlours.

DETECTION OF ABNORMAL UDDER TISSUE AND MILK BY NEAR INFRA-RED SPECTROSCOPY (COW SIDE)

R. Tsenkova[1], H. Morita[1], H. Shinzawa[1] & J.E. Hillerton[2]
[1]Graduate school of Science and Technology, Kobe University, Kobe, Japan
[2]Institute for Animal Health, Compton, United Kingdom

Quality control of milk at cow side has become very important issue. Recently, near infrared (NIR) technology has been introduced (Tsenkova *et al.*, 2000) for rapid and non-destructive milk quality evaluation and mastitis diagnosis, as well as for bio-fluid composition measurement. In this study, we present a new method for non-invasive mastitis diagnosis based on NIR spectra of udder tissue.

Holstein cows' udders were scanned in interactance mode, with portable NIR spectrophotometer FQA-NIRGUN. Five scans were acquired at various locations on each udder quarter making a data set of 20 spectra per cow. Milk electrical conductivity (MEC) and somatic cell count (SCC) of respective quarter milk sample were measured as reference.

Spectral data were analyzed by Soft Independent Modeling of Class Analogy (SIMCA). It is a well known classification method which builds a distinct confidence region around each class based on assumption that each class spans different space (Wold *et al.*, 1977). Classification was performed on each cow's data set where 4 times, consecutively, spectra from respective udder quarter were considered as one class and the rest of the spectra of the same cow were considered as another class. Four interclass distances were calculated and compared as indicator for mastitis diagnosis. High correlations between the reference data, MEC or SCC, and the respective interclass distances were obtained (r > 0.90) for each cow's udder quarter.

In-vivo mastitis diagnosis based on NIR spectra of udder tissue analyzed by SIMCA classificator proved to be a rapid non-invasive method especially suitable for robot milking where teats' locations are found by optical scanning of the udder.

References

R. Tsenkova, S. Atanassova, Y. Ozaki, K. Itoh, and K. Toyoda, 2000. Near infrared spectroscopy for bio monitoring: Cow milk composition measurement in a spectral region from 1,100 to 2,400 nanometers. J. Animal Science, 78: pp. 515-522.

S. Wold, M. Sjostrom, 1977. SIMCA: A Method for Analysing Chemical Data in Terms of Similarity and Analogy in 'Chemometrics: Theory and Application', B. R. Kowalski (Ed.), Am. Chem. Soc., Washington, D. C., pp. 243

GRAZING

AUTOMATIC MILKING AND GRAZING

Eva Spörndly[1], Christian Krohn[2], Hendrik Jan van Dooren[3] & Hans Wiktorsson[1]
[1]Department of Animal Nutrition and Management, Swedish Univ. of Agric. Sciences, Uppsala, Sweden
[2]Department of Animal Health and Welfare, Danish Institute of Agricultural Sciences, Foulum, Denmark
[3]Applied Research, Animal Sciences Group, Wageningen UR, Lelystad, The Netherlands

Abstract

Recent research on how to combine automatic milking (AM) systems and grazing is presented with the objective to summarise the results obtained in an EU-project and discuss these results in relation to present knowledge in the field.

Experiments with 24-hour grazing conducted in Sweden showed no difference in water intake between cows with drinking water indoors, and cows with water both indoors and on pasture. *Ad libitum* supply of silage did not increase milk yield or lower the milking interval compared with 3 kg dry matter*cow*day^{-1}. Free cow traffic (access to feed supplements) gave a significantly lower number of voluntary milkings compared with controlled access. Grazing near the barn (50m) gave a higher milk yield compared with grazing further away (260m).

Dutch surveys and studies on AM-farms with grazing showed that selection gates contributed to a lower number of manually fetched animals. In an experiment, the effect of pasture distance (<150m versus >500m) and supplementation level (6 versus 10 kg DM) on milking interval, pasturing time and milk yield was studied. No significant effects of distance or level of roughage supplements on milk yield and milking interval were observed.

Case studies on farms in Denmark showed that a relatively high milking frequency was obtained (variation 2.1-2.8) but most cows had to be fetched from the pasture. In periods when many cows were at pasture a decrease in number of milkings per hour often occurred. Cow behaviour was more synchronised at pasture than when in the barn, and only few cows entered the barn alone.

In conclusion, automatic milking can successfully be combined with pasture and grazing. High levels of supplementary feeding did not lead to higher production when pasture was available in sufficient amounts but supplementation can be used strategically as a management tool to obtain smooth cow traffic.

Introduction

During the last ten years automatic milking (AM) has developed from a new interesting invention to a well-functioning management system that is being adopted by more and more farmers. One of the major concerns with the new system is whether it can be successfully combined with grazing and there are several reasons for that concern. The main reason is that pasture is an important feed source for cows in countries where conditions for pasture growth are favourable. Well-managed pasture can offer a feed with high nutrient content at low cost. During recent years the interest for animal welfare and health has also led to an increased interest in pasture and grazing. At pasture, animals can move more freely and pasture offers the cows an opportunity to indulge in outdoor foraging. The role of grazing animals for the amenity of the landscape and for biodiversity must also be considered, as well as the need to find management systems that include grazing to facilitate for organic farmers interested in automatic milking

To address the question of how to combine automatic milking with grazing, the EU-project "Implication of the introduction of automatic milking on dairy farms" included the work package "Automatic milking and grazing" in its research activities. This paper will present the main results of the research activities performed in Denmark, The Netherlands and Sweden within the EU-project, but results from other experiments will also be included and discussed.

The objective of grazing research within the EU-project has been to identify the management factors that can contribute to a successful and well functioning AM system at pasture and to make recommendations that will improve management on all farms that wish to combine grazing with automatic milking. In this paper, a number of important factors will be discussed in relation to their effects mainly on milk yield, milking frequency/milking interval and animal behaviour.

Grazing hours

On-farm studies and surveys performed on AM farms with grazing have shown that there is a large variation among farms in the number of grazing hours. In a survey on 25 farms performed in The Netherlands, approximately half of the farms (13) allowed the animals to graze for more than 15 hours per day, 5 farms had grazing 11-15 hours per day and the remaining 7 farms practiced grazing 5-10 hours per day (van Dooren et al., 2002).

In an on-farm study on seven Danish farms with AM and grazing, animals were studied on three different occasions during the season, May, July and September (Munksgaard & Krohn, 2004). The data showed that a relatively high milking frequency was obtained on most farms (variation 2.1-2.8) and there was a large range in pasturing time (5.4-20.6 hours). No significant effect between the time cows spent on pasture and the total number of milkings per 24 hours was found.

Contrasting results were obtained in an experiment by Ketelaar-de Lauwere et al. (1999) where zero-grazing was compared with 12 and 24 hours of grazing in an automatic milking system with 24 lactating cows. The results showed that animals with 24-hour access to pasture had a significantly lower average milking frequency (2.3 milkings/day) compared with the other treatments (2.5-2.8 milkings/day).

In the Danish study it was also observed that on farms where the pasture was available only a few hours per day, animals made the most of these hours and were outdoors up to 85% of the available time. When pasture was available for more hours, average time spent outdoors was approximately 40-60% of the available outdoor time (Munksgaard and Krohn, 2004). On farms where the cows had free cow traffic out to the pasture area most of the cows went to the pasture immediately after the door was opened in the morning, showing that the cows were eager to use the opportunity to graze. There were fairly large differences between the amount of time that was spent fetching cows late for milking, a range of 0-55 minutes per day was reported, but farms normally used around 10-20 minutes daily. Most of the animals had to be fetched from the pasture on the Danish farms, while in the Dutch survey (van Dooren et al., 2002) the percentage of cows fetched varied from 0 to 90%

between farms. It was also observed that the time of day cows were fetched influenced the number that needed to be fetched; earlier fetching resulted in a higher percentage of cows to be fetched.

Cows synchronised their passage out to the pasture area and their return to the barn to a high degree (Munksgaard and Krohn, 2004). Similar observations were made in experiments of Ketelaar-de Lauwere *et al.* (1999; 2000).

Effects of distance to pasture and level of supplementary roughage

The effects of distance to the pasture area and the level of supplements offered in the barn have been studied in several experiments (Ketelaar-de Lauwere *et al.*, 2000; van Dooren, 2004; Spörndly and Wredle, 2004). In two-factor experiments with approximately 64 cows, the effects of distance between the pasture and barn and level of roughage supplements (maize silage) were studied during 2002 and 2003 (van Dooren, 2004). Two different distances between barn and pasture were compared (less than 150 m versus more than 500 m) and two levels of roughage supplements (6 versus 10 kg dry matter, DM) giving a total of four combinations. Treatments were applied after each other in time with four-day adaptation periods followed by registration periods of four days and two replicates for each combination.

During 2002, forced cow traffic was practiced and cows could only reach the feeding table and the door to the pasture by passing the automatic milking unit, whereas during 2003 cow traffic in the barn was free. A selection gate in front of the door to the pasture was open for 9.5 hours. In the evening (after 13.5 hours), animals that had not returned to the barn were fetched.

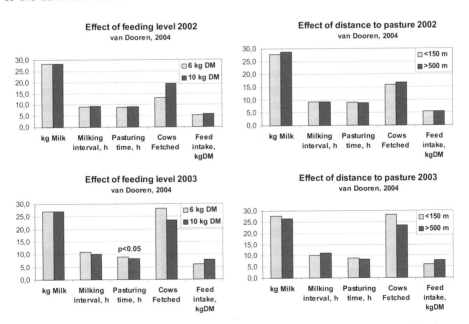

Figure 1. Effect of feeding level and distance to pasture on milk production, milking interval, pasturing time, no. of cows fetched and intake of feed supplements (van Dooren, 2004).

barn and on pasture. No significant effects of treatments on milking frequency were obtained in 2001 or 2003 apart from a somewhat higher milking frequency (+0.2; p<0.05) for animals with drinking water in the barn only during a sub-period in 2001 when the animals grazed at a longer distance from the barn. Weather conditions varied during the experiments but during very hot sunny days all animals spent more time lying indoors during the daytime and grazed mainly in the evening and at night and no treatment effects were observed, even during these hotter days.

Cow traffic during the grazing period

In the survey of 25 Dutch farms some farms reported improved cow traffic during the grazing season when supplementary roughages were available for the animals immediately after entering the barn and before, instead of after, milking (van Dooren et al., 2002). With this system, cows went to the pasture to graze immediately after milking, and they returned more often because the cow was rewarded with forage supplements directly after entering the barn.

In an experiment at the Swedish research station two different cow traffic systems were compared during the grazing season, free and controlled cow traffic (Spörndly and Wredle, 2002). At the start of the grazing season the herd was divided into two groups. The feeding system in the barn with transponder controlled feed troughs and feeding stations made it possible to offer half the herd free cow traffic while cow traffic for the remaining cows in the herd was controlled. Cows in the treatment group with free cow traffic had access to their daily ration of supplementary feeds directly upon entering the barn, whereas cows in the group with controlled cow traffic had to pass the milking unit or the selection gates before reaching the feeding area where the supplementary silage (approximately 4 kg $DM*cow^{-1}*day^{-1}$) and concentrates were offered. All cows were offered a small portion of concentrates (0.7 kg) at each milking and the remaining ration was given in the concentrate feeders. Minimum milking interval was six hours and selection gates were only open to cows during these first six hours after milking. Cows in both treatment groups had to pass the milking unit or selection gates to reach the exit gate leading to the cow lane and grazing area. The experiment lasted from mid-May until the end of July and was performed with 29 experimental cows in an AM barn with 45-50 cows. The gate to the pasture area was open 24 hours per day and cows alternated between two grazing areas with different distances to the barn (50 or 260 m). Animals that had not been milked during the last 12 hours were fetched twice daily, at 5:30 and 17:30 hours.

The results of the experiment, presented in figure 3, showed that free cow traffic gave a significantly lower number of milkings (per cow and day) and a significantly lower number of voluntary milkings (total milkings minus fechings) compared with controlled cow traffic.

The average number of animals that had to be fetched manually increased as the season progressed and was higher for the cows with free compared with controlled cow traffic, an average of 0.4 and 0.1 fetched animals per cow and day, respectively during July. When cows were fetched to be milked, cows in the free traffic group were in most cases found in the barn (82%), whereas cows late for milking in the controlled group were generally (67%) fetched out on the pasture. A trend for decreasing visiting and milking frequencies was seen over time.

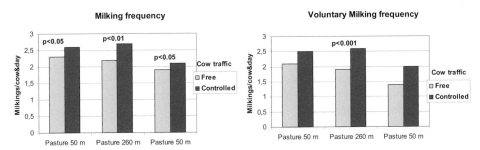

Figure 3. The effects of free and controlled cow traffic on milking frequency (milkings per cow and day) and voluntary milking frequency (excluding fetchings) during three periods at pasture.

Sward height

The effect of declining grass supply during the four days the animals spent in each paddock in a rotational grazing system was also studied in the experiment of van Dooren (2004) described earlier. When the data were analysed statistically including all the three variables (distance, level of feeding and effect over days), the effect of change over days (the slope) was not significant for any of the variables studied. Although the main model of statistical analysis showed that there were no significant effects of time (days 1-4) on the variables studied, the complete statistical analyses of the effects have not yet been completed. The non-significant averages seemed to indicate a certain effect of time for several variables as exemplified in figure 4. The number of cows that needed to be fetched to the barn in the evening decreased over the four days that the cows were in the same paddock, and the number of total visits to the milking unit was higher on days 3-4 compared with days 1-2. Also, the milking interval was lower on the last two days in the paddock compared with the first two days, especially in 2002. A decrease in pasturing time was also seen over the four-day period, whereas intake of maize silage supplements increased (Figure 4).

These results are similar to what was observed by Ketelaar-de Lauwere et al. (2000) who reported an increase in the total number of AM visits per cow and day from 4.4 to 7.3 and in the number of milkings per cow and day from 2.6 to 3.0, when sward height decreased in the paddock in a rotational grazing system. In another behaviour study where sward height decreased from 20 to 5 cm over a period of 11 days, a large difference in the effect

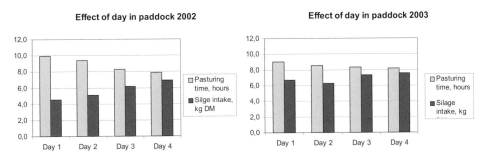

Figure 4. The effect of day in paddock, i.e. decreasing amount of pasture in a four-day rotational grazing system (van Dooren, 2004).

of decreasing sward height was seen between cows on different levels of supplementation (Salomonsson and Spörndly, 2000). Cows that

were offered only pasture and concentrates spent approximately 32% of the observation time grazing, with no difference between the first and last day in the paddock. whereas the cows that were offered *ad libitum* supplements of a mixed feed in the barn decreased their grazing time from approximately 22% during the first four days to 10% during the last four days in the paddock.

Behaviour, training and adaptation

During the EU-project, a number of experiments were performed to explore the possibilities of using individual auditory signals to stimulate cows due to be milked to go to the milking unit.

The aim of the first experiment (Wredle *et al.*, unpublished) was to study how to train cattle to approach a feed source in response to an auditory signal. In a longer perspective, the objective was to study if auditory signals could be used when cows were at pasture, in order to increase their motivation to return to the milking unit. Four different combinations of training methods and evaluation situations were studied in four experiments using classical and operant conditioning. The results were successful with regard to operant conditioning technique but the animals had difficulties in transferring what they had learnt in the experiment to a new place.

In the next experiment, ten animals in an AM barn were trained to respond to an individual auditory signal by going to the milking unit. The objective was to study if automatically milked cows could be taught to visit the milking unit in response to an individual auditory signal (emitted from their collar) during the pasture season.

Evaluation took place during the grazing season when the ten experimental animals were subjected to the normal management routines of the herd. During the evaluation period, animals were automatically given the signal with two reminders when eight and ten hours had passed from their latest milking. Many younger animals seemed to respond to the signal given by returning from the pasture to the barn or moving towards the milking unit when the signal was given in the barn. However, the animals were less successful in giving a clear response by entering the milking unit (Wredle, pers. comm.). A queue of cows waiting to be milked or a long distance between pasture and barn are examples of factors that may be obstacles to a full response to the stimuli. Preliminary results thus indicate that animals need a longer training period and perhaps a better reward system if auditory signals are to be used to entice the animals to go to the milking unit.

Discussion

Many farmers have shown that it is possible successfully to combine automatic milking and grazing, and the results of the experiments performed confirm this picture. There are, however, certain factors that should be considered.

Compared with the indoor season, the higher degree of synchronisation of behaviour of cows at pasture often leads to a more uneven distribution of milkings over the 24-hour period, with few or no milkings during certain hours when a majority of the cows are out on the pasture (Munksgaard and Krohn, 2004; van Dooren *et al.*, 2002). Thus, the utilization of the milking unit may be slightly lower during the grazing season, especially when animals are allowed to graze many hours per day. To maintain smooth cow traffic it is therefore advisable not to maximise the number of lactating animals in the barn during the grazing months compared with the indoor season.

Experience from farms indicated that in the long run, it appeared beneficial in terms of labour not to be too eager to fetch the cows but to give them the chance to return voluntarily. A general opinion was that habituation and training were important to minimize the number of cows that needed to be fetched. A selection gate out to the pasture area might also decrease the number of animals that need to be fetched (van Dooren et al., 2002).

The effects of distance between barn and pasture in different experiments were somewhat contradictory. In the experiment of Spörndly and Wredle (2004), animals were subjected to the same treatment throughout the whole grazing season, which enabled an evaluation of the effects over a longer time period compared with the other experiments. The experiments also differed with regard to grazing hours, levels of supplements and grazing system (rotational versus continuous). Another possible explanation of the different responses in the experiments may be found in the localisation of the pasture area in relation to the barn. At the Swedish experimental farm the pasture areas situated at a longer distance from the barn were localized "round the corner" from the barn exit. Thus, animals that left the barn on their way to the distant pasture could not see the paddock to which they were heading. The analysis of the data from the farm surveys showed that good production results could be obtained on farms with longer distances between barn and pasture but that distance in certain cases had an effect on factors such as number of cows that needed to be fetched.

The level of supplements offered in the barn did not have a significant effect on milk yield or milking interval in the experiments presented here, and when pasture allowance is normal, high levels of supplements may not be necessary. Naturally, in cases of pasture shortage, supplementation must be available in sufficient amounts. However, the results from several experiments have shown that the supplementation level and how supplements are offered (free or controlled cow traffic, time of feeding), have a strong influence on animal behaviour (van Dooren, 2004; Spörndly and Wredle, 2002). Although experience in this field is still somewhat limited, the results obtained indicate that supplementation strategy can thus be used as a management tool to achieve a well-functioning cow traffic during the grazing season.

Conclusions

Farm surveys and station experiments have shown that automatic milking can successfully be combined with pasture and grazing. The farm surveys found that there was a great variation in production results achieved and in the amount of labour used in the daily routines between AM-farms during the grazing season. This indicates that there is large scope for improvement on farms where results obtained during the pasture season are not satisfactory. Experiments performed on research stations showed that high levels of supplementary feeding did not lead to higher production. However, it seems possible that the strategy for supplementation in the barn can be used as a management tool to regulate cow traffic during the pasture season. Research and surveys have also shown that a well-functioning system can be obtained at longer distances between barn and pasture but due to negative effects of longer distances observed in some cases, it is an advantage if the pasture area is situated as close as possible.

References

Dooren, H.J.C. van, 2004. Results of partner 1 in: Automatic milking and grazing, Final report. Deliverable D27 within the EU-project Implications of the introduction of automatic milking on dairy farms (QLK5 - 2000 - 31006).

Dooren, H.J.C. van, Spörndly, E. & Wiktorsson, H. 2002. Automatic milking and grazing. Applied grazing strategies. Deliverable D25 within the EU-project Implications of the introduction of automatic milking on dairy farms (QLK5 - 2000 - 31006). May 2002, 28 pp. Available at http://www.automaticmilking.nl

Ketelaar-de Lauwere, C.C., A.H. Ipema, E.N.J. van Ouwerkerk, M.M.W.B. Hendriks, J.H.M. Metz, J.P.T.M. Noordhuizen & W.G.P. Schouten, 1999. Voluntary automatic milking in combination with grazing of dairy cows. Milking frequency and effects on behaviour. Applied Animal Behaviour Science 64:2, 91-109.

Ketelaar-de Lauwere, C.C., A.H. Ipema, C. Lokhorst, J.H.M. Metz, J.P.T.M. Noordhuizen, W.G.P. Schouten & A.C. Smits, (2000). Effect of sward height and distance between pasture and barn on cows' visits to an automatic milking system and other behaviour. Livestock Production Science 65:1-2, 131-142.

Munksgaard, L. and C. Krohn, 2004. Results of partner 3 in: Automatic milking and grazing, Final report. Deliverable D27 within the EU-project Implications of the introduction of automatic milking on dairy farms (QLK5 - 2000 - 31006).

Salomonsson, M. and E. Spörndly, 2000. Cow behaviour at pasture with or without supplementary roughage in automatic milking systems. In: Robotic Milking, eds. H. Hogeveen & A. Meijering. Proceedings of the International Symposium, Lelystad, The Netherlands, 17-19 August 2000, p. 192.

Spörndly, E. and H. Wiktorsson, 2004. Results of partner 6 in: Automatic milking and grazing, Final report. Deliverable D27 within the EU-project Implications of the introduction of automatic milking on dairy farms (QLK5 - 2000 - 31006).

Spörndly, E. and E. Wredle, 2004. Automatic milking and grazing - Effects of distance to pasture and level of supplements on milk yield and cow behavior. Journal of Dairy Science, in press.

Spörndly, E. and E. Wredle, 2002. Automatic milking and grazing - Motivation of cows to visit the milking robot. Deliverable D26 within the EU-project Implications of the introduction of automatic milking on dairy farms (QLK5 - 2000 - 31006). November 2002, 30 pp. Available at http://www.automaticmilking.nl

INTEGRATING AUTOMATIC MILKING INSTALLATIONS (AMIS) INTO GRAZING SYSTEMS - LESSONS FROM AUSTRALIA

R.K. Greenall[1], E. Warren[2] & M. Warren[2]
[1]National Milk Harvesting Centre, University of Melbourne, Ellinbank, Australia
[2]Winnindoo, Australia

Abstract

Australia's low-cost dairy production systems rely heavily on a diet of year-round pasture (grazing) supplemented with conserved fodder and concentrates. Currently only one commercial farm in Australia utilises AMIs - milking up to 220 cows through 4 units on 85 grazing hectares. This paper details five key lessons the farm owners consider are critical to the management of pasture-based AMI systems based on their experience. The lessons include: *feed adlib silage; don't remove silage from the AMI shed in an attempt to force cows to increase grazing pressure, cows need tracks in good condition; providing adequate pasture will reduce the rate of refusals*; and *changing the concentrate feed does not substantially change the cows' motivation to use the AMIs*. Balancing the quantity and quality of the diet and the location and time that the various components of the diet are fed is the key to motivating the cows to use the AMI. This requires considerable flexibility in management - constantly assessing and manipulating the cows' access to the various dietary sources. The challenge is to balance the quantity, quality and time of feeding pasture; with the quantity, quality, time and location of feeding silage; and the quantity and time of feeding concentrate. Further detailed monitoring begins in 2004.

Introduction

Max and Evelyn Warren are Australia's first AMI farmers and one of the first in the world to use AMI technology in a year-round, pasture-based dairy system. In the Macalister Irrigation District (MID) of SE Australia, pasture growth is seasonal (September through to June) and conserved fodder (mainly silage) and concentrates (usually grain) are routinely used to fill feed deficits (and also to increase per cow production) over the year.

Since 2001 the Warrens have found many answers to the myriad of unknowns that challenged them as they integrated AMI technology into this Australian farming system and are keen to share some of the lessons that they have learnt - many by trial and error. The most difficult to solve were the issues of balancing the various feed sources to keep the cows adequately fed but still motivated to move between the pasture and the automatic milking units (AMUs). Although there is still much to learn, the Warrens have tried a great range of different feeding strategies over the past 2.5 lactations. Their skills were especially tested recently as they battled to manage the system through the worst drought in over 50 years.

Cows enter the holding yards through Texas gates (one way) where they can access feed and water. After passing through an AMU, the cows exit via a single separation (drafting) gate to the appropriate section of the farm. The shed is cleaned by a flood wash system delivering 140 litres per second from a 90,000 litre water tank. Supplementary feed is stored in two 20 tonne silos and a 20,000 litre tank. Fresh water to the dairy is supplied from a 90,000 litre tank.

Getting started

In March 2001, 200 cows were moved from the home farm to the 'robot' farm over a 3 day period. Cows were selected on basic udder conformation and included a range of age and production levels. It was a massive undertaking by the Warrens - helped by two Lely technicians, a few family members, friends & neighbours. Very little was known about how to go about introducing the cows at that time - especially in Australia on a pasture-based system.

Cow selection was the most important lesson learnt. A small number of cows with poor teat placement complicated the start-up, as did the lack of experience of both the Warrens and their helpers. The use of a manual-backing gate to assist the entry of new cows to the AMUs has since been implemented and would have been a great help in those early days. Max and Evelyn also believe that some time spent on an AMI farm before starting up a new farm would be very beneficial.

The operation

For the first full lactation (2001-2) the Warrens milked 200 cows at the farm producing close to 8,000 litres per cow (Figure 3). The cows' health was better than that seen on their other (conventional) farm, especially in terms of mastitis and lameness.

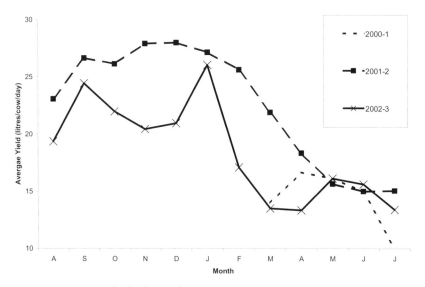

Figure 3. Average per cow production by month.

Automatic milking – A better understanding

Lush pasture growth and ample supplies of silage caused their first problems in balancing pasture and supplementary feeding. The challenge was to balance the different dietary components to ensure that cow flow between the robots and the pasture was acceptable. High milk prices encouraged a feeding regime targeting high per cow production (to the possible detriment of pasture utilisation). Cows produced up to 60L from up to 4 milkings per day. Public and industry interest was also intense - impacting on the time available for the finer management issues.

In the 2002-3 lactation the milk price plummeted by up to 30% and the MID experienced the worst drought conditions for over 50 years. Production fell to 6,500 litres per cow and condition score suffered dramatically in the latter half of lactation. The average cow milkings per day dropped from 2.6 to 2 (or below in some cases), resulting in changes to the feeding program. These changes were required to encourage the cows to rotate between the shed and the limited available pasture and have been detailed in Lesson 4.

The current year (Aug 03 onwards) started with cattle in poor condition, milk price down to A$0.26/litre and uncertainty about breaking of the drought. The spring break brought good pasture growth and a return to normal conditions. Milkings per cow peaked at 2.6 and production was slightly below the average level for this time of year, probably due to the poorer body condition of cows at the start of the season.

Lessons learnt from the Warren's experiences

These following 'lessons' perhaps raise more questions than answers. The Warrens have plenty of experience in managing their commercial AMI farm but have not scientifically tested their theories to date. It is hoped that by sharing their experiences AMI researchers can improve their understanding of AMI pasture-based systems and target their research to suit. The Warrens hope to undertake some initial research work with the National Milk Harvesting Centre in 2004.

Lesson 1 - Feed adlib silage

Cows will still move to pasture if given access to adlib silage as long as the pasture quality is better than the silage. When pasture quantity or quality is lacking, the quality of silage can be manipulated by mixing it with straw, or the silage can be fed in the paddock rather than the dairy. After eating silage the cows move through the AMU (and top up on pellets). They then seem keen to head back out to pasture (if it is available). Having some cows moving out to pasture encourages the other cows to follow one another back to the paddock.

The Warrens have tried removing silage from the shed at times of abundant pasture and feel that some roughage (straw) should still be available in the shed at this time. They believe that production is improved and AMU visits increase by 25-30%. Having some roughage available in the holding yard prior to the AMU is also useful when a number of cows enter the dairy at the same time and are forced to wait. They try to minimise the amount of time cows stand on concrete with nothing to eat and feel that a bit of roughage prior to a feed of pellets is likely to be beneficial. A further refinement would be to preselect cows, so that only cows ready to be milked are allowed access to the shed (and to the silage).

The reality is that no matter what quality of feed is available in the holding yard, the cows will still eat it while waiting to be milked.

Lesson 2 - Don't remove silage from the dairy in an attempt to force cows to increase grazing pressure

Cows forced to graze close to the pasture base will come out of the paddock looking for silage (if available). Feeding cows some silage (mixed with roughage) in the shed helps to motivate them to go back out and graze.

The amount of pasture allocated is critical to regulate cow visits to the dairy. The Warrens have found that giving the cows too little pasture will ensure that they return too soon to the shed. They will stand around at the dairy waiting for a gate change rather than returning to the paddock. Allocating too much pasture can make some cows lazy (especially when late in lactation). It will reduce visits to the AMUs and result in more cows being collected from the paddock.

During the drought the Warrens had very little pasture available. Silage was fed in the paddocks twice a day to encourage the cows through the dairy. Some cows (with low production) reverted back to a routine of twice a day milking to coincide with the feed rations. During this difficult time there were no benefits in having a pasture based system - the farm was really managed like a feedlot.

Lesson 3 - Cows need tracks in good condition

Cows that need to walk through mud after milking will prefer to lie down on the track rather than walk back to the paddock, especially if the cows are heavily pregnant (late lactation) or have free access to silage at the dairy.

The Warrens encountered a clinical mastitis problem due to cows sitting in muddy laneways in March 2002. Physically hunting the cows out was labour intensive and did not fix the problem. The cow flow however did improve once the tracks were resurfaced to remove the muddy areas. The Warrens now anticipate and modify the grazing rotation to reduce the distance that cows have to walk to their new paddock if the environmental conditions are likely to impact on cow flow.

Lesson 4 - Providing adequate pasture will reduce the rate of refusals

Cows that are hungry and do not have adequate pasture in paddock will continuously visit AMI in an attempt to get more grain & silage.

The Warrens have tried several different feeding strategies for their cows, especially during the recent drought. They aim to keep the cows moving through the AMUs and out of the shed at times of pasture restriction by:

- feeding silage out in the paddock (as well as the shed),
- strategically feeding silage (in the new paddock) as the gate changes to encourage cows to move through the AMU to the new paddock, and
- manipulating the timing of gate changes to regulate cow traffic through the AMI.

The Warrens have also recently tried 3 times (per 24 hours) grazing (A-B-C) system to keep cows stimulated, motivated and interested in moving through the AMI and out to the pasture. This strategy resulted in an increase in how many times the cows are milked per day and their production. However, the extra labour required was substantial. The Warrens intend to move to this system in the future by using more automation to reduce the labour commitment (see farm plan in figure 1).

Lesson 5 - Changing the concentrate feed does not substantially change the cows' motivation to use the AMIs

There is no benefit gained by adding sweeteners to the concentrate to motivate cows to visit the AMI more frequently. Different concentrates tried include pellets, triticale, wheat and Molafos (molasses).

The Warrens found that any benefit was minor and transient, with the cows getting used to the change quickly and returning to their previous routine. The Warrens now routinely feed high quality pellets as concentrate feed both in the AMUs and at a parlour (auto) feed station located just after the AMUs.

Discussion and conclusion

Constantly assessing and manipulating the cows' access to the various dietary components is the key to motivating the cows to visit the AMUs in pasture-based systems. The challenge is to balance the quantity, quality and time of feeding pasture; with the quantity, quality, time and location of feeding silage; and the quantity and time of feeding concentrate.

The Warrens have found that considerable flexibility in management is required, especially if pasture growth is unpredictable due to seasonal factors or climatic extremes. Managing an AMI system which is vulnerable to such environmental factors requires close monitoring of cow flow around the farm and quick actions to mitigate the factors as they arise.

Being a commercial farm, the scope for undertaking controlled experiments to investigate these theories at the Warren's farm has been limited. It is important to recognise that these principles are also difficult to weigh up in isolation - they seem to be sound when applied in the Warren's whole farm system but they may be reliant on one another, or on other unidentified (environmental or management) factors.

Further detailed monitoring at the Warren's farm is anticipated to begin in 2004 as a part of a wider program to investigate the potential for AMI technology in the Australian dairy industry.

Acknowledgments

The authors would like to thank Mr. Jürgen Steen and Lely Australia Pty Ltd for their technical advice and financial support. Without the support of Lely, this paper and our participation at this Symposium would not have been possible.

References

Dairy Australia (2003) Australian Dairy Industry In Focus 2003. Level 5 IBM Tower, 60 City Rd, Southbank, Victoria, 3006, Australia. ISSN 1448-9392.
International Symposium - A Better Understanding of Automatic Milking

AUTOMATIC DAIRY FARMING IN NEW ZEALAND USING EXTENSIVE GRAZING SYSTEMS

M.W. Woolford[1], R.W. Claycomb[2], J. Jago[1], K. Davis[1], I. Ohnstad[3], R. Wieliczko[1], P.J.A. Copeman[1] & K. Bright[1]
[1]Dexcel Ltd, Hamilton, New Zealand
[2]Sensortec Ltd, Hamilton, New Zealand
[3]ADAS, United Kingdom

The New Zealand dairy farm scenario

The development of AMS technology in Europe has been almost entirely within a milk production scenario where cows are housed for much of their lactation and there is limited access to grazing on fresh pasture.

New Zealand dairy farms operate with cows are on pasture all year round and housing of cows does not exist. Farms are geographically extensive, the national average herd of 285 cows being maintained on 111 ha of pasture. It is not uncommon in larger operations for cows to be walked 2 km or more from the distant parts of the grazing area to milking. Herd sizes are large relative to labour levels. This system enables a low cost of production being based on extensive pasture feeding and minimal labour input. Appropriate labour, however, is increasingly expensive and difficult to source.

Over the past 30 years, New Zealand dairy farmers have refined batch-milking to extreme levels of efficiency with large herringbone and rotary dairy systems currently accepted as best practice. Rotary milking systems vary in size over the range 30-100 bails, the median being approximately 50 bails. These milk harvesting systems achieve extremely high cow throughputs, operating with minimal operator time per cow, often as little as 5-6 s/cow, just enough time to attach the cluster. This high intensity batch-milking scenario is now mapped into the genes of New Zealand dairy farmers and they are arguably the best in the world at milking large numbers of cows very quickly with minimal labour while at the same time achieving excellent levels of milk quality. This has been an essential strategy given that 10% of New Zealand herds have over 500 cows.

The time has now come to step back and re-think the way we have come to do things in New Zealand, because another reality of the New Zealand system is that the commitment to batch milking large numbers of cows twice a day creates burn-out for many of the people involved and impacts dramatically on general lifestyle in the on-farm sector. Current practice is probably unsustainable in the long term due to the nature of the task, the environmental working conditions, and the decreasing pool of labour interested in pursuing this area of work as a career path.

A key element of AMS systems as they have evolved in Europe, is the shift in mindset away from "batch milking" to "distributed milking". Such a concept is light-years removed from what currently happens in New Zealand herds where large expensive milking systems that sit idle for 18-20 h/day are accepted practice and where milking is still very much a high pressure, hands-on, routine manual operation.

Could distributed milking work in the New Zealand pasture based scenario? Could a system be devised to generate cow traffic through an AMS unit from remote pasture on a 24 h basis and to what extent would it modify labour requirements on New Zealand dairy farms (Jago & Woolford, 2002)?

Addressing these questions is the mission of the Dexcel Greenfield Project.

The "Greenfield Project"

In early 2001 a joint venture agreement was set up in Hamilton, New Zealand between Dexcel Ltd and Sensortec Ltd (NZ based subsidiary of DEC International Ltd) to establish an AMS based research farm for joint purposes. The research objectives of Sensortec Ltd are to develop new sensing technologies for milk components and the research objective of Dexcel Ltd is to devise new farm systems that can capitalise on AMS technology and improve the productivity of New Zealand dairy farms (Woolford, 2001).

The Greenfield project started from the position that successful commercial AMS technology had been developed in Europe and that the question to be addressed was:

"What should an extensive pasture driven New Zealand dairy farm look like if AMS technology is to improve on existing productivity and labour utilisation?"

The over-arching research objective of the project has been to devise a system of generating continuous cow traffic from remote pasture sites through an AMS unit, so creating New Zealand's first "distributed" milking farm system. This is a farm production systems approach to the issue of automated milk harvesting. For that reason the end objective is an automatic farm, rather than an automated milking system.

Introducing robotic milking as a serious dairying research objective to be funded in New Zealand was a major problem given the industry mindsets previously discussed. However, in late 2000, visionary initial seed-funding was acquired from the Agricultural Marketing Research and Development Trust (Agmardt; www.agmardt.org.nz) and co-funded by the Dairying Research Corporation (now Dexcel Ltd) and also by Sensortec Ltd.

This funding scenario has now shifted to funding from the Public Good Science Fund administered by the New Zealand Foundation for Research Science and Technology, and by Dairy Insight from the new dairy producer levy (Dexcel, 2002a). The project has from its first inception had the benefit of a group of farmer mentors that meet on a 3-monthly basis to review issues and developments.

Design concept

The Greenfield Project started up on a very small scale. Adopting a "clean-slate" approach, all existing fences and infra-structure on the initial 10 ha experimental farm were removed. The farm was re-configured without any internal subdivision, with a central selection unit for the cows that channelled them either via an arterial dual-lane "cow-way" system leading to the AMS unit, directly back to the pasture, or to a new pasture break. Pasture feeding was by mobile electric fences that set up pie-shaped breaks of pasture that rotate around the central selection unit. The cow traffic across the total geography of the

Table 1. Incidence of new intramammary infections (IMI) due to major pathogens (Strep. uberis, E. coli, Nocardia) *over 865 days of operation of the Greenfield Project.*

	Clinical	Sub-clinical
New IMI (major paths)	9	34
Cow-days/new IMI	4748	1257
New IMI (minor paths)	1	109
Cow-days/new IMI	42 729	392
Total cow-days	42 729	

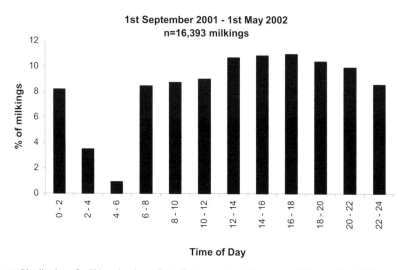

Figure 1. Distribution of milkings by time of day for the period 1 September 2001 - 1 May 2002.

from pasture to AMS was 400 m and the diet was grazed pasture + 2 kg barley/cow/day fed during milking.

The key finding to date from these and other data sets is that few milkings occur between 0200 h and 0600 h. Cow traffic to the dairy begins around sunrise and remains relatively constant until 0200 the next morning, although is dependent on the time at which a fresh pasture break is made available, and on weather conditions. Identifying ways to maintain cow traffic to the dairy during this apparent early morning "quiet period" will be a key factor in maximising cow throughput in a pasture-based AMS unit.

Key Greenfield Project outcomes

- Automatic milking within a pasture-based dairying system can be made to work with up to at least, 1.8 km return walking distances

Automatic milking – A better understanding

- There appears to be a lactational effect in terms of cows movement around the farm system. Cows in early lactation appear to be more active in their geographical movements and visit the remote selection units more often than later in lactation
- There is a perception that the New Zealand cows used in this project have quite different behavioural characteristics from European/North American cows in AMS situations. In particular there appears a much higher level of motivation in terms of foraging for feed.
- The rejection rate for new cows entering the herd has averaged 13%. Overall 8% were rejected for poor conformation, 2.4% for temperament, and 2.4% for other reasons.
- Automatic milking is as much about human behaviour factors, as it is about cow behaviour. Not all herd owners or managers will be able to make the change to a set of quite different philosophies.
- Viability of the concept in New Zealand probably depends heavily on the development of management and training techniques that are relevant to herd sizes of 200+ cows.

The future

Progress and learning to date from the prototype Greenfield system suggest real potential for implementing AMS technology in New Zealand extensive grazing systems. The project will continue to upsize in terms of cow numbers until the limits of cow capacity per AMS are established. Economic studies of the Greenfield system will be carried out and further development work will be pursued in the areas of reproductive strategies, cow and milk monitoring, and wider automation of the total farm system.

References

Dexcel. 2002a. Govt funding widens *Greenfield* horizon. New Zealand Dairy Exporter, September.

Dexcel. 2002b. Dexcel Annual Report 2002.

Dexcel. 2003. Automated milking project progresses to another level. New Zealand Dairy Exporter, January.

Jago, J., Copeman, P. J. A., Bright, K., McLean, D., Ohnstad, I., Woolford, M. W. 2002a. An innovative farm system combining automated milking with grazing. Proc. New Zealand Society of Animal Production. 62:115-119.

Jago, J., Davis, K. 2002. How it's done: cow training at automated milking farmlet. New Zealand Dairy Exporter, July.

Jago and Woolford 2002, Automatic Milling Systems: an option to address the labour shortage on New Zealand dairy farms ? Proc. New Zealand Grasslands Assn 64: 39-42.

Livestock Improvement Corporation. 2003. Business Information Unit, Private Bag 3016, Hamilton, New Zealand.

McEldowney, L. 2003. Huge potential benefits from robotic milkers. Editorial; New Zealand Dairy Exporter, February.

Mountfort, M. 2002. Robotic trial reveals exciting potential. New Zealand Dairy Exporter, March.

Mountfort, M. 2003. Automated system less stressful. New Zealand Dairy Exporter, February.

Woolford, M. W. 2001. Researching robotics. New Zealand Dairy Exporter, February.

Woolford, M. W., Jago, J. 2002. The use of robotic milking machines in pastoral dairy systems. Proc. Massey Dairy Farmers Conference, Palmerston North, New Zealand.

Woolford, M. W., Jago, J., Davis, K. 2003a. The Dexcel "Greenfield Project": A prototype for automatic dairyfarming on pasture. Proc. Australian Dairyfarmers Conference, Shepparton, Victoria.

TWO CASE STUDIES ON FARMS COMBINING AUTOMATIC MILKING WITH GRAZING - TIME BUDGETS, SYNCHRONISATION OF BEHAVIOUR AND VISITS TO THE ROBOT

Lene Munksgaard[1] & Mette Søndergaard[2]
[1]Danish Institute of Agricultural Science, Dept. of Animal Health and Welfare, Foulum, Denmark
[2]Zoological Institute, University of Copenhagen, Copenhagen, Denmark

Abstract

We investigated the behaviour of the cows and visits to the robot at two private farms, which differed considerable in their management routines.

At both farms there was one robot and approximately 65 Danish Friesian cows, which had access to pasture during the day. Herd 1 had free traffic and cows were only fetched from the pasture once a day; Herd 2 had forced traffic and cows were fetched from pasture more times a day. Each farm was visited three times in the grazing season. Occurrence and synchronization of "standing/walking", "lying" and "eating" was observed at pasture and in the barn in the fall. Furthermore cow traffic in and out of the barn was observed.

In herd 1 most cows left the barn immediately and in 79% of the observations 75% or more of the cows were at pasture. In herd 2 it took around three hours before most of the cows had left the barn, and only in 31% of the observations 75% or more cows were at pasture. The behaviour of the cows were more synchronised when being at pasture than in the barn (P<0.001). On both farms the cows almost always left/returned to the barn in groups, only in 4% (Herd 1) and 7% (Herd 2) a cow was leaving/returning to the barn with an interval of more than 9 minutes to the cow leaving/arriving before of after.

Milking frequency were 2.5 times (herd 1) and 2.7 times per day (herd 2). When the cows had access to pasture the number of milkings per hour decreased in herd 1 compared to the remaining part of the day (2.6 versus 8.0), but not in herd 2 (7.0 versus 7.3). In conclusion, the behaviour of the cows were more synchronised at pasture and cows rarely went to the barn alone. Differences in management routines had a severe effect of the use of the robot.

Introduction

The success of a system with automatic milking presupposes that the cows regularly, voluntarily and individually visit the milking unit, so that a high milking frequency is obtained and fetching of cows is kept at a minimum. The necessity of a de-synchronisation of the behaviour to achieve fully utilisation of the automatic milking system might results in difficulties of combining grazing with automatic milking. Previous studies suggest that the behaviour of cows is more synchronised at pasture than when kept in a barn (Krohn *et al.*, 1992, Miller & Wood-Gush, 1991). Furthermore, cows are gregarious animals, and may be reluctant to leave the group. Only few studies with rather few cows per robot have focused on the behaviour of cows in systems combining automatic milking and grazing (Ketelaar-de Lauwere *et al.*, 1999), therefore we investigated the behaviour and use of the robot on two farms with different management routines.

Methods and materials

Observations were made on two farms, each with one milking unit and approximately 65 Danish Friesians. In herd 1 there was free traffic, and cows were only fetched from pasture once a day. In herd 2 there was forced traffic, and smaller groups of cows were fetched from pasture 1-4 times a day. At both farms approximately 20 cows were separated in the morning and kept in the barn until milking was completed. Furthermore, in herd 2 approximately 15 cows were led to a smaller field in the morning and fetched again (usually after 2-3hours). Thereafter the small field was no longer in use that day. At both farms water was only available in the barn, and supplementary roughage was offered in the barn and concentrate in the robot.

Each farm was visited three times in the grazing season with an interval of 5 -6 weeks. Each visit had duration of 5 days. Furthermore, the farms were visited once in the fall, after the cows had been indoors all day for at least 2 weeks.

Behavioural observations

Sixteen cows were chosen according to the following criteria: 8 individuals with the highest milking frequency and 8 individuals with the lowest milking frequency. However, in each group of eight animals at least three should be in first lactation and at least three should be in second or later lactation. If this criteria was not fulfilled within the eight with the highest and eight with lowest milking frequency, the next cows on the list of high/low milking frequency was chosen. These 16 animals were marked with spray paint and coloured collars for individual observations (day 1). On the following two days (day 2 and 3) the behaviour of these animals was observed at pasture from the barn door was opened until the door was closed. The 16 cows were scanned with 15 minutes intervals in the following categories: standing/walking, lying or eating. Furthermore, the total number of cows' at pasture in the following categories: standing/walking, lying and eating was recorded with 15 minutes intervals.

On day 4 and day 5 observations were done from the barn door was opened until the door was closed. Every time a cow passed the door, the number of the cow; direction (in/out) and time of day were recorded either by direct observation or if a large group of cows were passing by analysis of video recordings.

In the fall the cows were observed on four succeeding days in the barn for the same time interval as when on pasture and the same observation method was used.

Results and discussion

Milking frequency

The milking frequency per cow per day was on average 2.5 in herd 1 and 2.7 in herd 2 during the grazing period, and 2.7 (herd 1) and 2.6 (herd 2) in the fall. During the time of the day where the cows did not have access to the pasture, the milking frequency per hour was 8.0 in herd 1 and 7.3 in herd 2. However, during the period of the day, where the cows had access to the pasture the milking frequency per hour decreased to 2.6 in herd 1, while it was almost at the same level as during the night time in herd 2 (7.0). Krohn (2004) also observed a decrease in the use of the robot during periods where cows have access to pasture. Unvisited periods of up to two hours during the period where cows had access to pasture were observed in herd 1, but not in herd 2, where cows were fetched from pasture more times a day. The milking frequency was rather high in both herds also during the grazing season, but it should be taken into consideration that the cows only had restricted access to the pasture (Figure 1). In agreement, Ketelaar-de Lauwere et al. (1999) only found a reduction in the milking frequency per cow, when the cows had 24-hour access to pasture.

Behaviour

In herd 1 with free cow traffic most of the cows moved outside when the door was opened in the morning (Figure1), and on average 75% of the cows were outside at 79% of the scan samplings, and 95% of the cows at 41% of the scan samplings. In herd 2, where cows had to go through the robot to get access to the pasture, it took approximately three hours before most cows had left the barn. On average 75% of the cows were at pasture at 31% of the observations (scan sampling), but only in 3% of the scan sampling were 95% of the cows at pasture. There were no differences in the time budget and percentage of time at pasture between cows with high (3,5 in both herds) and low (1,8 herd 1 and 2.4 in herd 2) milking frequency. Thus the results suggest, that cows prefer being at pasture rather than in the barn, and that cows with the highest milking frequency did not spend more time in the barn than cows with the lowest milking frequency.

In all observations periods the cows spent a third or more of their time at pasture lying down even though they had restricted access to pasture. At both farms the cows spent more time feeding and less time lying and standing/walking when at pasture than when indoors in the fall (Table 1). Day to day variation was considerable larger in the summer than during the winter observations.

Feeding and lying patterns of one randomly chosen day from summer and fall at each farm are shown in figure 2. They are representative of the difference between seasons and farms. The proportion of cows lying or feeding at each scan sampling varied a lot, especially in the summer. At pasture, lying behaviour alternated with feeding behaviour. In the barn

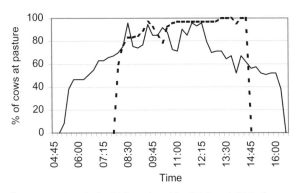

Figure 1. Percentage of cows at pasture in herd 1(– – –) and herd 2 (——) (data from one day in July).

Automatic milking – A better understanding

Table 1. Percentage of time cows spent on different activities estimated using data from 10 and 15 min scan sampling (Mean ± std.).

Behaviour	Herd 1		Herd 2	
	Summer	Winter	Summer	Winter
Eating	44.5±7.0	18.3±0.8	33.7±15.2	14.4±2.1
Stand/walking	15.5±3.2	35.9±1.3	25.3±14.0	35.9±1.6
Lying	40.2±6.2	45.7±6.2	41.0±4.9	49.7±1.9

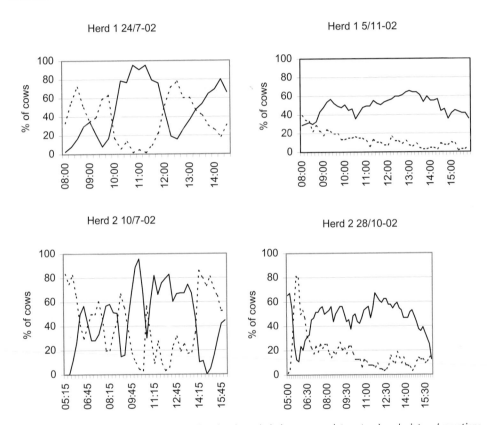

Figure 2. Proportion of cows lying (——) and eating (– – –) during summer (at pasture) and winter observations.

this pattern was only seen after the morning delivery of forage. However, during the fall observations feed was delivered approximately one hour before the observation started in herd 1, while food was delivered during the first hour of observation in herd 2. In fall, feeding was concentrated in the hours following the morning delivery of forage. The percentage of cows lying peaked around midday both during the summer and the fall. In summer this pattern was clearer in Herd 1 with a long period, where almost all cows were lying.

Cows arriving at pasture alone or in small groups during the day were more often performing the same behaviours as the majority of the herd than other behaviours half an hour after arrival at pasture (Herd 1- 65%, Herd 2 - 59%).

In both herds' the pattern of lying and eating showed a higher degree of synchrony when at pasture than when indoors (summer: Kappa coefficient (K)=0.34 herd 1, K=0.19 herd 2; fall: K=0.04 herd 1 and K=0.09 herd 2). Furthermore, at pasture the synchrony was higher in herd 1 than in herd 2, where the farmer controlled cow traffic in/out of the barn more.

At both farms cows mostly arrived at pasture or in the barn in close successions (Figure 3). When cows were leaving/entering the barn, another cow was leaving/entering the barn within an interval of 4 min or less in between, in 90% of the cases in herd 1 and in 75% in herd 2. Within a limit of 9 min, 96% and 93% went together with another cow.

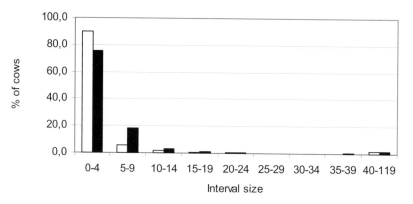

Figure 3. Distribution of cows leaving/entering the barn with various time intervals to another cow leaving/entering the barn in Herd 1 (□) and Herd 2 (■).

Conclusion

This study demonstrate that differences in management routines on farms combining grazing and automatic milking can have great impact on both the behaviour of the cows as well as the use of the robot. The behaviour of the cows was more synchronised at pasture than in the barn, and cows rarely went to the barn alone. This suggest, that it is necessary to fetch cows from pasture or to develop other management strategies to attract cows one by one or in smaller groups to go from the pasture to the barn. However, even though the number of milkings per hour decreased considerable when cows were at pasture in herd 1 with completely free traffic, the average number of milkings per cow per day was rather high.

References

Ketelaar-de Lauwere, C.C., Ipema, A.H., van Ouwerkerk, E.N.J., Hendriks, M.M.W.B., Metz, J.H.M., Noordhuizen, J.P.T.M., Schouten, W.G.P., 1999. Voluntary automatic milking in combination with grazing of dairy cows. Milking frequency and effects on behaviour. Applied Animal Behaviour Science 64, 91-109.

Krohn, C.C. 2004. Seven case studies about automatic milking and grazing in private herds. Proceedings of international Symposium: A better understanding - Automatic milking. Lelystad, The Nederlands March 24-26. Poster presentation.

Krohn, C.C., Munksgaard, L., Jonasen, B., 1992. Behaviour of dairy cows kept in extensive (loose housing/pasture) or intensive (tie stall) environments I. Experimental procedure, facilities, time budgets - diurnal and seasonal conditions. Applied Animal Behaviour Science 34, 37-47.

Miller, K., Wood-Gush, D.G.M., 1991. Some effects of housing on the social behaviour of dairy cows. Animal Production 53, 271-278.

THE INFLUENCE OF THREE GRAZING SYSTEMS ON AMS PERFORMANCE

H.J.C. van Dooren, L.F.M. Heutinck, G. Biewenga & J.L. Zonderland
Applied Research, Animal Sciences Group, Wageningen UR, Lelystad, The Netherlands

Abstract

Pasturing is common practice in The Netherlands and has several advantages. Feeding costs are lower compared to zero-grazing, animal health and welfare benefits from an outdoor period and public opinion toward dairy farming is positively influenced. However dairy farmers increasingly incline to reduce the grazing period in terms of days per year and hours per day. Among them are relatively many farmers with an automatic milking system (AMS). For an efficient use of the AMS and a maximum milk yield per unit they aim to maximize the AMS occupation. Pasturing is believed to increase the peaks and off-peaks in visiting pattern of the AMS. However, pasturing can be organized in many different ways resulting in different cow behaviour and consequently different visiting patterns. This papers aims to investigate the effect of different ways of pasturing on AMS performances. Based on a rotational grazing system, three alternatives were tested: a) Daytime grazing - one group (BE); b) Day and night time grazing - one group (OT) and c) Day and night time grazing - two groups (OG).

AMS performance was highest for the BE system and lowest for the OT system. The fact that half of the herd stayed indoor in the OG system could not compensate the loss of daily milk production per AMS. Aiming at optimising AMS occupation to maximise AMS milk yield the BE system is most appropriate. Moreover the other two systems also require a higher labour input to achieve good results. However, if available labour is not limited and if maximum grazing opportunities are positively appraised both OT and OG are possible.

Introduction

Grazing is common practise among dairy farmers in The Netherlands and has several advantages. Feeding costs are lower compared to zero-grazing, animal health and welfare benefits from an outdoor period and public opinion towards dairy farming is positively influenced. Despite these advantages an increasing number of farmers reduce the grazing period in terms of days per season and hours per day in pasture (Pol, 2002), decreasing the advantages of grazing proportionally. Among them are a relatively high number of dairy farmers with an automatic milking system (AMS). Only around 50% of these farmers offer grazing to their herds (Dooren et al., 2002) while this percentage among farmers with a conventional milking parlour is believed to be around 90%. Farmers generally aim at optimising the use of the AMS in order to maximize the milk yield per unit. This requires a more or less constant occupation of the AMS without much queuing of cows. Pasturing influences the visiting pattern of the herd to the AMS, which may lead to peaks and off-peaks in the occupation, queuing and therefore a sub-optimal use of the AMS. However, pasturing can be organized in many different ways resulting in different cow behaviour and consequently different visiting patterns. This papers aims to investigate the effect of different ways of pasturing on AMS performances. The most commonly used pasturing system among AMS farmers in The Netherlands is the rotational system (Dooren *et al.*, 2002). In this system a field with enough grass for a number of days is offered to the herd, after which a new fresh field is available. The number of days per field can range from 3 to more than 10. Within this system a farmer can choose for restricted or unrestricted access to the pasture in terms of time (daytime grazing or day and night time grazing) or in terms of grouping of the herd (only a part of the herd has access to the field). Based on a rotational pasturing system three of these alternatives were tested in this research.

Material and methods

On Nij Bosma Zathe, one of the experimental dairy farms of Applied Research situated in the province of Fryslân, around sixty cows are milked with a two-units-in-row Galaxy AMS. Figure 1 gives the layout of the section of the barn where this group is housed.

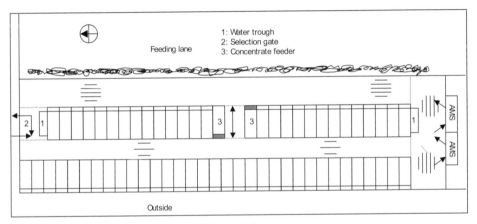

Figure 1. Lay-out of barn section with AMS at Nij Bosma Zathe.

Pastures were situated around the east side of the barn on a distance between 50 and 800 meter from the entrance of the barn. Pasturing was regulated by a selection gate always refusing access to pasture for cows within less than 2 hours before the end of their minimum milking interval. Additional restrictions on herd level depended on grazing system (see section 0). The length of the minimum milking interval was 6, 8 or 10 hours depending on production level and lactation stage. Concentrate was supplied in the AMS and in a concentrate feeder. Water was available at two sides of the barn and at the entrance of the field. Roughage (a mixture of grass and mais silage) was supplied twice a day. On average an amount 7,5 kg DM*cow^{-1}*day^{-1} was available. Cows were fetched from pasture one or two times a day depending on grazing system (see section 0), meaning that all the cows out in pasture were fetched to be milked.

Grazing systems

Based on a rotational grazing system, three alternatives were tested:

a. Daytime grazing - one group (BE)
    In addition to the individual criteria described above, access to pasture was restricted between 4:30 pm and 6 am for the whole herd. Fetching occurred one time a day at dawn (around 8 pm).

b. Day and night time grazing - one group (OT)
    Given the individual criteria, access to pasture was unrestricted in time for the whole herd. Fetching occurred twice a day at 6 am and at 4 pm.

c. Day and night time grazing - two groups (OG)
    The herd, ranked on days in lactation, was divided in two equal sized groups. Access to pasture was refused for the early lactation group and unrestricted in time (given the individual criteria) for the other, late lactation, group. Fetching occurred twice a day at 6 am and at 4 pm.

Table 1 gives a summary of the three grazing systems used in this experiment.

Table 1. Summery of three grazing systems.

Pasturing system	Start date	Number of groups	Number of fetchings	Number of days in field
BE (day)	18-8-2003	1	1	15
OT (night & day)	2-9-2003	1	2	31
OG (night & day)	2-10-2003	2	2	21

Recorded data

AMS recordings of each visit (cow-id, time and if relevant milk yield) allowed calculating daily number of visits (milkings, refusals and failures), milking intervals and daily milk yield. Number of animals fetched from the field was recorded as was the total roughage intake per day.

Analyses of visiting pattern

As stated above the grazing system probably influences the visiting pattern of the AMS. To gain insight into this visiting pattern and make it comparable between the different grazing systems, an evenness ratio was introduced. Therefore, in each of the 24 hours of a day the average milkings and visits per hour was calculated for each grazing system. The ratio was calculated as the 90 percentile divided by the 10 percentile over these 24 hours. In case the visiting pattern is complete even this ratio will be 1 and it will increase with an increasing unevenness of the visiting pattern. Together with the total daily milk production this evenness ratio expresses the AMS performances.

Results and discussion

Table 2 and figure 2 give an overview of the AMS performance indicators.

Automatic milking – A better understanding

Table 2. AMS performance indicators at three different grazing systems.

Grazing system	Visits (#)	Milkings (#)	Milkings (% of visits)	Refusals (#)	Refusals (% of milkings)	Failures (#)	Fetched cows (#)
BE	335	171	51%	142	83%	22	26
OT	269	147	55%	106	72%	17	88
OG	348	162	47%	169	104%	17	43

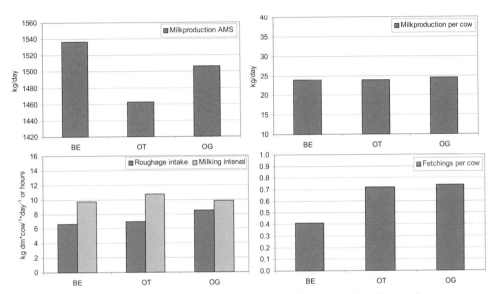

Figure 2. Milking interval, milk production, roughage intake and fetching results of the three grazing systems.

The average numbers of visits per day for the BE and OG system are comparable, whereas the number of visits at the OT system is considerably lower. The same can be said about the number of milkings. However, the number of milkings related to the total number of visits is the highest during the OT system. The use of the AMS, in terms of 'relevant visits', seems therefore more efficient.

The milking interval was for both the BE and OG system 9.8 hours on average. The interval during the OT system was almost one hour longer (10.7). This higher milking interval did not lead to lower daily milk yields per cow. Differences between the three systems were minimal: 24 kg per cow for BE and OT system and 24.5 for the OG system. Combining the results of milking interval, milk yield and visits to AMS, leads to the milk production per AMS (Figure 2). This was highest for the BE system (1537 kg per day) and lowest for the OT system (1462 kg per day). During the OG system an amount of 1506 kg milk per day was produced.

Roughage intake seems not to be influenced by length of grazing period. The total daily intake for BE and OT system was comparable (6.6 and 6.9 kg*day^{-1}*cow^{-1} respectively). The intake during the OG system was somewhat higher (8.5 kg*day^{-1}*cow^{-1}), presumably caused by the longer indoor period of half of the herd. The number of fetched cows (Table 2) gives an unclear picture. Related to the number of fetchings per day and the number of cows with access to pasture (Figure 2), the OT and OG systems appear to have a considerably higher fetching rate (number of fetched cows per cow) than the BE system so that both systems requires more labour input. Both the lower number of milkings and the longer milking interval during the OT system seems in accordance with expectations of cow behaviour in this grazing system. The more the herd is outside, the longer the milking interval will be, and the more cows have to be fetched to keep the milking frequency at an acceptable level. Despite this extra input of labour the lower daily milk yield per AMS was reduced. Keeping half of the herd inside did increase the total daily milk production but not the level accomplished during the BE system. Considering the number of visits (including milkings) per day during the OT system compared to the other systems it seems reasonable to assume that in the OT system a higher number of animals per AMS is possible. Based on a estimated, fixed duration of a milking, a refusal and a failure (6, 1 and 6 minutes respectively) the amount of time per day the AMS is occupied can be calculated. Maximum is 22 hours as two hours are needed for cleaning procedures. Together with the average visiting frequency the number of extra cows and new total daily milk production can be estimated (Table 3).

Table 3. Estimation of extra cows and extra daily milk production.

Grazing system	Occupation (hours)	Frequency per cow per day			Extra cows	New AMS production
		Milking	Refusal	Failures		
BE	21.7	2.7	2.2	0.01	1	1561
OT	18.2	2.4	1.7	0.01	14	1798
OG	20.7	2.6	5.8	0.01	4	1604

Visiting pattern

Figure 3 gives the visiting patterns and the evenness ratio of the three grazing systems.

During the period the pasture is available for the herd the number of visits to the AMS decreases in all three systems. In the OT and OG system the moments of fetching can be taken easily from the figures as the number of visits increases sharply afterwards. In the BE system the autonomous return seems higher than in the other systems as the number of visits is already increasing before cows are actually fetched (around 8 pm). The lowest level at the OG system is higher than the OT system caused by the restricted access to pasture for the early lactation group.

These effects become also clear from the evenness ratios. Taking only milkings into account the OG systems shows the most even visiting pattern, the OT system the most uneven. This is somewhat different compared with the total AMS visits, caused by a different proportion of milkings and refusals over the day.

At all three systems the visiting pattern shows a dip at 9 am and 11 pm caused by the extensive cleaning procedure of the AMS during these hours.

Based on the same procedure as mentioned above the AMS occupation per hour can be calculated. During the BE system in 13 of the 24 hours queuing will occur. During OT and

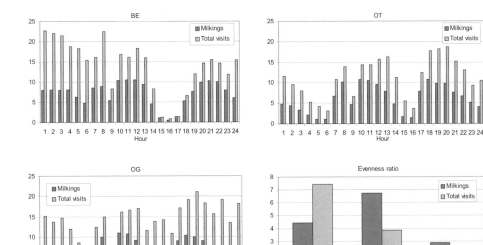

Figure 3. Visiting patterns (total visits and milkings) and evenness ratio (bottom right) of the three grazing systems.

OG system in respectively 7 and 9 of 24 hours this will happen. Although based on daily averages there seems to be enough capacity to increase the number of animals per AMS. However this will most likely lead to more queuing as well due to group wise visits to the AMS.

Despite this extra input of labour the lower daily milk yield per AMS was reduced.

Conclusions

AMS performance was highest for the BE system and lowest for the OT system. Despite the more even visiting pattern of the OG system the loss of daily milk production could not be compensated. Aiming at optimising AMS use to maximise AMS milk yield the BE system is most appropriate. The other two systems also require a higher labour input to achieve current results. However, if available labour is not limited and if maximum grazing opportunities are positively appraised both OT and OG are possible. Although AMS capacity seems available an increasing number of animals in the OT system will lead to more queuing.

References

Pol-van Dasselaar, A. van der, W.J. Corré, H. Hopster, G.C.P.M. van Laarhoven, C.W. Rougoor, 2002 . Belang van weidegang, PraktijkRapport 14, Praktijkonderzoek Veehouderij, Lelystad, 82p.

Dooren, H.J.C. van, E. Spörndly and H. Wiktorsson, 2002. Automatic milking and grazing, Deliverable 26, EU-project Implications of introduction of automatic milking on dairy farms, May 2002, 38 pp. See: http://www.automaticmilking.nl

HOW PASTURING INFLUENCES THE USE OF AMS: A SURVEY AMONG 15 DAIRY FARMS IN THE NETHERLANDS

H.J.C. van Dooren[1], M.E. Haarman[2], J.H.M. Metz[2] & L.F.M. Heutinck[1]
[1]Applied Research, Animal Sciences Group, Wageningen UR, Lelystad, The Netherlands
[2]Farm Technology Group, Agrotechnology and Food Sciences, Wageningen UR, Wageningen, The Netherlands

Abstract

One of the changes that comes with the introduction of automatic milking is that milk frequency does not depend on farmer's initiative only, but is mainly determined by cow behaviour. As milk yield per cow is dependant on milking frequency and interval and since the introduction of an AMS is a major investment, maximizing milk yield per AMS is important for farmers. Whether maximum milk yield can be achieved, largely depends on cow visiting pattern. Unevenness of the visiting pattern in terms of visits per hour will also lead to queuing in front of the AMS during peak hours and idle time during off-peaks. Pasturing is believed to be one of the factors that influence the visiting pattern. The objective of this research was to study the effects of pasturing on the use of AMS capacity on commercial dairy farms in The Netherlands.

Data of fourteen farms were collected during two visits, one in the winter period and one in the grazing period. During these visits the following data were collected: AMS data, general farm characteristics and observation of cow behaviour. Data analysis implied analysis of differences between winter and pasture observations *within* each farm. For each farm and observation period milking frequency, milking interval, number of milkings per day, distribution of milkings per hour and milk yield per cow and per day were calculated. Differences between the observation periods were analysed with a t-test.

Results did not give evidence to assume that use of AMS capacity is generally decreasing during the pasture period compared to the winter period. Total daily milk production decreased with only 2.5%, number of milkings with 6% and milking frequency decreased from 2.68 to 2.61. Also visiting pattern expressed in an evenness ratio increased moderately. However, differences between farms were great. Apparently other factors than recorded in this research had bigger influence on the outcomes. It can be concluded that grazing does not have to be disadvantageous for AMS performances. Management skills of individual farmers are likely to be more decisive.

Introduction

In modern dairy farming milk is the most important source of income. The farmer as entrepreneur has the task to maximize milk yield under economical, environmental and legal conditions. This is also valid with automatic milking, although with the introduction of the AMS a new way of milking has emerged.

One of the main changes that comes with the introduction of automatic milking is that milk frequency does not depend directly on farmer's initiative only, but is mainly determined by cow behaviour. Milking frequency is not longer fixed (two of three) but will vary from 1.5 to over 4 milkings per day for individual cows. As milk yield per cow is dependent on milking frequency and interval and since the introduction of an AMS is a major investment, maximum use of its capacity, or in other words increasing capacity by maximizing milk yield per AMS is an important goal for farmers. The maximum capacity in kg milk per day for an AMS can be calculated by multiplying the yield per milking with the number of milkings per day. The maximum number of milkings per day can be calculated with handling time of the AMS (identification, udder preparation, teat attachment, cluster cleaning) and machine on time (milk yield divided by flow rate of individual cows) with the assumption that apart from cleaning cycles the AMS is more or less constantly occupied for about 80% (Koning and Ouweltjes, 2000).

Whether calculated maximum milk yield can be achieved largely depends on cow visiting pattern. Unevenness of the visiting pattern in terms of visits per hour will lead to queuing in front of the AMS during peak hours and idle time during off-peaks. Pasturing is believed to be one of the factors that influence the visiting pattern. Pasturing has always been practice on Dutch dairy farms. Despite the benefits in terms of animal health and welfare and public acceptance, more and more dairy farmers, under which relatively many farmers with an AMS, decide to keep their cattle indoors during the summer. Only 50% of AMS farmers offered grazing to their herd. (Dooren et al., 2002).

Much of the research up till now on combining automatic milking and grazing has focused on specific aspects of pasturing and their effect on milking frequencies, milking intervals and milk production. Examples are the effect of level of roughage feeding indoors (Jagtenberg and Van Lent, 2000a), sward height (Ketelaar-de Lauwere, 1999), field distance (Ruis-Heutinck et al.,2001; Ketelaar-de Lauwere, 1999; Wredle, 2001) water supply in field (Jagtenberg and Van Lent, 2000b) and number of fetchings per day (Jagtenberg and Van Lent, 2000b). These studies have been conducted with relatively small groups of cows under controlled circumstances. Objective of this research was to study the effects of pasturing on the use of AMS capacity on practical dairy farms in The Netherlands.

Materials and methods

From a database with around 120 dairy farmers a selection was made based on three criteria: a) at least 2 years experience with automatic milking and grazing, b) only one AMS unit installed and c) having a Lely Astronaut AMS. The third last criterion was only chosen for reasons of uniformity in data collection. This led to a group of 60 farmers that met these criteria. From this group fifteen farmers, selected randomly, were asked to participate in the project. Fourteen of them agreed.

Data were collected during two observation periods. The first observation was during the winter period, the second during the grazing period. In each observation the farms were visited. During these visits the following data were collected:

a. AMS data. AMS log files of fifteen days preceding the visits were collected containing information of cows (id, lactation number and lactation stage) and milkings (date, time, cow-id, duration and yield).

b. General farm characteristics. During an interview with the farmer information about grazing system, grazing time, access to pasture, frequency of fetching, number of cows, water supply, system of roughage and concentrate feeding and data about quotum and average farm production levels were collected. Additionally, observations like location of barn and pasture, cow traffic and barn lay out were gathered.
c. Observation of cow behaviour. A short behaviour analysis resulted in an indication of hierarchy and other behaviour like imitation and autonomous return from pasture.

Data analyses

Data analysis implied analysis of differences between winter and pasture observations *within* each farm (Figure 1).

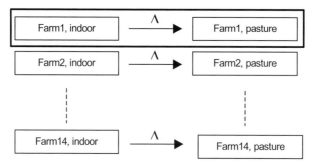

Figure 1. Analysis of differences between observation periods within each farm.

For each farm and observation period the following characteristics were calculated:
a. Milking frequency
b. Milking interval
c. Number of milkings per day
d. Distribution of milkings per hour. For comparison of the evenness of the distribution of milkings per hour between the two observation periods an evenness ratio was calculated as the 90% percentile divided by the 10% percentile of the number of milkings per hour. A complete even pattern of milkings over 24 hour will give a evenness ratio of 1, a more uneven pattern will give a higher ratio.
e. Milk yield per cow and per day with $MY_{pasture} = MY_{winter} * (LF_{pasture}/LF_{winter})$ where the Lactation Factor (LF) from Coolman (1981) is used to correct daily milk yield per cow for lactation stage and age at last calving.

Differences in number of milkings, milk yield per cow and per day and milk frequency between the observation periods were analysed with a t-test.

Results and discussion

A summary of recorded farm characteristics is given in table 1
The data show a broad variety in farm characteristics. Especially Farm 8 is somewhat an outlier. This farmer did have another job as an alternative source of income. The majority

Table 1. Summery of some farm characteristics.

Farm	1	2	3	4	5	6	7	8	9	10	11	12	13	14
Lactating cows (#)	85	56	76	52	65	80	70	39	75	56	60	59	59	56
305 day production (ton)	6.6	8.2	8.2	9.2	7.7	8.8	9.5	5.0	9.2	8.3	9.2	7.3	8.0	9.1
Intensity [ton/ha]	12	7.7	14.1	13.5	n.a.	9.1	7.5	3.0	13.5	10.2	11.9	4.9	6.5	14.1
Calving season[1]	3	4	4	4	4	4	4	3	4	3	4	0	4	3
Grazing time [hours/day]	9	9	15	9	7	14	9	23	7	9	n.a	6	7	11
Fetching frequency pasture	2	2	1	1	1	1	3	0	3	2	1	1	1	n.a
winter	2	3	2	2	3	3	2	2	3	4	2	2	3	n.a
AMS idle time [%]	6	13	7	17	32	26	16	64	n.a.	21	12	n.a.	11	9
Total milkings [/day]	170	150	176	166	144	139	186	63	n.a.	157	154	126	161	176
Auto return rating[2]	2	2	2	3	1	3	3	1	3	n.a.	1	3	1	2
Hierarchy rating[3]	3	4	4	3	2	n.a.	4	2	3	3	4	2	3	3
Cow traffic[4]	1	1	1	1	2	1	3	1	1	1	1	1	1	1
Concentrates[5]	2	1	2	1	2	2	2	1	2	2	2	1	1	1
Water supply[6]	2	2	2	1	2	2	1	2	1	n.a.	2	1	1	1
Stable exit[7]	2	2	3	2	2	1	3	1	3	n.a.	2	2	3	3
Sight line pasture[8]	2	1	3	3	2	3	1	2	3	n.a.	3	1	3	1

[1] 0=flat, 1=spring, 2=summer, 3=fall, 4=winter; [2] 1=minimal,2=moderate,3=good; [3] 1=no...5=much; [4] 1=free, 2=semi-free, 3=forced; [5] 1=AMS only, 2=AMS+feeder; [6] 1=barn, 2=barn+pasture; [7] 1=free, 2=free after milking, 3=selection gate; [8] 1=blocked sight... 3=open view

of the farmers had a peak in calving season during the winter, resulting in high producing cows in an early stage of lactation during the firsts half of the grazing period. The average grazing time was 10.4 hours per day. The majority of the farmers apparently had a restricted grazing pattern allowing the cows to pasture only during daytime. Average number of cows per farm was 63, the average intensity was 8,950 kg/hectare and the average 305-days production was 8,166 kg.

Table 2 gives an overview of most important results concerning the differences between the winter and the pasture period.

Average daily milk yield in pasture period was 0.86 kg lower than during the winter period. Average milking frequency decreased from 2.68 during winter period to 2.61 during pasture period. Average evenness ratio in the winter period was 2.01 and in the pasture period 2.31. This average however was highly influenced by the results of Farm 8. Excluding this farm gives an average evenness ratio during the winter period of 1.53 and during the pasture period of 2.03. In both cases (Farm 8 included and excluded) these differences were not significant ($p=0.377$ and $p=0.107$ respectively). Average milking interval during winter period was 8:04 hours during winter period and 8:30 hours during pasture period. Average number of milkings per day during winter period was 166.0 and during pasture period 156.4. Average milk production per day was during the winter period 1580 kg and during the pasture period 1541 kg. This decrease of milk production is only 2,5%, corrected for lactation stage.

Due to intensive use of the AMS the evenness ratio was close to one. Although it did rise in the pasture period this increase was not significant. Both the number of milkings

Table 2. Overview of the differences between winter and grazing period.

Farm	Milk yield[1] [kg/cow]	Milk yield[1] [kg/day]	Milking Frequency[1]	Milking interval	Milkings[1] [/day]	Evenness ratio Winter	Grazing
1	0.59	-263***	-0.10	0:37	-45 (-23%)***	1.19	1.59
2	-1.69**	-230***	0.47***	-1:15	9 (6%)***	1.37	1.64
3	-1.23*	-31	-0.09*	0:09	-5 (-3%)~	1.31	1.82
4	3.14***	165*	0.03	1:10	0 (0%)	1.34	1.67
5	-3.00***	-225***	-0.07~	-0:22	-14 (-9%)***	2.41	2.68
6	-2.35**	-139***	0.15*	-0:45	-12 (-8%)**	1.59	1.79
7	-1.86***	148***	-0.13	0:14	-1 (-1%)	1.69	1.69
8	-1.31*	3	-0.45*	2:14	-14 (-19%)***	8.24	6.02
9	0.42	194***	-0.34***	0:37	-11 (-4%)**	1.50	1.62
10	-1.11*	59	-0.31*	1:21	-21 (-13%)***	1.53	5.37
11	-2.32*	-78**	n.a.	1:33	-17 (-10%)***	1.30	1.55
12	-3.10***	-95***	-0.29***	0:46	-17 (-13%)***	1.67	1.80
13	1.10*	46~	n.a.	-0:21	2 (1%)	1.58	1.56
14	0.75	5	-0.11**	0:19	-8 (-4%)**	1.43	1.54

[1]~ $0.1 < p < 0.05$; * $0.05 < p < 0.01$; ** $0.01 < p < 0.001$; *** $p < 0.001$

and the milk production per day however decreased in the pasture period. For the majority of the farms this decrease was significant. Increase of milking interval and decrease of number of milkings per day, milk frequency per cow, milk production per cow and milk production per day have been reported by others (see introduction). The change in visiting pattern did not seem to have an effect on these changes.

Graphs of number of cows, farm intensity, number of fetching and grazing time on one hand and variables of table 2 on the other hand didn't reveal any meaningful relationships.

Conclusions

Results described above do not give evidence to assume that use of AMS capacity is generally decreasing during the pasture period compared to the winter period. Total daily milk production decreased with only 2.5%, number of milkings with 6% and milk frequency decrease from 2.68 to 2.61. Also visiting pattern expressed in evenness ratio increased moderately. However, differences between farms were big. Apparently other factors than recorded in this research had greater influence on the outcomes. Generally can be concluded that grazing does not have to be disadvantageous for AMS performances. Management skills of individual farmers are likely to be more decisive.

References

Coolman, F., L.H. Huisman, W. Sybesma, 1981. Automatisering ten behoeve van het bedrijfsbeheer op melkveebedrijven, IMAG, Wageningen

Dooren, H.J.C. van, E. Spörndly and H. Wiktorsson, 2002. Automatic milking and grazing, Deliverable 26, EU-project Implications of introduction of automatic milking on dairy farms, May 2002, 38 pp. See: http://www.automaticmilking.nl

Jagtenberg, K., J. van Lent, 2000a. Access to pasture and the milking robot, Praktijkonderzoek 13(2), Research Institute for Animal Husbandry, Lelystad

Jagtenberg, K. & J. van Lent, 2000b. Cows go to and from the milking robot. Praktijkonderzoek 13 (6), Research Institute for Animal Husbandry, Lelystad, pp.1-5.

Ketelaar-de Lauwere, C.C., 1999. Cow behaviour and managerial aspects of fully automatic milking in loose housing systems. Wageningen Agricultural University, Wageningen.

Koning, C. de , W. Ouweltjes, 2000. Maximizing the milking capacity of an automatic milking system, In: Hogeveen, H., A. Meijering (eds.), Robotic milking: proceedings of the international symposium held in Lelystad, 17-19 August 2000., Research Institute for Animal Husbandry, Lelystad.

Pol-van Dasselaar, A. van der, W.J. Corré, H. Hopster, G.C.P.M. van Laarhoven, C.W. Rougoor, 2002. Belang van weidegang, PraktijkRapport 14, Praktijkonderzoek Veehouderij, Lelystad, 82p.

Ruis-Heutinck, L.F.M., H.J.C. van Dooren, A.J.H. van Lent, C.J. Jagtenberg & H. Hogeveen, 2001. Automatic milking in combination with grazing on dairy farms in The Netherlands. Proceedings of the 35th Congress of the International Society of Applied Ethology, 4-8 August 2001, Davis, USA.

Wredle, E., 2001. Grazing in automatic milking system - Effect of distance between pasture and barn and different levels of supplementary feeding on cow's behaviour. MSc report 143, Department of Animal Nutrition and Management, Swedish University of Agricultural Sciences, Uppsala, Sweden.

SEVEN CASE STUDIES ABOUT AUTOMATIC MILKING AND GRAZING IN PRIVATE HERDS

Christian C. Krohn
Danish Institute of Agricultural Sciences, Foulum, Denmark

In Denmark the first automatic milking systems were established in 1998. Since then the number has increased significantly - a fact resulting in more than 250 herds being milked automatically today. Due to various reasons, many of these herds will be interested in combining automatic milking with daily grazing. The present experiment was an initial screening of the conditions in 7 private farms with automatic milking systems in combination with different grazing strategies.

The herds were visited 3 times during the summer period at an interval of approx. 2 months. During each visit the cows were videorecorded for a period of 3 days when they went into and out of the barn, thus making it possible to calculate the average number of hours the cows spent outdoors. During the same 3 days a back up was made from the automatic milking system, thus allowing the actual milking data to be compared to the cows' grazing rhythm.

The conditions varied significantly both between herds and within herds. Free as well as forced cow traffic was used. The cows were either let out to graze in the morning (from 3-4 AM) via the robot, or they were examined manually in the morning and then either let out to graze or milked. After 10 AM all cows were free to enter or leave the cowshed.

There were 43 - 69 cows per robot with an average milk yield of 19.8 - 29.1 kg milk/cow/herd. The number of milkings per cow was 2.1-3.0, or 107-177 milkings per robot. The average number of hours spent grazing varied between 4-10 hours in May/June, 3.9-9.6 hours in July/August and 4.6-6.7 hours in September. In herds were the cows have access to the pasture via the robot, the number of cows on pasture increased gradually during the morning, and most of the cows were first outdoors about noon. In the other herds the number of cows on pasture increased more rapidly after the door was opened. When more than 70% of the cows were outdoors the frequency of milkings per hour often dropped to 2-4 cows. There was no significant correlation between the total number of milkings per 24 h and the time the cows spent on pasture during the same period.

An average of 5-52 minutes per herd was spent every day fetching the cows from pasture. The interviews of the farmers indicate that it was often a few difficult cows, which increased the fetching time on pasture. The distance from the barn to the beginning of the pasture varied between 30 and 400 m, whereas the distance to the outer corner of the pasture was more than 8-900 m. The distance between barn and pasture had no significant effect on numbers of cows outside. Daily indoor feeding with roughage/concentrates varied between 6 and 13 SFU (Scandinavian Feed Unit), whereas an amount of 3.5-7.0 kg concentrates was supplied in the milking robot. No significant correlation was found between the average number of hours spent grazing and the milking frequency across herds.

The investigation shows that by 4-10 hours of grazing it was possible to achieve a relatively high milking frequency even with considerably long distances between pasture and cowshed. However, 10-80% of the cows always have to be fetched from pasture.

Automatic milking – A better understanding

COMBINING AUTOMATIC MILKING AND GRAZING. PRACTICE IN THE NETHERLANDS

H.J.C. van Dooren, L.F.M. Heutinck & G. Biewenga
Applied Research, Animal Sciences Group, Wageningen UR, Lelystad, The Netherlands

A wide majority of the farmers offer grazing to their herd although the number of days per season and the number of hours per day is under pressure. In 2001 90% of Dutch dairy cows were put out on pasture for some time (Pol *et al.*, 2002). The introduction of an automatic milking system (AMS) influences almost every aspect of a dairy farm. It is obvious that also grazing is affected. Where with conventional milking the grazing period is well defined between the two milkings, the use of an AMS requires free cow traffic between the barn and the pasture. This may lead to a reduction of the occupation of the AMS. These uncertainties about the performance of the AMS during grazing, lead to an increasing number of farmers choosing for a zero-grazing system. In 2001 a number of dairy farmers in The Netherlands have been interviewed about management, housing and grazing system as part of a broader questionnaire on milk quality issues. The objective was to document the existing practices of grazing in herds with automatic milking systems and to identify strengths and weaknesses. Results have been reported in (Dooren, *et al.*,2002). A total of 121 dairy farmers, evenly spread over The Netherlands, were interviewed between October 2001 and January, 2002. As four were excluded for several reasons, data of 117 farmers remained for further analyses. A single unit system was used by 86% of the farmers, 27% of them having two or more units. Among the 14% farmers with a multi stall system, 63% had two units and 37% three or more units. The majority of the farmers (83%) had less than 3 years experience with automatic milking. Almost 53% (61) of the farmers offered grazing to their herd, 41% did not and 6% only offered a paddock of less than 1 ha. From the group of farmers applying grazing, 53 answered further questions about applied grazing systems and grazing time. Table 1 gives an overview of these results. The average overall grazing time was 9.6 hours per day.

More than half of the AMS farmers combined automatic milking with grazing. Although this is much less than among non-AMS farmers, it is a positive result regarding the sometimes negative attitude of farmers toward the combination of automatic milking and grazing.

Table 1. Overview of grazing time per grazing system.

Grazing system	# farms	% farms	Hours (min-max)
Rotational	28	52.8	10.7 (4-24)
Strip	5	9.4	13.0 (10-16)
Siësta	6	11.3	8.5 (6-15)
Continuous	10	18.9	6.8 (3-14)
Other	2	3.7	5.5 (3-8)

References

Pol-van Dasselaar, A. van der, W.J. Corré, H. Hopster, G.C.P.M. van Laarhoven, C.W. Rougoor, 2002 . Belang van weidegang, PraktijkRapport 14, Praktijkonderzoek Veehouderij, Lelystad, 82p.

Dooren, H.J.C. van, E. Spörndly and H. Wiktorsson, 2002. Automatic milking and grazing, Deliverable 26, EU-project Implications of introduction of automatic milking on dairy farms, May 2002, 38 pp. See: http://www.automaticmilking.nl

IS AUTOMATIC MILKING POSSIBLE WITH A 100% PASTURE DIET?

J. Jago[1], A. Jackson[1], K. Davis[1], R. Wieliczko[1], P. Copeman[1], I. Ohnstad[2], R. Claycomb[3] & M. Woolford[1]
[1]Dexcel Limited, Hamilton, New Zealand
[2]ADAS Bridgets, United Kingdom
[3]Sensortec Ltd, Hamilton, New Zealand

In New Zealand dairy herds graze 12 months of the year with the majority fed a 100% pasture or conserved pasture diet. The economic viability of automatic milking in New Zealand will depend partly upon the amount of supplementary feed necessary to encourage cows to walk from the pasture to the milking unit. Maximising the level of fresh pasture in the diet and milking cows less frequently are two approaches to automatic milking being researched in New Zealand, which contrast with common practice in most other countries. This paper will report on a study that aimed to determine the importance of offering concentrate in the milking unit and the effect of milking frequency settings on cow movement, milking behaviour and performance in a pasture-based automatic milking system.

The effects of feeding level (FL0=0kg or FL1=1kg crushed barley/day) and minimum milking interval (MI6=6h or MI12=12h) on cow movement and milking behaviour were studied in a multi-factorial cross-over (feeding level only) design. The study involved 27 mixed breed cows run as part of a larger herd (mean herd size=31, min=27, max=41) milked through a single automatic milking system. Diet was 100% fresh or conserved pasture (fed at pasture) with the exception of the crushed barley fed in the AMS for the FL1 treatment. The maximum walking distance from the furthermost pasture to the dairy was 400m. A pre-selection unit where cows accessed water, was located 180m from the dairy in the center of the farm. Cows had unlimited 24hour access to the pre-selection unit but were prevented from visiting the milking unit until a minimum of either 6 or 12h had elapsed since their last milking. Each feeding level treatment lasted 4 weeks.

Observations showed that on 5.8% of FL0 milking visits significant residual barley was left from a preceding FL1 cow, usually following a failed attachment. Feeding 1kg barley in the milking unit resulted in a higher visiting frequency to the pre-selection unit (FL0=4.6visits/d, FL1=5.4visits/d, sed=0.35, P<0.05) and a higher yield (FL0=22.5kg/d, FL1=23.6kg/d, sed=0.385, p<0.01) but had no effect on milking frequency (FL0=1.6 milkings/d, FL1=1.7 milkings/d, sed=0.04, ns). Minimum milking interval settings were the major factor influencing milking frequency (MI6=1.9, MI12=1.4 milkings/d, sed=0.15, p<0.01). The absence of feeding in the milking unit had no negative effect on behaviour during milking or the number of cows that had to be manually driven from the paddock. The results show that automatic milking can be combined with a near 100% pasture diet. Factors other than concentrate fed in the AMS, for example availability of fresh pasture, were also significant in controlling cow movement.

needs to be managed in another way. Several sources of information such as conductivity, temperature and color of the milk, yield and machine on time figures are integrated and inform the farmer on the status of the milk and cows.

However, previous research has shown that the milk quality of farms with an automatic milking system was significantly lower both when compared to the milk quality of the period before the introduction of the AM-system and when compared with the quality of conventional milking parlors (Billon, 2001; Klungel et al., 2000; Justesen and Rasmussen, 2000; Pomies and Bony, 2000; Van der Vorst and Hogeveen, 2000).

In the EU project Automatic Milking (Meijering et al., 2002), within work package 4 "Milk quality and automatic milking", much attention is paid to milk quality and automatic milking. In order to obtain more information regarding automatic milking and milk quality within this research, three projects were carried out. First milk quality results from farms with AM-systems from three countries were analyzed. Secondly the technical and management factors affecting the milk quality on farms with an AM-system were identified and quantified in an epidemiological study. In the final phase the issue of free fatty acids (FFA) was studied more in depth. This paper describes the main results of the research on milk quality within work package 4 of the EU project.

Study 1: Milk quality of farms with an AM-system in comparison with conventional milking

All farms in Denmark, Germany and The Netherlands, using an AM-system (AM-farms) between February 2001 and October 2001 were selected. The farms were divided in groups (AM-groups) as presented below, based on the installation dates of the AM-system:
1. Before January 1, 1998 (AM1)
2. January 1, 1998 - March 31, 1999 (AM2)
3. April 1, 1999 - June 30, 2000 (AM3)
4. July 1, 2000 - December, 2000 (AM4)

In total 99 Danish, 33 German and 262 Dutch farms were included in the study.

For the period of January 1997 until December 2000, for every selected farm, bulk milk quality data were collected. The collected milk quality parameters were: Total plate count (TPC), bulk milk somatic cell count (BMSCC), freezing point (FP) and free fatty acids (FFA). Total Plate Count is mainly a measure for the bacteria present in the milk, BMSCC is a measure for the inflammatory cells in the udders of the cows, FP provides an indication on the amount of water in the milk and FFA is a measure of damage to the fat globules and of susceptibility of cows for lipolysis. FFA data were not available for Denmark and Germany. Other quality parameters were not taken into account due to low sampling frequencies and not enough data available.

Furthermore, two control groups were used. For The Netherlands, one group of 295 farms, milking two times a day (C2), was randomly selected. The second control group consisted of all farms (n=40) milking three times a day (C3) with a conventional milking parlor. Data of such control groups for Denmark and Germany were not available. Therefore, it was chosen to present the national averages of all dairy farms in these countries as a reference to the outcomes of the AM farms.

Results study 1: Milk quality of farms with an AM-system in comparison with conventional milking

Before introduction of the AM-system all groups of farms had similar milk quality, independent of the date of installation of the AM-system. The only exception was that the Dutch conventional farms that milked three times daily had a higher level of FFA in the milk then all other Dutch farms.

After the introduction of the AM-system significant differences were found. The results (predicted means and recalculated geometric means) are presented in table 1. The average figures of conventional farms (two and three times milking per day) are presented in table I as a reference to the AM-farms. For all three countries and for most milk quality parameters (TPC, BMSCC, FP, FFA) the milk quality was slightly negatively affected after introduction of the AM-system in comparison to the period before. One exception was seen for the BMSCC in Germany, where no significant difference was found between before and after introduction. Before introduction of the AM-system the TPC was comparable to conventional farms in Denmark and The Netherlands. For Germany a lower TPC was found before the introduction of the AM-system than the national average.

Both in Denmark and Germany the BMSCC of the AM-farms before introduction seemed to be slightly higher than the national average BMSCC. Furthermore, the FP increased significantly after introduction of the AM-system both in Germany and The Netherlands by 0.005 °C. The FFA-levels in The Netherlands show an increase from 0.39 to 0.57 meq/100 g fat after introduction.

For all countries the variation in the levels of TPC, BMSCC, FP and FFA could generally be explained by similar variables. Differences between the AM-brands explained 32% percent of the variation in TPC, 30% was explained by installation period and 22% by farm effect. Regarding BMSCC, 56% of the variation could be explained by the farm differences. For the

Table 1. Predicted Log means (PM) and recalculated geometric means (GM) before versus after introduction of the AM system, with figures of conventional farms as reference (Van der Vorst et al., 2002).

Country	Group	No. of farms	TPC [Cfu/ml]		BMSCC [Cells/ml]		FP [°C]	FFA [Meq/100 g fat]	
			PM	GM	PM	GM	PM	PM	GM
DK	C2*	All	-	9,000	-	246,000	-	-	-
	Before	99	2.080[x]	8,000	5.558[x]	259,000	-	-	-
	After	99	2.633[y]	14,000	5.633[y]	279,000	-	-	-
D	C2*	All	-	21,000	-	181,000	-	-	-
	Before	33	2.835[x]	17,000	5.302[x]	201,000	-0.521[x]	-	-
	After	33	3.033[y]	21,000	5.313[x]	203,000	-0.516[y]	-	-
NL	C2*	295	1.995	7,000	5.169	176,000	-0.521	-0.8142	0.44
	C3*	40	2.016	8,000	5.214	184,000	-0.522	-0.5870	0.56
	Before	262	2.006[x]	7,000	5.138[x]	170,000	-0.522[x]	-0.9310[x]	0.39
	After	262	2.559[y]	13,000	5.320[y]	204,000	-0.517[y]	-0.5569[y]	0.57

*shaded sections not included in model / x and y = averages with different superscripts within each column and country differ significantly (p<0,05)

FP 62% could be explained by differences in AM-brand and the last parameter, FFA, was mainly influenced by the interaction of time and the farm (58%) but also for 26% by differences in brand. Therefore, farm and time differences and the make of AM-system seem to play a great role in the variation of the outcomes of milk quality results.

Course of TPC

No significant differences were found in the average TPC between the different AM-groups (AM1-AM4) after introduction of the AM-system for all countries. However, differences in tendencies can be seen. To illustrate this, the course of TPC for the Dutch dairy farms milking with an AM system is presented in figure 1. As a reference the mean values of the conventional farms, C2 and C3, are also given. All TPC levels for the AM groups are higher than the reference values of the conventional farms. During the first 45 days after introduction in all groups a quick increase of TPC is noted. After this time the TPC seems to stabilize for all four groups. The same result can be found on Danish and German farms.

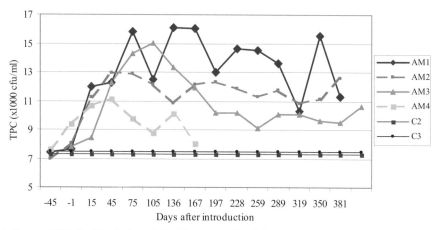

Figure 1. Course of TPC after introduction of the AM-system on Dutch farms.

Course of BMSCC

Regarding the course of BMSCC after introduction again similar patterns were found for all three countries. During and just after introduction BMSCC increased slightly. However, after some time the BMSCC decreased in all countries to the same level as the conventional farms. The strongest decrease was found in Denmark (see figure 2) where the AM3 group (largest number of farms) already reached the conventional level in about 110 days after introduction of the AM system.

Courses of FP and FFA

Regarding the course of FP, an increase was seen immediately after introduction of the AM-system and the level remained substantially higher. Little fluctuation was found after introduction. Regarding FFA the increase is less impulsive after introduction. However it appears to rise slowly (Figure 3).

After 6 months the FFA-level of the AM2 and AM3 group was significantly higher than on the conventional farms that milk twice daily (C2). No difference was found when

Automatic milking – A better understanding

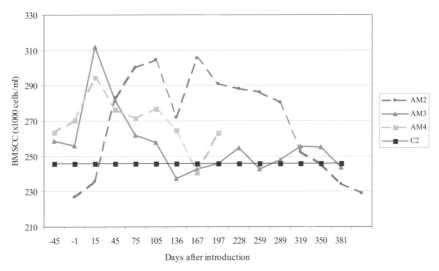

Figure 2. Course of BMSCC after introduction of the AM-system on Danish farms.

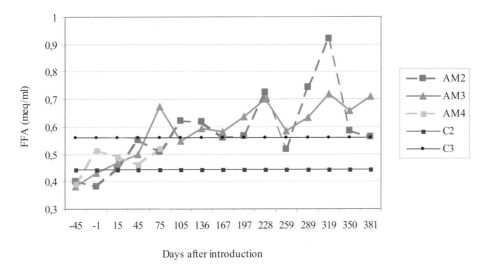

Figure 3. Course of FFA of four groups of AM farms after introduction of the AM system.

comparing with the C3 group during 1,5 years after introduction. The significant difference with the C2 group, on the other hand, was still found at the end of year 2 after introduction of the AM system. As mentioned before, it is known that the milking frequency has an increasing effect on FFA (Klei *et al.*, 1997). However, on the AM-farms it appears as if the FFA keeps increasing, although average milking frequency is less than 3 milkings per day.

As shown before, the introduction of the AM-system has a slightly negative effect on the milk quality (Billon, 2001; Klungel *et al.*, 2000; Justesen and Rasmussen, 2000; Pomies and Bony, 2000; Van der Vorst and Hogeveen, 2000, Rasmussen *et al.*, 2002). However, in this study no significant differences were found between the different groups of farms (AM1-AM4) according to their date of installation in contradiction to earlier results where difference were shown between the AM1 and AM2 group (Van der Vorst and Hogeveen, 2000). In comparison to that study, study 1 had more data available over a longer period before and after introduction of the AM-system. As can be seen in figure 1 and 2 TPC and BMSCC increase directly after introduction and then stabilize after some time. The point of stabilization for the four groups is around the same level for TPC, BMSCC, FP and FFA. Therefore, the longer the period after introduction is taken into account, the fewer differences will be found between the overall averages of the different groups.

The changes in the four milk quality parameters after introduction of the AM-system are relatively small, especially when compared with the European penalty limits. Besides substantial farm effects, for several milk quality parameters most variation could also be explained by differences over time and between brands. Farm effects are mainly based on management, time effects on changes over time such as seasonal effects and the brand effects are mainly based on the technique. Most probably interactions between these factors play an important role in affecting milk quality.

Study 2 - Identification and quantification affecting milk quality on AM-farms

The second study aimed to identify possible risk factors that affect milk quality on farms with AM-systems. A preliminary study was started in 2001 in The Netherlands to identify the risk factors on a larger scale (Van der Vorst & Ouweltjes, 2003). In that study 124 AM-farms were visited. The farms were analyzed for risk factors focusing on four milk quality parameters: TPC, BMSCC, FP and FFA. For study 2, out of these 124 farms, a selection was made based on the bulk milk quality performance during month 7 through 18 after introduction. The milk quality performance levels used resulted in the following groups (Table 2).

Table 2. Milk quality performance levels.

	Low	Medium	High
Total plate count (TPC) (x1000 CFU/ml)	<8	8-14	>14
Bulk milk somatic cell count (BMSCC) (x1000 cells/ml)	<170	170-265	>265
Free fatty acids (FFA) (mMol/100 g fat)	<0.55	0,55-0.78	>0.78

The freezing point was omitted, because the preliminary study showed a clear explanation for the involved risk factors. Using the above criteria, 28 farms were selected for this study. Automatic milking was introduced on these farms between November 1996 and December 1999. Milk quality data over the period January 1997 to November 2002 were provided by the national Milk Control Station. For studying the possible risk factors on the farms six sources were used; a questionnaire, a scoring list on management items, the periodic test report of the AM-system, milk recording data, a hygiene check list and available data from the AM management program (Van der Vorst et al., 2003). For the final analysis farms were classified for TPC, BMSCC and FFA as described in table II, based on their milk quality during the 12 months before the farm was visited. All farms were visited once.

Results study 2: Identification and quantification affecting milk quality on AM-farms

The analysis of the data of the 28 farms in the study (Van der Vorst *et al.*, 2003) showed that size of milk quota and numbers of cows milked were not related to milk quality. The results showed a significant relationship between the production level after introduction with TPC and BMSCC. Both were lower for higher yielding herds. The results largely agree with those of a Dutch study on 124 farms in The Netherlands (Van der Vorst & Ouweltjes, 2003). Free Fatty Acid level was not significantly related to production level, but on average herds with higher yielding cows, had higher FFA levels in the bulk milk.

Mastitis incidence was on average estimated at 22%, with a range from 0 to 71%. The estimated incidence was not related to milk quality. The farmers were asked about their opinion on conductivity as a tool to control mastitis. Only 2 farmers considered it useless, most farmers (25 out of 28) regarded it as necessary. Five of them judged it as very useful. There was no significant difference in milk quality between farmers with different judgments. Half of the farmers used synthetic rubber liners, the other half used silicon liners. The type of liner used was not related to milk quality. Earlier replacement of liners tended to relate to a lower BMSCC. On average TPC and FFA levels were also lower with earlier replacement, but these relationships were not significant. The maximum interval between main cleaning cycles of the milking unit varied from 8 to 15 hours. The length of this interval seemed not to be related to milk quality. Most farmers spent more time between the cows and on management programs, and less time on cleaning and other activities around milking. This however, was not related to milk quality. On average the farmers reported to have more than 2 hours of labour time saved each day, with a range from 0 to 8 hours. The amount of saved labour was not related to milk quality.

Hygiene appeared to be important for TPC as well as for BMSCC. Both milk quality parameters showed increased values on farms with a poor overall hygiene. The TPC value was related to hygiene in the clean part of the AM-system and the waiting area. Also the overall impression of the farm was related to TPC. The relationship with hygiene is logical because with a poor hygiene, the chance for bacterial growth and the chance of contamination of milk increase. However, a direct relationship between hygiene on the farm and TPC is difficult to prove under experimental conditions (Slaghuis *et al.*, 1991). It was also found that TPC and BMSCC are lower on higher producing farms. It may be that these farms have better overall management, resulting in a good milk production, animal health and hygiene (De Koning *et al.*, 2002).

It was remarkable that on average farmers replace their teat cup liners (rubber as well as silicon) much too late. Rubber liners were replaced after on average 4636 milkings, silicon liners after 9579 milkings. The analyses showed that a high number of milkings is a risk for increased BMSCC. A technical factor that was also found to be important for BMSCC was the level of air inlet in the teat cups. However, there was an unexpected relationship between air inlet and BMSCC. The results showed a higher air inlet on farms with low BMSCC. Perhaps a higher air inlet might cause less teat washing in the liner.

There are big differences among and within farms in milk quality. Some farms have excellent milk quality results after introduction of the AM-system, some have a decreased milk quality, while others show incidental peaks in for example bacterial plate count.

Study 3: An study on Free Fatty Acids and automatic milking

Increased milking frequencies in general will result in an increase in yield and a small decrease in fat and protein content. However, also increases in FFA levels have been reported (Ipema and Schuiling, 1992, Jellema (1986),Klei et al., 1997).

Study 2 paid extra attention to the FFA issue. That study showed both technical and management factors influencing FFA. It appeared that a regular check of the attention lists and check of the cleaning system was related to a lower FFA value. This may indicate a better overall management of the farmers who do this regularly. Farms that had more alerts from the cooling system had a higher FFA value than other farms. Unfortunately, the type of alert was not registered. Especially cows in late lactation seemed to be at risk. The explanation is that this milk is more susceptible to lipolysis than milk from cows in early lactation (Jellema, 1986). A calving pattern spread evenly around the year may decrease the risk of a rise in FFA in a certain period because of the herd being in late lactation. If cows are in late lactation it is important that the milking frequency is about 2 milking per day with > 5-7 kg per milking.

The technical factors related to FFA, mainly concerned air inlet in the system, bubbling (excessive air inlet) of the milk and a too long post run time of the milk pump, which increases the lipolysis. An interesting aspect is that air inlet in the teat cup seemed to have a positive effect on BMSCC and a negative effect on FFA levels. A remarkable result was that FFA seemed to be related with the number of feeding places and the accessibility of the water trough. The more places at the feeding fence and the better the accessibility of the water troughs, the lower the FFA. The exact relationships are unknown, but it may be related to some stress on the cows. It might be that stress induces the susceptibility of the milk to lipolysis, but more research is needed.

Results study 3: free fatty acids and automatic milking

Also in this study the relation with milking frequencies en free fatty acids was found. Short milking intervals contribute to increased FFA-levels (Table 3).

Lipolysis results from the enzymatic hydrolysis of milk fat, causing an accumulation of free fatty acids (FFA) which are responsible for the rancid flavour of milk. The milk fat present in milk fat globules is protected against the action of the endogenous milk lipoprotein lipase enzyme (mLPL) by the milk fat globule membrane. If this membrane is disrupted by e.g. agitation, fatty acids can be splitted up from the glycerol part. But in untreated milk, lipolysis also occurs due to spontaneous lipolysis (Jellema, 1975, Cartier & Chilliard, 1990).

Table 3. Effect of milking interval on FFA contents (meq/100 g fat) and standard deviations (between brackets) in raw milk after 0 hours and 24 hours storage after sampling.

Interval (h)	4		8		12	
0 hours at 5°C	0.20[a]	(0.04)	0.19[ab]	(0.07)	0.15[b]	(0.07)
24 hours at 5°C	1.23[a]	(0.98)	0.71[b]	(0.47)	0.42[c]	(0.31)
increase in 24hours	0.97[a]	(0.98)	0.49[b]	(0.45)	0.25[c]	(0.26)

[a,b,c] statistically significant difference on the same row P<0.05

Late lactation milk and milk form cows milked 3x per day or more is more susceptible to spontaneous lipolysis (Jellema,1986, Cartier *et al.,* 1990).

Variation of FFA was determined on 4 farms milking conventionally, on 4 farms milking three times a day conventionally and on 4 farms milking with an AM-system. During one year samples were taken monthly and during two periods of two weeks every delivered bulk tank was sampled. Selection between high and low FFA levels was based on two previous results. The high FFA levels showed more variation and previous selection for farms milking three times a day was not adequate, indicating that management strategies may affect FFA levels. More variation within a year was also found for the farms milking 3x with low previous FFA levels. For the other type of farms (milking conventionally and AM), previous selection was adequate. Technical aspects of the milking technology used and management strategies including feeding routines will be further investigated.

Increased FFA-levels are related with increased milking frequencies and the construction of the milking machine in the AM-system. Precautionary measures like adjusted milking frequencies for cows in late lactation and preventive maintenance of the AM-system, can help to avoid high FFA-levels resulting in taste abnormalities of the milk. The preliminary results indicate that further fundamental research is necessary to understand the mechanism of lipolysis.

Conclusions

Milk quality, to a certain extent, is negatively affected when milking with an automatic milking system. The highest levels for TPC and BMSCC are found in the first six months after introduction. After this period the milk quality slightly improves and all farms more or less stabilize their levels around the average of conventional farms. Differences between farms are seen both in averages and in variation. Possible risk factors related with milk quality concern several general farm characteristics, animal health, AM-system, cleaning and cooling, housing, management of the farmer and the hygiene on the farm. An increased milking frequency is not the only explanation of increased FFA levels. Especially cows in late lactation seem to be at risk. Technical factors related to FFA mainly concern the air inlet in the teat cups, bubbling (excessive air inlet) and a too long post run time of the milk pump.

References

Billon, P., 2001. Les robots de traite en France; impact sur la qualité du lait en le système de production. Proceedings Symposium Il Robot di Mungitura in Lombardia; October 26, 2001, Cremona, Italy

Cartier, P., Chilliard, Y., 1990. Spontaneous lipolysis in bovine milk: combined effects of nine characteristics in native milk. J. Dairy Sci. 73:1178-1186.

Cartier, P., Chilliard, Y., Paquet, D., 1990. Inhibiting and activating effcts of skim milks and proteose-peptone fractions on spontaneous lipolysis and purefied lipoprotein lipase activity in bovine milk. J. Dairy Sci. 73:1173-1177.

De Koning, K., van der Vorst, Y., Meijering, A., 2002. Automatic milking experience and development in Europe, Proc. 1st North Am. Conf. on Robot Milking, Toronto, Canada. Wageningen Academic Publishers, Section 1, pp1-11

Ipema, A.H., Schuiling,E., 1992. Free fatty acids; influence of milking frequency. in Proceedings of the Sumposium Prospects for Automatic Milking November 23-25, 1992, EAAP Publ. 65, Wageningen, The Netherlands, pp 491-496

Jellema, A., 1975. Susceptibility of bovine milk to lipolysis. Neth. Milk Dairy J. 29:145-152.

Jellema, A., 1986. Some factors affecting the susceptibility of raw cow milk to lipolysis. Milchwissenschaft. 41:553-558.

Justesen, P., Rasmussen, M.D., 2000. Improvement of milk quality by the Danish AMS self-monitoring program. pp 83-88 in Proc. Int. Symp. on Robotic Milking, Lelystad, Wageningen Academic Publishers, The Netherlands.

Klei, L. R., Lynch, J. M., Barbano, D.M., Oltenacu, P.A., Lednor, A.J., Bandler, D.K., 1997. Influence of milking three times a day on milk quality, J. Dairy Sci. 80: 427-436.

Klungel, G.H.,. Slaghuis, B.A., Hogeveen, H., 2000. The effect of the introduction of automatic milking on milk quality, J. Dairy Sci. 83:1998-2003.

Meijering, A., van der Vorst, Y., de Koning, K., 2002. Implications of the introduction of automatic milking on dairy farms, an extended integrated EU project, Proc. 1st North American Conference on Robot Milking, Toronto, Canada. Wageningen Academic Publishers, pp Section 1, 29-38

Pomies, D., Bony, J., 2000. Comparison of hygienic quality of milk collected with a milking robot vs. with a conventional milking parlor. pp 122-123 in Proc. Int. Symp. on Robotic Milking, Lelystad, The Netherlands.

Rasmussen, M.D., Bjerring, M., Justesen, P., Jepsen, L., 2002. Milk quality on Danish farms with automatic milking systems. J. Dairy Sci. 85: 2869-2878.

Slaghuis, B.A., de Vries, T.,.Verheij, J.G.P, 1991. Bacterial load of different materials which can contaminate milk during production. Milchwissenschaft. 46: 574-578.

Van der Vorst, Y., Hogeveen, H., 2000. Automatic milking systems and milk quality in The Netherlands, pp 73-82 in Proc. Int. Symp. on Robotic Milking, Lelystad, Wageningen Academic Publishers, The Netherlands.

Van der Vorst, Y., Knappstein, K., Rasmussen, M.D., 2002. Milk quality on farms with an automatic milking system. Effects of automatic milking on the quality of produced milk. Report D8 of the EU Project Implications of the introduction of automatic milking on dairy farms (QLK5-2000-31006).

Van der Vorst, Y, Ouweltjes, W., 2003. Milk quality and automatic milking, a risk inventory. Report 28 Research Institute for Animal Husbandry, Lelystad, The Netherlands.

Van der Vorst, Y.,. Bos, K., Ouweltjes, W., Poelarends, J., 2003. Milk quality on farms with an automatic milking system; Farm and management factors affecting milk quality. Report D9 of the EU Project Implications of the introduction of automatic milking on dairy farms (QLK5-2000-31006)

QUARTER MILKING - A POSSIBILITY FOR DETECTION OF UDDER QUARTERS WITH ELEVATED SCC

I. Berglund[1], G. Pettersson[1], K. Östensson[2] & K. Svennersten-Sjaunja[2]
[1]Dept. Animal Nutrition and Management, Swedish University of Agricultural Sciences, Kungsängen Research Centre, Uppsala, Sweden
[2]Dept. Obstetrics and Gynaecology, Swedish University of Agricultural Sciences, Uppsala, Sweden

Abstract

One of the most common diseases in dairy production is mastitis. Clinical mastitis is easily detected and usually milk composition is changed. Subclinical mastitis often remains undetected, since there are no clinical signs of inflammation and the milk appears normal when inspected. The aim of this study was to find out whether a moderate increase in somatic cell count (SCC) is associated with measurable changes in milk composition and milk yield when analysed on individual udder quarters and comparisions are made with the opposite healthy quarter. During 13 weeks, 4158 bulk quarter milk samples from 68 cows were collected twice weekly and analysed for milk composition and SCC. Milk yield was registered at udder quarter level. For calculations, three groups of cows were formed according to their SCC value. Group 1 cows, where all quarters had a SCC <100 000 cells/ml, were considered to be unaffected. Group 2 cows had one udder quarter with a SCC >100 000 cells/ml and 1.5-fold higher than the opposite quarter at one sampling occasion. For group 3 cows, the increase remained for more than one consecutive sampling occasion. Data from group 1 cows revealed that front and rear quarters were similar when compared to each other. For both groups 2 and 3 cows, the lactose content in milk decreased statistically significant simultaneously with the increase in SCC in the affected quarter. For group 3 cows, the levels remained for two sampling occasions after the initial increase in SCC. It was concluded that deviations in lactose content within pairs of front and rear quarters, respectively, may be a useful tool for detection of a moderate increase in SCC in separate udder quarters.

Introduction

Fully automated milking systems are the latest progress in milking management. In some of these systems it is possibile to milk each udder quarter separately with consideration to milk yield and milk flow, and with the possibility measure and register quantities of data at quarter level at each milking. However, in AM systems it is not possible to visually inspect the milk before the milking cluster is attached, whereby sensors for detection of abnormal milk, such as mastitic milk, is needed.

Mastitis, which is one of the most costly diseases in dairy production, can be either clinical or subclinical. Clinical mastitis are easily detected, while subclinical mastitis often remains undetected (Schalm, *et al.*, 1971), since there are no clinical signs of inflammation and the milk appears normal when inspected. Both clinical and subclinical mastitis causes an increase in milk somatic cell count (SCC). The SCC in bulk tank milk is analysed routinely by the dairies for quality control and monthly the SCC is analysed at the cow level by the dairy-herd improvement organisations for breeding purposes. The results are also used by the farmers for individual udder health control and management decisions. Whether monthly registrations of cow composite milk is enough for those purposes can be questioned. The udder consists of four separate quarters and when milk sampling for analyses of SCC is performed at whole udder level, an increase in SCC in one quarter may be masked by the dilution effect from the other, healthy quarters.

During clinical mastitis milk yield decreases and milk composition change; content of fat, casein and lactose decrease while whey protein increase (Claesson 1965; Harmon, 1994). If similar changes can be observed during a period of slight or moderate increase in SCC (during subclinical mastitis) is not fully evaluated. If so, frequent analysis of the milk and comparisons between the quarters within the udder, could be a way to detect udder quarters with disturbances that could be interpreted as being due to inflammatory reactions.

The aim of this study was to find out if a moderate increase in milk SCC was associated with measurable changes in milk composition and milk yield when analysed on individual quarter basis and when using deviations from the values of an opposite healthy quarter.

Material and methods

The study took place at Kungsängen's Research Centre, Swedish University of Agricultural Sciences, Uppsala, Sweden. During 13 weeks, 4158 bulk quarter milk samples were taken twice a week from 68 Swedish Red and White Breed cows. The cows were kept in the barn provided with an AM system. The herd records were 9500 kg Energy Corrected Milk, [ECM (Sjaunja, et al., 1990)]/ year. The means from 19 analyses of bulk tank milk were: SCC 102 000 cells/ml, fat 4.7% and protein 3.3%. Mean milking frequency was 2.5 times/day and varied from 1.7 to 3.1.

The cows were milked at the udder quarter level in the milking unit VMS^{TM} (Voluntary Milking System, DeLaval, Sweden). The system vacuum was 42 kPa, pulsation rate 60 cycles/minute and pulsation ratio 35/65 massage/suction. Detachment of teat cup was done when the milk flow at quarter level was <100 g/min, with vacuum detachment at 15-16 kPa. The teat cups were flushed with water after every milking.

For analyses of milk composition (fat, protein, and lactose) and SCC, spectroscopic mid infrared technique (Dairylab2 Multispec, A/S N. Foss Electric, Denmark) and fluorescence based electronic cell count (Fossomatic 5000, A/S N. Foss Electric, Denmark) were used, respectively. Whole udder milk yield was registered by Flomaster Pro (DeLaval, Sweden) and udder quarter yield was registered by Free Flo (SCR Engineers Ltd. Israel). Sampling was done at Mondays and Thursdays. The day where the increase in SCC was detected was set to sampling occasion 0. For calculations three groups of cows were formed according to their series of numerical values from the SCC analyses.

Group 1 (n=22). Cows that were considered as having non-affected quarters. All quarters had a SCC ≤100 000 cells/ml at nine consecutive sampling occasions. Only the last seven occasions were used for further calculations.

Group 2 (n=16). Cows that, during one sampling occasion (0), had one udder quarter considered as affected with a SCC > 100 000 cells/ml and ≥ 1.5 times higher than the opposite quarter. The later quarter should have a SCC ≤ 100 000 cells/ml during all samplings. The affected quarter should have been proceeded of and followed by at least two sampling occasions with a SCC ≤ 100 000 cells/ml.

Group 3 (n=12). Cows that, during more than one sampling occasion in a row had one udder quarter considered as affected with a SCC > 100 000 cells/ml and ≥ 1.5 times higher than the opposite quarter. The later quarter should have a SCC ≤ 100 000 cells/ml during all samplings. The sampling (0) of the affected quarter should have been proceeded of at least two sampling occasions with a SCC ≤ 100 000 cells/ml.

For statistical evaluation, the general linear model (GLM) in SAS (SAS Institute, 1996) was used. Differences within pairs of udder quarters (non-affected - affected) in the parameters milk yield and composition were calculated. It was then tested if the calculated difference was significantly differed from 0. All differences between udder quarters were calculated within rear and front quarters respectively. For calculations of LS-means ±SE in differences within pairs of udder quarters the following model was used:

$$Y_{ijk} = \mu + \alpha_i + \beta_j + e_{ijk}$$

Where μ is the mean of all observations
α_i is the effect of the i:th cow (n= 16 group 2, n = 12 group 3)
β_j is the effect of the j:th sampling occasion (n=6)
e_{ijk} is the effect of the ijk:th random errors.

Results

Group 1. All quarters non-affected

These cows were considered as having non-affected udder quarters based on results from milk analyses of SCC. Front and rear quarters had the same protein and lactose content while the rear quarters had a higher milk production/hour and a lower fat content than front quarters. However, the both front quarters and the both rear quarters had the same production respectively (Table 1).

Table 1. Group 1 (22 cows). SCC, milk production/hour and contents of fat, protein and lactose at udder quarter level. Data are expressed as LS means (±SE).

	Log$_{10}$SCC (cells/ml x1000)	SCC[1] (x 1000 cells/ml)	Milk prod/h (g)	Fat (%)	Protein (%)	Lactose (%)
RR[2]	1.14±0.6	18±11	388±52	4.63±0.6	3.33±0.1	4.95±0.1
LR	1.05±0.5	14±9.3	399±44	4.60±0.7	3.33±0.1	4.95±0.1
RF	1.13±0.4	18±7.8	265±28	4.85±0.7	3.33±0.1	4.91±0.1
LF	1.10±0.4	16±6.5	265±29	4.81±0.6	3.33±0.1	4.91±0.1

[1]Anti-logarithmic value
[2]RR - Right rear; LR - Left rear; RF - Right front; LF - Left front.

Table 5. Group 3 (12 cows), increase in SCC at more than one sampling occasion. Milk production and contents of fat, protein and lactose in non-affected (N) and in affected (A) udder quarters. Differences between N and A udder quarters within pairs in milk production and in contents of fat, protein and lactose. The pairs consist of rear and front quarters respectively. Data are expressed as LS means ± SE.

Variable		Sampling occasion[1]					
		-3	-2	-1	0	+1	+2
Milk Prod.	N	398±16	370±16	388±16	412±14	381±18	389±18
(g/h)	A	403±16	389±16	378±16	401±14	384±18	374±18
Diff N-A		9.3±14.7	-1.9±12.7	22.4±12.7	17.0±12.7	-2.7±12.7	12.4±16.3
Sign level[2]		ns	ns	(*)	ns	ns	ns
Fat	N	4.69±.21	5.39±.21	4.95±.21	4.92±.18	5.27±.23	5.01±.23
(%)	A	4.80±.21	5.34±.21	4.95±.21	4.75±.18	5.23±.23	4.94±.23
Diff N-A		0.10±.12	0.01±.12	0.03±.11	0.20±.10	0.08±.13	0.07±.13
Sign level		ns	ns	ns	*	ns	ns
Protein	N	3.41±.04	3.44±.04	3.37±.04	3.37±.04	3.40±.05	3.32±.05
(%)	A	3.41±.04	3.42±.04	3.33±.04	3.43±.04	3.43±.05	3.33±.05
Diff N-A		0.01±.02	0.02±.02	0.05±.02	-0.06±.02	-0.04±.02	-0.03±.02
Sign level		ns	ns	*	***	(*)	ns
Lactose	N	4.87±.03	4.86±.03	4.88±.03	4.86±.02	4.87±.03	4.89±.03
(%)	A	4.81±.03	4.83±.03	4.86±.03	4.69±.02	4.73±.03	4.71±.03
Diff N-A		0.05±.04	0.02±.04	0.04±.04	0.17±.03	0.14±.04	0.17±.04
Sign level		ns	ns	ns	***	***	***

[1]Sampling occasion -3 = 10 or 11 days before sampling occasion 0; -2 = 7 days before sampling occasion 0; -1 = 3 or 4 days before sampling occasion 0. +1 = 3 or 4 days after sampling occasion 0; +2 = 7 days after sampling occasion 0.
[2]ns=not statistically significant, (*) P<010; *P<0.05; *** P<0.001

Discussion

In this study it was shown that non-affected quarters are similar when respective rear and front quarters were compared to each other regarding milk production and content of fat, protein, and lactose, which agrees with earlier findings (Linzell & Peaker, 1972). The rear quarters produced more milk of a lower fat content than the front quarters, but the contents of protein and lactose were the same in front and rear quarters. Therefore, the hypothesis that disturbances in one quarter, for instance caused by elevated SCC, could be detected by within udder comparisons, seems to be useful.

In the group 2 cows with an increased SCC at one occasion, it was found that the SCC was, or tended to be, slightly higher in the affected quarters at all sampling occasions. The milk production/hour was also lower in the affected quarters before the sampling occasion

0. This indicates that the affected quarters had some udder health disturbance already from the beginning. Previous records of the SCC levels in these cows would have been useful for more accurate interpretation of the results. The only statistically significant difference related to the increased SCC at sampling occasion 0 was the decrease in lactose content in the affected quarter (Table 3).

In group 3 cows, with an increase in SCC at more than one sampling occasion, the SCC differed between non-affected and affected udder quarters with the affected quarters having a slightly higher SCC already from sampling occasion -3 (Table 4). As for the group 2 cows, this may indicate a minor disturbance in the udder health in the affected quarters. However, there were no differences in the milk production/hour in the affected quarters until sampling occasion -1 when milk production/hour tended to be lower in the affected quarter. That there was no decrease at all in milk production/hour, at sampling occasion when SCC was increased, is contradictory to generally accepted findings in several previous studies (for a review, see Deluyker, 1991). The protein content showed a tendency to decrease at sampling occasion -1 while it increased at sampling occasion 0 in the affected quarters. It has been suggested that increased milk protein content is due to an influx of proteins from the blood and extracellular fluids. Thereby the whey content in the milk increases (Korhonen & Kaartinen, 1995) as opposed to the content of casein, which has been observed to decrease during mastitis (Claesson, 1965). The lactose content decreased significantly from sampling occasion -1 to sampling occasion 0 in the affected quarters and continued to be statistically significantly lower from sampling occasion 0 and during sampling occasions +1 and +2, which agrees with Claesson (1965). The change in lactose content indicates that the milk synthesis may be altered when the SCC was increased. Lactose has osmotic regulating functions (Kaartinen, 1995), and the content in milk during normal circumstances is therefore very stable.

In conclusion, it was shown that in cases where one quarter was affected, there was a difference in lactose content when SCC was increased. The non affected and affected quarters usually differed in SCC (groups 2 and 3) already before the day when the SCC increased, and to a certain extent in milk production in group 2. The fact that milk production/hour in this study was not unequivocally decreased when the SCC was increased may be due that the quarters were already slightly affected already at the start of the study. However, the most promising indicator for detection of increased SCC in this study, was lactose content.

Acknowledgements

This study was funded by SLF (Swedish Farmers Foundation for Agricultural Research) and DeLaval, Tumba, Sweden.

References

Claesson, O. 1965. Variation in the rennin coagulation time in milk. Annals of the Agricultural College of Sweden. 31. 237-332

Deluyker, H. A. 1991. Milk yield fluctuations associated with mastitis. Flemishe Veterinary Journal, Suppl 1, 62:207-216.

Harmon, R. J. 1994. Physiology of mastitis and factors affecting somatic cell counts. Journal of Dairy Science. 77: 2103-2112

Kaartinen, L. Physiology of the bovine udder. 1995. In: The bovine udder and mastitis. (eds. Sandholm, M., Honkanene-Buzalski, T., Kaartinen, L. & Pyöräälä,S.) University of Helsinki, Faculty of Medicine, Helsinki, 14-23.

Korhonen, H. & Kaartinen, L. 1995. Changes in the composition of milk induced by mastitis. In: The bovine udder and mastitis. (eds. Sandholm, M., Honkanene-Buzalski, T., Kaartinen, L. & Pyöräälä,S.) University of Helsinki, Faculty of Medicine, Helsinki, 76-82.

Linzell, J. L. & Peaker, M. 1972. Day-to-day variation in milk composition in the goat and cow as a guide to the detection of subclinical mastitis. British Veterinary Journal. 128: 284-295.

SAS Institute. 1996. SAS/Stat Software. Changes and enhancement through release 6.12. SAS Institute. Cary. North Carolina. USA.

Schalm, O.W., Carroll, E.J. & Jain, N.C.1971. Bovine Mastitis. LEA & Febiger. Philadelphia, USA.

Sjaunja, L.-O., Baevre, L., Junkarinen, L., Pedersen, J. & Setälä, J. 1990. A Nordic proposal for an energy corrected milk (ECM) formula. ICAR, 27th session. July 2 - 6, Paris, France. EAAP Publication N°. 50. 1991. 156-157.

PATTERN OF SOMATIC CELL COUNT IN MILK UNDER AUTOMATIC MILKING CONDITIONS (VMS) AND INTERACTIONS WITH MILK CONSTITUENTS

J. Hamann
Department of Hygiene and Technology of Milk, School of Veterinary Medicine Hannover, Hannover, Germany

Abstract

A group of approx. 40 high-yielding German Holstein Frisian cows were milked robotically (VMS®, DeLaval) throughout 400 days. During 20 session days (20-day intervals), samples (quarter foremilk [QFM], quarter composite milk [QCM] and cow composite milk [CCM]) were drawn continuously for 24 hours. Cell count (SCC) determination (Fossomatic) was done in all milk fractions, culturing (CULT) in QFM. Milk NAGase activity, lactate, fat, protein and lactose were analysed in QCM. The milk secretion rate was determined as yield/quarter/hour. With a mean milking frequency of 2.72 milkings/day/cow, mean daily yield was 26.87 ± 10.59 kg/cow/day.

Data analyses (SAS, PROC GLM) evaluated the influence of VMS® on udder health. The latter was categorized (cyto-bacteriological results of QFM, using the SCC threshold of 100,000 cells/ml) into "normal secretion (NS)": CULT (-), SCC < 100,000; "latent infection (LI)": CULT (+), SCC < 100,000; "unspecific mastitis (UM)": CULT (-), SCC > 100,000 and "mastitis (M)": CULT (+), SCC > 100,000.

35% of all quarter samples were CULT positive (n = 919). Of those, *Corynebacterium* spp. were isolated in 788 cases (= 86%). Overall SCC in QCM was 4.75 lg (= 56,234 cells/ml). The mean SCC values (incl. sd) were below 100,000 cells/ml for the health categories NS and LI and the fractions QFM and QCM. These data justify the statement that an automated milking system is not causing an increase in SCC *per se*. Milk constituents in QCM differed significantly ($p < 0.05$ at least) between all health categories and between the classes 50,000, 100,000 and 100,00 - 200,000 cells/ml. In addition, the secretion rate was highest (347 g) at a SCC of 10,000 (QFM). A reduction by 22% (266 g) occurred at the SCC of 100,000 (QFM). This re-opens the debate on the physiological thresholds of milk constituents.

Introduction

There are a variety of influences on the SCC in milk. The milking interval may be considered the most important non-microbiological factor, in addition to feeding, milking performance etc.

It has been reported several times that the application of automatic milking systems (AMS) under field conditions increased SCC. Until now, very little has been published indicating that AMS can keep the cows' mammary glands at a stable health status. This study was performed to evaluate the influence of an AMS on SCC and related changes in milk constituents of quarter composite milk (QCM) samples. Moreover, the secretion rate per udder quarter was estimated.

Material and methods

A group of 40 high-yielding German Holstein Frisian cows at different lactation stages and numbers were milked robotically (VMS®: voluntary milking system, DeLaval; vacuum 42 kPa, pulsation rate 60 cycles/min, pulsation ratio 65%). Over 400 days, survey was carried out every 20 days, and each session included 24 hours of continuous sampling. Udder quarter health was categorized in QFM in four groups - normal secretion (NS), latent infection (LI), unspecific mastitis (UM) and mastitis (M) - applying the definition provided by DVG (1994), i.e. SCC threshold of 100,000 /ml and CULT according to the NMC (1999). The cow udder health was always defined based on the quarter with the severest category within the udder. SCC, NAGase activity, lactate, fat, protein and lactose were determined in all QCM samples (n = 6194). Means were calculated for six different SCC classes at udder foremilk level (QFM) in a range of < 50,000 to > 1 million cells/ml milk by using GLM for repeated measurements (SAS, Version 6.12). Significances were calculated using Student-Newman-Keuls-tests. The secretion rate per udder quarter [g/hour] was calculated in QCM samples by using cell classes between 10,000 to > 1 million cells /ml.

Results

Udder health categorization

Table 1 and 2 detail the distribution of all QFM results regarding udder health and the percentages of the most important pathogens isolated.

Table 1. Mean distribution [%] of udder health categories (quarter foremilk samples) at quarter and cow level throughout the study (400 days; 20 session days; quarters: 130 ± 12; cows: 33 ± 3).

Evaluation level	Normal secretion	Latent infection	Unspecific mastitis	Mastitis
Quarter	52 ± 7	23 ± 4	13 ± 3	12 ± 3
Cow	20 ± 9	28 ± 5	20 ± 7	32 ± 9

Table 2. Comparison of infections with Corynebacterium spp. or other pathogens (919 quarters [= 35%] out of a total of 2461 were bacteriologically positive).

Pathogen	Health category	Number (%)	Xg SCC/ml
Corynebacterium spp.	Latent Infection	558 (71)	24,000
n = 788 (86%)	Mastitis	230 (29)	182,000
Other pathogens	Latent Infection	51 (39)	34,500
n = 131 (14%)	Mastitis	80 (61)	388,500

It is obvious that the main pathogen found in this herd with an overall percentage of 35% culturally positive results was *Corynebacterium* spp. (86%).

SCC in relation to udder health categories and milk fractions

Table 3 shows the level of somatic cells in QFM in relation to different health categories and in corresponding different milk fractions (QFM, QCM, CCM) and table 4 gives the SCC level in relation to different SCC classes in QFM.

Table 3. Comparison of SCC in different milk fractions with regard to 4 udder health categories (culturing of quarter foremilk samples) throughout the study (400 days; 20 session days; quarters: 130 ± 12; cows: 33 ± 3).

Daily diagnosis	Normal secretion	Latent infection	Unspecific mastitis	Mastitis
QFM: n =	3212	1592	703	687
SCC [lg]*	4.24 ± 0.35^c	4.41 ± 0.31^b	5.31 ± 0.47^a	5.34 ± 0.50^a
QCM: n =	3212	1592	703	687
SCC [lg]	4.49 ± 0.35^c	4.63 ± 0.32^b	5.46 ± 0.47^a	5.49 ± 0.50^a
CCM: n =	312	499	309	562
SCC [lg]	4.39 ± 0.30^d	4.56 ± 0.26^c	5.17 ± 0.44^b	5.23 ± 0.44^a

* = different letters within lines indicate significant differences (p ≤ 0.05)

Table 4. Comparison of SCC in different milk fractions with regard to six different cell count classes throughout the study (400 days; 20 session days; quarters: 130 ± 12; cows: 33 ± 3).

SCC [1000/ml]	< 50	51 - 100	101 - 200	201 - 400	401 - 1 Mio	> 1 Mio
QFM: n =	4238	886	484	233	205	148
SCC [lg]	4.22^f	4.84^e	5.13^d	5.44^c	5.79^b	6.28^b
QCM: n =	4238	886	484	233	205	148
SCC [lg]	4.48^f	5.01^e	5.26^d	5.55^c	5.91^b	6.37^a
CCM: n =	710	366	280	177	103	46
SCC [lg]	4.40^f	4.85^e	5.14^d	5.43^c	5.78^b	6.24^a

* = different letters within lines indicate significant differences (p ≤ 0.05)
Mio = million

SCC and related milk constituents

Table 5 describes the concentration of milk constituents in QCM in relation to udder quarter health categories, whereas Table 6 deals with the corresponding data in relation to six different SCC classes.

Table 5. Comparison of cell counts in quarter composite milk (QCM) samples considering the four udder health categories and other milk constituents (400 days; 20 session days; quarters: 130 ± 12; cows: 33 ± 3).

Daily diagnosis	Normal secretion	Latent infection	Unspecific mastitis	Mastitis
QCM: n =	3212	592	703	687
SCC [$\lg_{cells/ml}$]*	4.49 ± 0.35^c	4.63 ± 0.32^b	5.46 ± 0.47^a	5.49 ± 0.50^a
NAGase [$\lg_U$]	0.17 ± 0.23^c	0.23 ± 0.23^b	0.56 ± 0.31^a	0.57 ± 0.33^a
Lactate [$\lg_{\mu mol/l}$]	1.32 ± 0.47^c	1.37 ± 0.43^b	1.76 ± 0.45^a	1.79 ± 0.41^a
Fat [%]	4.13 ± 1.08^c	4.29 ± 1.07^b	4.41 ± 1.10^a	4.48 ± 1.11^a
Protein [%]	3.32 ± 0.32^c	3.33 ± 0.32^c	3.36 ± 0.34^b	3.47 ± 0.41^a
Lactose [%]	4.85 ± 0.19^a	4.82 ± 0.21^b	4.58 ± 0.37^c	4.61 ± 0.38^d

* = different letters within lines indicate significant differences (p ≤ 0.05)

Table 6. Means of selected milk constituents in relation to SCC classes.

SCC [1000/ml]	< 50	50 - 100	100 - 200	200 - 400	400 - 1000	> 1000
n = samples	4238	886	484	233	205	148
SCC [$\lg_{cells/ml}$]	4.48^{f*}	5.01^e	5.26^d	5.55^c	5.91^b	6.37^a
NAGase [$\lg_U$]	0.17^f	0.36^e	0.45^d	0.62^c	0.75^b	0.95^a
Lactate [$\lg_{\mu mol/ml}$]	1.31^f	1.55^e	1.64^d	1.82^c	2.00^b	2.28^a
Fat [%]	4.12^b	4.52^a	4.58^a	4.51^a	4.16^b	4.64^a
Protein [%]	3.31^c	3.43^b	3.43^b	3.39^b	3.37^b	3.52^a
Lactose [%]	4.85^a	4.74^b	4.68^c	4.59^d	4.47^e	4.25^f

* = different letters indicate significant differences (p < 0.05) within lines

SCC and milk secretion at quarter level (QCM)

Table 7 and table 8 present the interaction between SCC in QFM and the milk secretion rate at QCM level.

Table 7. Comparison of secretion rates (SR; [g/quarter/hour] in quarter composite samples (QCM) with regard to different cell count classes (1,000 - 50,000 cells/ml) throughout the study (400 days; 20 session days; quarters: 130 ± 12; cows: 33 ± 3).

SCC [1000/ml]	1 - 10	11 - 20	21 - 30	31 - 40	41 - 50	total
QCM; n =	1142	1331	887	521	357	4238
SR [g/h]*	343 ± 138^a	317 ± 119^b	303 ± 114^c	288 ± 118^d	287 ± 129^d	315 ± 126

* = different letters indicate significant differences (p < 0.05) within lines

Table 8. Comparison of secretion rates (SR; [g/quarter/hour] in quarter composite samples (QCM) with regard to different cell count (SCC) classes (< 50,000 - > 1 million cells/ml) throughout the study (400 days; 20 session days; quarters: 130 ± 12; cows: 33 ± 3).

SCC [1000/ml]	< 50	51 - 100	101 - 200	201 - 400	401 - 1 Mio	> 1 Mio
QCM; n =	4238	886	484	233	205	148
SR [g/h]*	315 ± 26[a]	266± 125[b]	251 ± 123[bc]	236 ± 114[c]	251 ± 133[bc]	234 ± 180[c]

* = different letters indicate significant differences (p < 0.05) within lines
Mio = million

Discussion

The udder quarter health categorization resulted in 75% of foremilk samples with a SCC of < 100,000 cells/ml. This can be regarded as an expression for a good udder health (Hamann and Reinecke 2000), even if the percentage of latent infection (25%) is relatively high compared to the corresponding mean value in Germany (i.e. 16%; Sobiraj et al., 1997). The majority (86%) of all bacteriologically positive foremilk samples displayed *Corynebacterium* spp. as causative agent. Published information has shown that insufficient post milking teat disinfection (e.g. spraying versus dipping) can markedly contribute to an increase in *C. bovis* infections (Bramley et al., 1976, Huxley et al., 2002). During this study, post milking automatic teat disinfection by spraying has been applied. Yet, a specific evaluation of the efficacy of this technique was not performed.

Several publications verify that in udder quarters that are healthy, i.e. remain free of pathogens for the whole lactation, the SCC does not exceed a threshold of 100,000 cells/ml (Schüttel 1999, Laevens et al., 1997, Doggweiler and Hess 1983). Table 3 indicates that 53% of all quarter samples presented a SCC level below 100,000 cells/ml in both QFM and QCM. This confirms earlier data, despite that the SCC level in QCM is significantly higher (p < 0.05) than in QFM, showing that these milk fractions have a SCC level < 100,000 cells/ml (Hamann 2002). Table 4 details the SCC in relation to different SCC classes and indicates that 82.7% of all quarter samples showed a SCC < 100,000 cells/ml.

Overall, the application of an automated milking system (e.g. VMS®) as such has proven not to cause an increase in milk SCC since the overall mean SCC value in QCM was 4.75 lg and 75% of all quarter milk samples had a SCC < 100,000 cells/ml.

Significant differences (p < 0.05) between milk constituents' concentration at different SCC classes < 100,000 cells/ml in QCM have been previously published (Hamann 2002). The results in table 5 show significant differences between the health categories NS and LI. Similar results are given in table 6 since the concentrations differ significantly at least between the classes < 50,000 and 50 to 100,000 cells/ml. It should be clear however that the concentration level of these parameters in the SCC class below 100,000 represents the physiological range and that changes towards the threshold reflect the commencing failure of the blood udder barrier.

The data in table 7 indicate that milk secretion rate at udder quarter (QCM) level is highest when SCC in QFM drops to < 10,000/ml. The comparison of this secretion rate with that at SCC between 51,000 and 100,000 cells/ml resulted in a reduction of 77 g (= -22%;

table 8). This shows clearly that a threshold being a limit between the physiological and the pathological stage is something different than an optimal SCC level (10,000 cells/ml) where secetion rate is highest.

These preliminary data should stimulate further research in defining physiological thresholds for milk constituents.

References

Bramley, A.J., R.G. Kingwill, T.K. Griffin and D.L. Simpkin, 1976. Prevalence of *Corynebacterium bovis* in bovine milk samples. Veterinary Record 99: 295

Doggweiler, R. and E. Hess, 1983. Zellgehalt in der Milch ungeschädigter Euter. Milchwissenschaft 38: 5-8

DVG, Deutsche Veterinärmedizinische Gesellschaft (editors), 1994. Leitlinien zur Bekämpfung der Mastitis des Rindes als Bestandsproblem. DVG, 126 pp.

Hamann, J., 2002. Relationships between somatic cell count and milk composition. Bulletin of the International Dairy Federation 372: 56-59

Hamann, J. and F. Reinecke, 2002. Machine milking effects on udder health - comparison of a conventional with a robotic milking system. In: Wageningen Pers, First North American Conference on Robotic Milking, Proc., 17-27

Huxley, J.N., M.J. Green and A.J. Bradley, 2002. The prevalence and significance of minor pathogen intramammary infections, in low somatic cell count herds, in the UK. In: Kaske, M., H. Scholz and M. Höltershinken (editors), 22th World Buiatrics Congress, CD-ROM

Laevens, H., H. Deluyker, Y.H. Schukken, L. de Meulemeester, R. Vandermeersch, E. de Muelenaere and A. de Kruif, 1997. Influence of parity and stage of lactation on the somatic cell count in bacteriologically negative cows. Journal of Dairy Science 80: 3219-3226

NMC, National Mastitis Council (editors), 1999. Laboratory Handbook on Bovine Mastitis. NMC, 222 pp.

SAS. Inst. Inc. 1996. SAS/Stat Software, Changes and Enhancement through release 6.12.

Schüttel, M., 1999. Vergleich der N-Acetyl-beta-D-glucosaminidase-Aktivitäten (NAGase) in Milch, Blut und Harn beim laktierenden Rind. School of Veterinary Medicine, Hannover (thesis)

Sobiraj, A., A. Kron, U. Schollmeyer and K. Failing, 1997. Bundesweite Untersuchung zur Erregerverteilung und in-vitro-Resistenz euterpathogener Bakterien in der Milch von Kühen mit subklinischer Mastitis. Tierärztliche Praxis 25: 108-115

INTRODUCTION OF AMS IN ITALIAN DAIRY HERDS: EFFECTS ON COW PERFORMANCES AND MILK QUALITY IN A HERD OF THE GRANA PADANO AREA

G. Pirlo[1], G. Bertoni[2] & R. Giangiacomo[3]
[1]Animal Production Research Institute, Rome, Italy
[2]Istituto di Zootecnica, Facoltà di Agraria, Piacenza, Italy
[3]Dairy Research Institute, Lodi, Italy

Abstract

One hundred and five Italian Friesian cows (69 of them were heifers) were allotted to two milking systems (AMS or MP). Milk was stored at 4-5 °C. From the experimental herd 38 heifers were selected for additional analyses concerning metabolism and milk characteristics. Twenty-one of them were milked with MP and 17 with AMS. In addition, the entire bulk milk of the two experimental groups was used for the production of 30 Grana Padano cheeses. AMS cows were milked 2.56 times a day on average. No significant difference was observed in overall milk production between AMS and MP. Slight differences between MP and AMS were observed for some blood parameters at the beginning of lactation and for cortisol since the first days after introduction into AMS, suggesting a light but prolonged stress-like situation. Milk fat was 3.61 and 3.33% ($p<0.05$) for MP and AMS respectively, however no difference was observed for milk protein and lactose content. SCC was higher in AMS than in MP, pH was 6.706 and 6.759 ($p=0.002$), FFA (meq/100 g of fat) was 0.531 and 0.700 ($p=0.001$) and titratable acidity (°SH/100 ml) was 7.143 and 6.759 ($p=0.002$) for MP and AMS respectively. No significant differences were found for other milk characteristics. Cheese of MP and AMS groups had very similar characteristics.

Introduction

Two are the main reasons why many farmers have installed a milking robot: to improve the quality of life and to increase economic efficiency of their firm. Furthermore, they expect to improve milk yield as a consequence of the increase of milk frequency. Number of milkings per day per cow has a large range of variation according to traffic system (Hermans *et al.*, 2003), season (Artmann and Bohlsen, 2000; Speroni *et al.*, 2003) and feeding (Rodenburg and Wheeler, 2002). It was demonstrated that the reduction of milking interval is related to an increase of milk yield (Artmann and Bohlsen, 2000), but in some trials milk yield was higher in AMS than in the traditional systems (Wagner-Storch and Palmer, 2003), in some others it was not (Hopster *et al.*, 2002). Field studies gave also different results: Billon and Tournaire (2002) found a reduction of milk yield as a consequence of a reduction of duration of lactation, whereas Løvendahl and Madsen (2001) found an increase of milk yield. Effect of milking interval depends on milk ability (Hogeveen *et al.*, 2001). Milking intervals have also a large variability per cow (de Koning and Ouweltjes, 2000) and per season (Speroni *et al.*, 2003). Most of these studies have been carried out in Northern Europe and

bleeding. A detailed description of these results is given in the poster "Introduction of AMS in Italian condition: welfare assessment based on metabolic and endocrine responses in primiparous dairy cows".

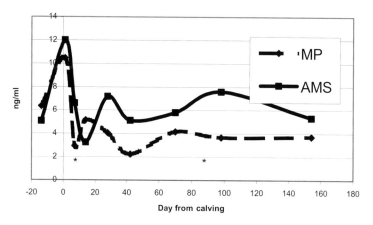

Figure 2. Plasma cortisol in primiparous cows milked with AMS or MP.

The milk of heifers milked with the AMS had a lower concentration of fat, a higher value of SCC and of FFA (Table 2) and significantly at 7, 42 and 98 days in milk. Freezing point was not affected, probably because samples were drawn with a automatic sampler and they did not include technical water that remains in the pipes after rinsing. The milk of AMS heifers had higher pH and lower titratable acidity than that of MP heifers. This results appear in contrast with some theoretical expectations. A possible explanation might be the higher value of SCC which could have affected mineral composition and casein phosphorilation. Fat globule size has not often been considered and a few papers are available to compare

Table 2. Least square means and standard error of milk parameters of primiparous cows in MP and AMS.

Item	MP	AMS	P
Fat, (%)	3.61±0.07	3.33±0.09	0.02
Protein, (%)	3.54±0.04	3.50±0.03	NS
Lactose, (%)	5.17±0.,03	5.17±0.02	NS
SCC, ln (10^3/ml)	4.37±0.15	4.81±0.11	0.02
Urea, mg/100 ml	24.69±0.46	25.12±0.34	NS
Freezing point, °C	-0.5293±0.0006	-0.5291±0.0004	NS
FFA, meq/100 g of fat	0.531±0.039	0.700±0.030	0.001
Natural creaming, % of fat	39.74±1.50	36.11±1.24	0.06
Median fat globule Ø, μm	4.55±0.11	4.61±0.08	NS
pH	6.706±0.010	6.745±0.007	0.002
Titratable acidity, °SH/100 ml	7.143±0.100	6.759±0.072	0.002
Casein N, % of N	75.73±0.39	75.62±0.25	NS
NPN, % of N	5.49±0.13	5.72±0.09	NS

Automatic milking – A better understanding

methods and results. We considered the median value on the fat globule size distribution as representative of the sample. As evidenced in table 2, there were no differences in the median value of fat globule size distribution between the two milking systems.

Concerning cheese production, technological parameters, cheese yield, fat content of cream, total solids in whey indicate no differences between the milk collected by the two systems. Also the expert cheese maker did not notice any difference during the whole process. Weight loss during the first 7 months of ripening proceed correctly in both cheeses and no blowing has been recorded. The aspect of cheeses were similar at 7 months (Figure 3). A detailed description of these results is reported in the poster "Introduction of AMS in Italian dairy herds. Long ripened cheese production from robotic milking".

Figure 3. Pictures of experimental Grana Padano cheese type (7 months aged): A. AMS; B. MP.

Conclusion

Throughout the year, AMS did not increase milk yield in comparison to MP. However, we observed that milk yield of AMS cows was higher than that of MP cows when the environmental condition was favourable. This is particularly important in the area of Grana Padano, where the summer is very hot and humid and systems to alleviate disconfort should be found, in order to increase presentations to the milking robot and to maintain milk yield. AMS seems to be a source of slight but prolonged stress situation, that apparently has no serious consequence, but requires further researches to be better understood (i.e. if the cause is the AMS box, the one-way traffic system or other reasons).

Chemical and technological parameters indicate that milk obtained by AMS is suitable for further processing into cheese. In particular, cheese-making process into Grana Padano type long ripening cheese did not show difficulties and ripening process up to 7 months was normal, creating the tipical structure and aroma. This result indicates that AMS is potentially suitable also for the production of cheese whose current production protocol strictly prescribes the use of milk obtained from two consecutive milkings.

References

Abeni, F., L. Degano, M. Capelletti, G. Pirlo, 2002. Effect of robotic milking on physiochemical and renneting properties on bovine millk: prliminary report from an Italian experimental farm. The First North American Conference on Robotic Milking - March 20-22, 2002. Toronto, Canada, pp. V 64-V 67.

Armann R., Bohlsen E. (2000). Results from the implementation of automatic milking system (AMS)-multi-box facilities. In: Hogeveen H., Meijering A. (editors), Robotic Milking, Internation Symposium., Lelistad, The Netherlands, 17-19 Aug. 2000, pp. 38-46.

Billon, P. and F. Tournaire, 2003. Impact of automatic milking systems on milk quality and farm management: the French experience. The First North American Conference on Robotic Milking - March 20-22, 2002. Toronto, Canada, pp. V 59-V 63.

Hermans G.G.N., A.H. Ipema, J. Stefanowska, J.H.M. Metz (2003). The effect of two traffic situation on the behavior and performance of cows in an automatic milking system. Journal of Dairy Science 86: 1997-2004.

Hogeveen, H., W. Ouweltjes, C.J.A.M. de Koning, K. Stelwagen, 2001. Milking interval, milk production and milk flow-rate in an automatic milking system. Livestock Production Science 72: 157-167.

Hopster, H., R.M. Bruckmaier, J.T.N. var der Werf, S.M. Korte, J. Macuhova, G. Korte-Bouws, C.G. van Reenen, 2002. Stress responses during milking; comparing conventional and automatic milking in primiparous dairy cows. Journal of Dairy Science 85: 3206-3216.

Justesen, P. and M. D. Rasmussen, 2000. Improvement of milk quality by the Danish AMS self-monitoring programme. In: Hogeveen H., Meijering A. (editors), Robotic Milking, International Symposium, Lelistad, The Netherlands, 17-19 Aug. 2000, pp. 83-88.

Klungel, G.H., B.A. Slaghuis, H. Hogevee, 2000. The effect of the introduction of automatic milking systems on milk quality. Journal of Dairy Science 83: 1998-2003.

de Koning, K. and W. Ouweltjes, 2000. Maximising the milking capacity of an automatic milking system. In: Hogeveen H., Meijering A. (editors)., Robotic Milking, International Symposium, Lelistad, The Netherlands, 17-19 Aug. 2000, pp 38-46.

Løvendahl, P., P. Madsen, 2001. Does automatic milking change the shape of the lactation curve and composition of milk? 52nd Annual Meeting of EAAP.

Luther, H., W. Junge, E. Kalm, 2002. Space requirements in feeding, resting and waiting areas of robotic milking facilities. The First North American Conference on Robotic Milking - March 20-22, 2002. Toronto, Canada, pp. II 15-II 25.

Rodenburg, J., B. Wheeler, 2002. Strategies for incorporating robotic milking into North American herd management. The First North American Conference on Robotic Milking - March 20-22, 2002. Toronto, Canada, pp. III 18-III 27.

Speroni, M., F. Abeni, M. Cappelletti, L. Migliorati, G. Pirlo, 2003. Two years of experience with an automatic milking system. 2. Milk yield, milk interval and frequency. Italian Journal of Animal Science 2 (Suppl. 1): 260-262.

Wagner-Storch, A.M., R.W. Palmer, 2003. Feeding behavior, milking behavior, and milk yields of cows milked in a parlor versus an automatic milking system. Journal of Dairy Science 86: 1494-1502.

ROBOTIC MILKING AND FREE FATTY ACIDS

B.A. Slaghuis[1], K. Bos[1], O. de Jong[1], A.J. Tudos [2], M.C. te Giffel [2] & K. de Koning[1]
[1]Applied Research, Animal Sciences Group, Wageningen UR, Lelystad, The Netherlands
[2]NIZO food research, Ede, The Netherlands

Abstract

With the introduction of automatic milking (AM) systems, increased levels of free fatty acids (FFA) in milk were observed, which might result in off-flavours in milk and dairy products.

The aim of this study was to investigate the factors contributing to elevated FFA levels: influence of the milking frequency, technical parameters of the milking system, and finally, farm management aspects.

Milking frequency was studied in a Latin square design with milking intervals of 4, 8 and 12 hours and showed increased FFA -levels for the shorter intervals.

Technical factors were studied in a laboratory study using milking machine components of AM-systems and of conventional systems. With susceptible milk, differences in increase in FFA levels were found, but results were difficult to interpret. Tests will be repeated.

Some FFA problems remained after solving technical and milking frequency problems. Indications were found that feed composition and feeding regime might influence susceptibility of milk for FFA-formation. These farm management aspects are subject of ongoing research.

Introduction

On the introduction of milk lines and farm milk tanks in the seventies, problems with elevated levels of free fatty acids (FFA) were observed, resulting in off-flavours in milk and dairy products. Therefore, measurement of FFA was integrated in the quality control system in The Netherlands (determination twice a year, in spring and autumn). The problems of elevated FFA levels were tackled practically by avoiding blind pumping, reducing air leakage and improving maintenance of the milking equipment. Especially the introduction of a national milking machine maintenance program contributed largely to overcome problems with FFA.

When milking three or more times a day instead of two times is applied, increased milk yields, decreased fat and protein contents and increased free fatty acid levels have been reported (Ipema and Schuiling, 1992, Jellema, 1986, Klei et al., 1997, Svennersten-Sjauna et al., 2002).

With the introduction of AM systems, the problem of increased FFA levels (also known from farms milking 3 times per day) occurred again (Klungel et al., 2000,Vorst and Koning, 2002). This problem seems due to the higher average milking frequency with AM systems and to technical features of this system compared to conventional milking systems.

As elevating FFA levels are a result of lipolysis, the possible mechanism of lipolysis is explained below.

Possible mechanism of lipolysis

Lipolysis is enzymatic hydrolysis of milk fat by milk lipoprotein lipase (mLPL), causing accumulation of FFA. The milk fat present in milk fat globules is protected against the action of mLPL by the milk fat globule membrane. If this membrane is disrupted by e.g. agitation, fatty acids can be split from the glycerol part by mLPL.

But in untreated milk, lipolysis also occurs due to spontaneous lipolysis. Jellema (1986) defined susceptibility as FFA level in untreated milk after 24 hours storage at 4°C. Late lactation milk and milk from cows milked three times per day or more is more susceptible to spontaneous lipolysis (Jellema,1986). The milk fat globule is not disrupted, but some factors (activators) present in milk may favour the interaction of mLPL with milk fat resulting in higher degree of FFA. Activators are related to blood serum components, as addition of blood serum to raw milk increases FFA levels (Jellema, 1975,1986). Inhibitors, that inhibit contact between mLPL and milk fat globule membrane, have been determined as proteose pepton component 3 fractions (Cartier et al., 1990).

Several treatments, referred to as activation treatments, enhance lipolysis (induced lipolysis) and are often used to study lipolysis of milk fat by endogenous mLPL. Treatments which cause activation include agitation, homogenisation, temperature changes and the addition of blood serum or heparin to milk. Milk from individual cows differs in susceptibility to these treatments. Correlations between spontaneous and induced lipolysis have been reported (Jellema, 1986; Chazal and Chilliard, 1987; Cartier and Chilliard, 1989).

The mechanisms that promote lipolysis are not fully understood.

The aim of this study was to investigate the factors contributing to elevated FFA concentrations: influence of the milking frequency, technical parameters of the milking system, and finally, farm management aspects.

Material and methods

Milking frequency

In the first study spontaneous lipolysis (Cartier and Chilliard, 1990) was determined in a Latin square design experiment with 12 cows, three periods of four days, three milking intervals (4, 8 and 12 hours corresponding to 6, 3 and 2 times milking a day) and 4 cows per group (Table 1). Cows were selected to form equal groups (parity and lactation stage). Cows were milked according to this scheme for 4 days and on day 4 all cows were milked conventionally and sampled.

Milk samples were divided into two sub samples: one was inactivated by hydrogen peroxide immediately (0,02%, Jellema, 1979) to stop lipolysis, the other one was stored

Table 1. Latin square design of spontaneous lipolysis study, with groups of cows defined as A, B and C.

Period	1	2	3
Milking interval			
2x per day	A	B	C
3x per day	B	C	A
6x per day	C	A	B

Automatic milking – A better understanding

at 5°C ±1°C for 24 hours and then inactivated with hydrogen peroxide. Difference between the sub samples is defined as spontaneous lipolysis (Cartier and Chilliard, 1990). Samples were analysed according to the BDI-method (IDF,1991).

Technical parameters of the AM-system

Fresh milk, susceptible for lipolysis (from 4-5 cows at the end of lactation), was divided in two parts. One part was passed through a reconstructed AM-system (two brands) and milk was sampled before and after passing the system. The other part of milk was passed through a conventional system and sampled before and after passing the system. The milk was mixed and used again for a repetition of the experiment.

Farm management

In a study on variation of FFA levels in farm tank milk of 12 farms (4 conventional (milking twice a day), 4 conventional milking (three times a day) , 4 AM; table 2.) were determined monthly and in two periods of 14 days from every bulk tank. Groups of farms were selected as high (>0.80 mmol/100 g fat during the last year) or low (<0.70 mmol/100 g fat).

Table 2. Number of farms in design of study on variation of FFA levels.

FFA level	high	low
Type of farm		
Conventional 2x per day	2	2
Conventional 3x per day	2	2
AM	2	2

The effect of farm management aspects was studied at 8 farms using the same brand of AM-system and the same cooling system and 6 farms milking three times a day (Table 3.). Half of the farms were classified as high (FFA >0.75 mmol/ 100 g fat during at least the last year) and the other half as low (<0.70 mmol/100 g fat).

Table 3. Number of farms in design of study on farm management.

FFA level	high	low	remarks
Type of farm			
Conventional 3x per day	3	3	different brands, same cooling system
AM	4	4	same brand, same cooling system

Bulk tank milk was sampled and analysed for FFA. Detailed farm (including milk frequencies), robotic and management information regarding feeding, housing conditions and animal health was obtained by a questionnaire during a visit at the farm and correlations among these factors with FFA levels were calculated.

Automatic milking – A better understanding   343

Results and discussion

Milking frequency

Analysis of variance of log transformed FFA showed slight differences between intervals after 0 hour storage and significant differences after 24 hours of storage (Table 4).

Table 4. Effect of milking interval on FFA contents (mmol/100 g fat) and standard deviations (between brackets) in raw milk after 0 hours and 24 hours storage after sampling.

Interval (times/day)	6		3		2	
0 hours at 5°C	0.20[a]	(0.04)	0.19[ab]	(0.07)	0.15[b]	(0.07)
24 hours at 5°C	1.23[a]	(0.98)	0.71[b]	(0.47)	0.42[c]	(0.31)
increase in 24hours	0.97[a]	(0.98)	0.49[b]	(0.45)	0.25[c]	(0.26)

[a,b,c] statistically significant difference on the same row P<0.05)

The increase after 24 hours storage, defined as spontaneous lipolysis (Cartier and Chilliard, 1990) was significant for the different milking intervals. Initial levels of FFA were rather low compared to other studies.

As all bulk milk samples contain milk of 3 days old and the increase in FFA is the highest in the first 24 hours (data not published), no hydrogen peroxide was used anymore during the rest of the experiments and no sub samples were taken anymore.

Technical parameters of the AM-system

The milk used in test 1 with brand 1 was not susceptible enough, because initial FFA level was low (mean FFA: 0.36 mmol/100 g fat). Selection of susceptible milk was not properly. The milk used for test 2 with brand 2 was susceptible enough to show difference in FFA increase (mean FFA: 0.52 mmol/100 g fat), but different initial FFA levels in milk passing the conventional and in milk passing the reconstructed AM system made results difficult to interpret. Tests will be repeated. In AM systems more air is needed to transport the same amount of milk in comparison with conventional milking, so FFA increase may be expected.

Farm management aspects

Preliminary results from the variation study indicate that previous selection, based on the FFA levels of the previous year, of farms in high (>0,80 mmol/100 g fat) and low (<0,70 mmol/100 g fat) FFA level groups was in accordance with the selection for farms milking conventionally and robotically (Figure 1). Twice a year determination of FFA was a good indication for these farms. All farms milking three times a day had high levels of FFA on a regular base (>0,80 mmol/100 g fat) and FFA levels were not in accordance to previous results. But the previous results were based on two FFA determinations: one in spring and one in autumn. So more variation in FFA during the year was found for these type of farms.

Two of the high 2x farms and one of the high 3x farms were tied stalls with long high milk lines, indicating that FFA levels are more elevated with elevation of milk and long milk lines.

Automatic milking – A better understanding

On one high and one low 3x farm, FFA increased when cows were pastured. Ferlay *et al.* (2002) found a higher lipolysis for milk of pasturing cows than for milk from cows feeding a diet rich in concentrate or a corn silage-based diet. The cows grazing on pasture were slightly underfed. Increased lipolysis due to underfeeding in cow milk were observed previously (Jellema, 1975). So for two farms there might be a feeding problem, especially in summer.

On one high AM farm, the software was changed, resulting in a decrease in the number of failed milkings. As a result FFA levels decreased. This may have been a technical problem related with excessive air inlet. For the other high AM farm no clear explanation could be found, but feeding experiments showed a decrease in FFA levels.

Selection of farms, based on FFA levels in the previous year, in the study on farm management aspects seemed to be adequate for 3x milking per day but not for AM-systems (Figure 2). Farms milking 3x with high FFA levels had a lower fat content in the milk than farms with low FFA levels. This has to be studied in more detail. For AM farms the selection

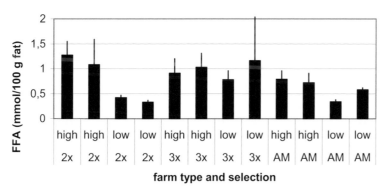

Figure 1. *Mean FFA and standard deviations on twelve farms based on monthly sampling and two times two weeks sampling of every farm milk tank (5-6 samples in two weeks). Results were based on 9 -21 samples per farm.*

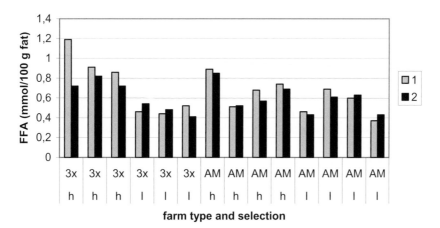

Figure 2. *FFA in farm tank milk from 6 farms milking 3x per day and from 8 farms with an AM-system. Results of two samplings (1 and 2) per farm within six weeks in autumn. Selection was based on high (h) and low (l) FFA-levels in at least the last year.*

was made on the same brand and cooling systems, all quite recently installed, but FFA results for at least one year were known. After solving some starting problems most of the FFA levels decreased and no significant difference was found between high and low FFA on AM farms. On the farm with the highest FFA levels, the milking frequency was not adequately adjusted. Cows producing less (7 kg per milking) were milked too often.

Conclusions

Increased FFA levels were due to increased milking frequencies (both for AM and conventional: 3x milking per day) and probably to milking machine components of AM systems, but tests must be repeated. Milking frequencies seemed to be of more importance than technical parameters of the AM system, because FFA levels for farms milking three times per day and AM systems were comparable. However, technical aspects cannot be excluded. Compared to conventional milking, the air/milk ratio is higher, probably resulting in more disruption of milk fat globule membranes.

Apart from milking frequencies and technical parameters, management aspects probably play a role. Feeding might be a factor of importance and could be studied in more detail. Also more fundamental research is needed regarding the susceptibility of cows.

Acknowledgements

The work was funded by the European Commission within the EU project "Implications of the Introduction of Automatic Milking on Dairy Farms (QLK5-2000-31006)" as part of the EU-programme "Quality of Life and Management of Living Resources".

The content of this publication is the sole responsibility of its publisher, and does not necessarily represent the views of the European Commission nor any of the other partners of the project. Neither the European Commission nor any person acting on behalf of the Commission is responsible for the use, which might be made of the information presented above.

The authors thank the farmers for their cooperation during this investigation.

References

Cartier, P., Chilliard, Y. 1989. Lipase redistribution in cows' milk during induced lipolysis. I. Activation by agitation, temperature change, blood serum and heparin. Journal of Dairy Research 56:699-709.

Cartier, P., Chilliard, Y., Paquet, D., 1990. Inhibiting and activating effcts of skim milks and proteose-peptone fractions on spontaneous lipolysis and purefied lipoprotein lipase activity in bovine milk. Journal of Dairy Science 73:1173-1177.

Chazal, M.P., Chilliard, Y., Coulon, J.B., 1987. Effect of nature of forage on spontaneous lipolysis in milk from cows in late lactation. Journal of Dairy Research 54: 13-18.

Ferlay, A., Martin, B., Pradel, Ph., Chilliard, Y., 2002. Effect of the nature of forages on lipolytic system in cow milk. In: Proceedings Congrilait 2002, 26th IDF World Dairy Congress, 24-27 september 2002, Paris, France. poster B4-38.

IDF, 1991. Determination of free fatty acids in milk and milk products. IDF Bulletin No. 265. Brussels, International Dairy Federation.

Ipema, A.H., Schuiling,E., 1992. Free fatty acids; influence of milking frequency. In: Ipema, A.H., A.C. Lippus, J.H.M. Metz and W. Rossing (editors) Proceedings of the International Symposium on Prospects for Automatic Milking, Wageningen, EAAP series 65: 491-496.

Jellema, A., 1975. Susceptibility of bovine milk to lipolysis. Netherlands Milk and Dairy Journal 29:145-152.

Jellema, A., 1986. Some factors affecting the susceptibility of raw cow milk to lipolysis. Milchwissenschaft 41:553-558.

Klei,L.R., Lynch, J.M., Barbano, D.M., Oltenacu, P.A., Lednor, J., Bandler, D.K., 1997. Influence of milking three times a day on milk quality. Journal of Dairy Science 80: 427- 436.

Klungel, G.H., Slaghuis, B.A., Hogeveen,H., 2000. The effect of the introduction of automatic milking systems on milk quality. Journal of Dairy Science 83:1998-2003.

Svennersten-Sjauna, K., Persson, S., Wiktorsson, H., 2002. The effect of milking interval on milk yield, milk composition an raw milk quality. In: Proceedings of the first North American Conference on Robotic Milking, Wageningen Pers: V43 -V48.

Vorst, Y., Koning, K. de, 2002. Automatic milking systems and milk quality in three European countries. In: Proceedings of the first North American Conference on Robotic Milking, Wageningen Pers: V1-V12.

IMPACT OF SIZE DISTRIBUTION OF MILK FAT GLOBULES ON MILK QUALITY AFFECTED BY PUMPING

Lars Wiking[1], Lennart Björck[2] & Jacob H. Nielsen[1]
[1]Dept of Food Science, Danish Institute of Agricultural Sciences, Foulum, Denmark
[2]Dept of Food Science, Swedish University of Agriculture Science, Uppsala, Sweden

Abstract

In the present study a correlation (r^2=0.54) was found between average diameter of the milk fat globule (MFG) and the diurnal fat yield of cows. The changes in size distribution of MFGs showed transition of medium size globules into large globules with increased fat yield. Furthermore, analysis of fatty acid composition demonstrated that the content of stearic and palmitoleic acids correlated positively with the average diameter of the MFG, while the content of polyunsaturated fatty acids correlated negatively. The importance of the size distribution of milk fat globules was demonstrated in an experiment where three groups of Holstein cows were fed concentrates with different fatty acid compositions; one high in saturated lipids, another high in unsaturated lipids and the last one simulating high de novo synthesis. The diets resulted in milk with fat contents of 5.0, 3.7 and 4.0%, respectively. The milk fat globules were significantly larger in the milk with the highest fat content. All three types of milk were pumped at various flow rates and temperatures. Afterwards, analysis of fat globule size distribution showed that the highest coalescence of milk fat globules in the milk occurred with the largest fat globules. Furthermore, the fat globules were less resistant against coalescence at a pumping temperature of 31°C compared with lower temperatures. Likewise, an increase was found in free fatty acids for milk with the largest milk fat globules, indicating that milk with a high fat content is more unstable when subjected to pumping.

Introduction

The present focus on increasing formation of free fatty acids in milk from automatic milking systems deals with milking intervals and pumping technology. However, the size of milk fat globules has crucial influence on the stability and technological properties of the milk. Recent research has shown that the size of native small milk fat globules affects moisture content, firmness and elasticity of camembert cheeses (Michalski *et al.*, 2003). Likewise, the size of milk fat globules affects the viscoelasticity of acid and rennet gels. Therefore, knowledge of the fat globule distribution is important already at the time of milking.

Material and Methods

Milk samples were obtained from 52 Danish Holstein cows on the farm at Research Centre Foulum. A true-tester was used to collect the milk samples during morning and evening milking. Composite milk samples for analysis of fat, protein, lactose and citrate were collected from each cow at the two daily milkings.

Determination of the particle size distribution (MFG) in milk

Particle size distributions were determined by integrated light scattering using a Mastersizer 2000 (Malvern Instruments Ltd.). The volume-based diameter,

$$d_{(4,3)} = \frac{\Sigma Nidi^4}{\Sigma Nidi^3}$$

was calculated by the software (where N_i is the number of globules in a size class of diameter d_i).

Fatty acid composition

Prior to GC separation and quantification, the lipid was trans esterified to methyl esters in a sodium methylate solution (2 g/L methanol). Analysis of the fatty acid methyl esters was carried out with a GC (Hewlett-Packard Co.) using a FFAP-column (polyethylene glycol TPA 25 m x 200 µm x 0.30 µm) and helium as carrier gas and a flame ionisation detector. Injection was splitless with an injector temperature of 250 °C. The detector temperature was 300 °C. The initial column temperature was 40 °C, which was held for 4 min. The temperature was then raised by 10 °C /min to 240 °C, which was held for 1 min.

Experimental feeding

Thirty Danish Holsteins were divided into three equal groups and were fed concentrates with different fatty acid compositions; one high in saturated lipids, another high in unsaturated lipids and the last one simulating high de novo synthesis. The animals used in the experiment were in mid-lactation. The milk used for experiments was collected between 15 and 30 days after introduction of the experimental feeding.

Pumping experiment

The milk was cooled to the pumping temperatures (5, 20 and 31 °C) immediately after milking. The pumping system contained a 9.5 m pipeline, one valve, a balance tank and a centrifugal pump (Alfa-Laval), which was regulated by an ABB ACS 140 frequency converter. The diameter of the pipes was 2.25 cm. In each treatment, eight litres of raw milk were pumped through the system over 450 seconds.

Analysis of free fatty acids in milk

The free fatty acids were analysed using the B.D.I method (IDF, 1991).

Statistical analyses

Data were analysed by use of the procedure GLM in the program package SAS version 8.1 (SAS institute, Cary, NC, USA).

Results

The average diameter of MFG in milk samples from the 52 investigated cows was found to correlate with the diurnal fat production (Figure 1). This indicates that when cows producing a high level of milk fat, the synthesis of membrane material covering the MFG is limited.

Figure 2 shows that it is particularly the large population that increases at the expense of the main population upon increasing fat yield. Meanwhile the population of small globules is nearly constant.

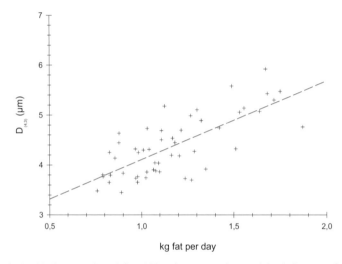

Figure 1. Relationship between diurnal fat yield and average volume-weighted diameter of MFG from 52 Holstein cows.

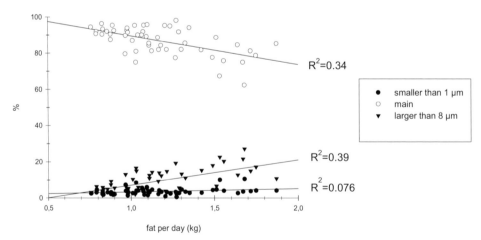

Figure 2. Changes in subpopulations by increasing diurnal fat yield. The subpopulations are divided in three groups: small (< 1 μm in diameter), main (1-8 μm) and large (> 8 μm).

Automatic milking – A better understanding

In the present study, the average diameter of MFG correlated positively with the concentration of C16:1 and C18 but negatively with the concentration of C18:2 +C18:3 (Table 1). No correlations were observed between the average diameter of MFG and concentration of C4-C14, C16 and C18:1.

Table 1. Pearson's correlation coefficients for the correlations between the average diameter of MFG and concentration (wt%) of fatty acids. The result is compared with literature. + indicates significant increase in fat globule size, and vice versa.

Fatty acid	Pearson's correlation coefficient Present study Spring milk	Briard et al. 2003 Winter milk	Briard et al. 2003 Spring milk	Timmen and Patton, 1988
C4-C14				+ (C4-10)
C16		-		
C16:1	0.30	-	-	
C18	0.33	+	+	+
C18:1		+	-	-
C18:2 + C18:3	-0.55	+ (only C18:2)		

Pumping stability of the milk fat

The three experimental diets used to feed the cows in the present study contained different fat sources with diverse fatty acid composition, and the diets resulted in milk with fat contents of 3.7, 4.0 and 5.0%, respectively (Table 2). The MFG were significantly larger in the milk with the highest fat content (Table 2). All three types of milk was pumped at various temperatures and flow rates in a system containing pipes, balance tank and a centrifugal pump to mimic the mechanical stress that milk is subjected to at the farm, from the udder to the bulk tank. The pumping temperatures of the milk chosen were 31, 20 and 5°C, ranging from the temperature of milk when leaving the udder to a cool storage temperature. The measurement of particle size distribution of the MFGs clearly showed the highest coalescence of the MFGs in the milk with the highest fat content at 31 °C (Figure 3). The influence of the pumping temperature was crucial for the stability of the MFGs containing 5.0% fat.

Table 2. The milk fat% and the volume weighted-diameter ($d_{(4,3)}$) of fat globules from the raw milk from the three different feedings.

Diet	% milk fat	$d_{(4,3)}$ (μm)
Unsaturated fat	3.7	3.54
High de novo	4.0	3.62
Saturated fat	5.0	4.33

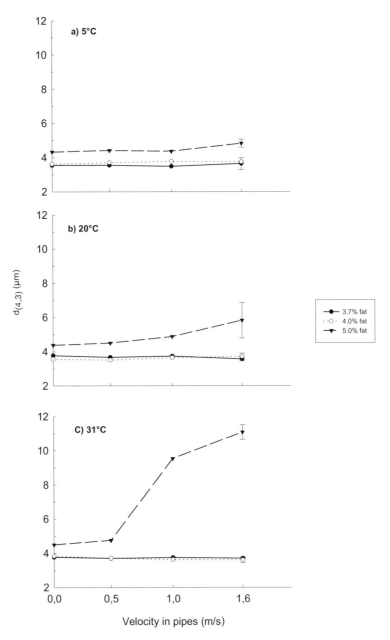

Figure 3. The volume weighted-diameter $(d_{(4,3)})$ of fat globules from the three types of raw milk. The milk is pumped for 450 s at 5, 20, 31°C, respectively.

The observations of coalescence of MFGs in milk with a high fat content generally agree with analyses of free fatty acids. The formation of FFA after pumping and cold storage was highest in the milk with 5.0% fat (Figure 4). However, increasing flow rates affected the lipolysis of all milk types, especially at pumping temperatures of 20 and 31°C.

Automatic milking – A better understanding

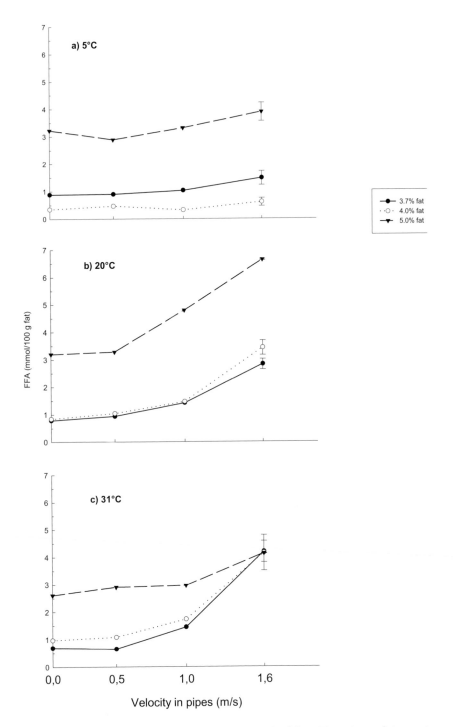

Figure 4. Free fatty acids content of the three types of raw milk after pumping followed by 24 hours of storage at 5°C. The milk is pumped for 450 s at 5, 20, 31°C, respectively.

Automatic milking – A better understanding

Discussion

The present work demonstrates for the first time a correlation between the milk fat output and the average diameter of MFG. In contrast to the present study, Walstra (1969) found no positive correlation between fat content and average globule size, neither within a lactation period, nor between different cows within a breed. However, that study did not take account of the diurnal yield as do the present study. The average MFG size declines with advancing lactation (Walstra, 1969). At the same time the milk yield decreases, and the fat percentage starts to increase approximately after the second month of lactation. In summary, both the yield of fat and the phospholipid secretion decrease during lactation, hence the phospholipid to fat ratio is fairly constant throughout lactation (Kinsella & Houghton, 1975). Bitman & Wood (1990), on the other hand, found decreasing content of phospholipid and cholesterol in the total lipid from day 42 to 180 of lactation. In the present study, the size of the fat globule size did decrease with advancing lactation (data not shown), but the diurnal fat yield explains the data better. Our observation suggests that it is the diurnal fat yield that mainly affects the average size of the MFG.

It has previously been suggested that MFGs are composed of three subpopulations: one containing the small globules with a diameter less than 1 μm which comprises approximate 80% of the total number of globules, the medium size population containing approximate 94% of the fat, and the last subpopulation includes the large globules (Walstra, 1969). The present finding that the main milk fat globules becoming larger, and the small globules are constant with constant fat to surface ratio when the fat synthesis increases, indicates that the enveloping of small and large fat globules in the epithelial cell is not similar.

The present findings of correlations of various fatty acids with the average diameter of the MFG are in table 1 compared with studies of Timmen & Patton (1988) and Briard et al. (2003). In the two mentioned studies the MFGs were fractionated by centrifugation or microfiltration into a small and a large size population within fatty acid composition was analysed. All three studies agreed that C18 is positively correlated to the milk fat globule size. The studies are conflicting considering all the other fatty acids. Our results indicate that cows producing high levels of fat and large fat globules have a high utilization of C18 from the diet. The observation of a negative correlation between MFG size and polyunsaturated fatty acids indicates that these cows have a high activity of biohydrogenation in the rumen. The observation that these fatty acids synthesized *de novo* do not correlate to MFG size, indicates that the size of MFGs is not dependent on the *de novo* fatty acid synthesis. Consequently, it is obvious that the lipid supplement in the diet has effect on the size of the MFG.

The present data from pumping raw milk with MFGs of various size clearly demonstrate that an understanding of the factors affecting the milk fat globule size is necessary. The current feeding experiment with dairy cows followed by pumping of the raw milk displays the relevant effects of diet on the stability of MFGs. The two milk types containing 3.7% and 4.0% fat with smaller MFGs were very stable when subjected to pumping flow rates regarding coalescence of MFG. The development of larger fat globules is caused by damage to the MFGM which leads to it that two fat globules stick to each other by a common membrane or with liquid fat as a sticking agent. The higher pumping stability of small MFGs compared with larger MFGs is supported by Mulder and Walstra (1974) describing the collision energy as being higher for large fat globules. Furthermore, the surface potential is higher for small globules than it is for large globules, therefore more energy is needed to coalesce small globules. In milk cooled to 5°C, the majority of the lipids are crystallized, resulting in more stable MFG when subjected to high pumping flow rates, as is shown by the smaller coalescence of MFG.

In the milk with 5.0% fat, the concentration of FFA is high in the samples without pumping treatment, which apparently may be by caused by the fact that the MFG in this milk is inherently unstable, making it more susceptible to damages during milking. Moreover and similar to the present study, Astrup et al. (1980) observed that addition of palmitic acid to the feed increased the level of FFA in the milk compared with a control diet. The present observation of maximum formation of FFA at a pumping temperature of 20°C at the two highest flow rates is comparable to older studies (Fitz-Gerald, 1974; Deeth & Fitz-Gerald,1977). A comparison between the present measurements of the particle size in the milk and the analysis of FFA indicates that damage to MFGM leads to casein absorption, and that the lipoprotein lipases associated with caseins induce lipolysis.

The use of automatic milking systems has caused an increased in the formation of free fatty acids (Justesen & Rasmussen, 2000; Klungel et al., 2000), probably due to a hasher mechanical treatment such as portions from each individual milking being continuously pumped and temperature fluctuations in the bulk tank. It is likely, that the physiological factors, such as change in milking interval also affect the content of free fatty acid. This study shows that the effect of these factors can be inhibited through manipulation of cow feeding.

References

Astrup, H.N., L. Vik-Mo, O. Skrøvseth, and A. Ekern, 1980. Milk lipolysis when feeding saturated fatty acids to the cow. Milchwissenschaft 35: 1-4.

Briard, V., N. Leconte, F. Michel and M.C. Michalski, 2003. The fatty acid composition of small and large naturally occurring milk fat globules. European journal of Lipid Science and Technology 105: 677-682.

Bitman, J. & D.L. Wood, 1990. Changes in milk fat phospholipids during lactation. Journal of Dairy Science, 73: 1208-16.

Deeth, H.C. and C.H. Fitz-Gerald, 1977. Some factors involved in milk lipase activation by agitation. Journal of Dairy Research 44: 569-583.

Fitz-Gerald, C.H., 1974. Milk lipase activation by agitation - influence of temperature. Australian Journal of Dairy Technolog, 29: 28-32.

IDF. 1991. Determination of free fatty acids in milk & milk products. International Dairy Federation. Bulletin 265.

Justesen, P. and M.D. Rasmussen, 2000. Improvement of milk quality by the Danish AMS self-monitoring programme. In: Hogeveen H. and A. Meijering (editors). Robotic Milking - proceedings of the international symposium held in Lelystad, The Netherland 17-19 August 2000. Wageningen Pers.: 83-88.

Kinsella, J.E. and G. Houghton, 1975. Phospholipids and fat secretion by cows on normal and low fiber diets: Lactational trends. Journal of Dairy Science 58: 1288-93.

Klungel, G.H., B.A. Slaghuis and H. Hogeveen, 2000. The effect of the introduction of automatic milking systems on milk quality. Journal of Dairy Science 83: 1998-2003.

Michalski, M.C., J.Y. Gassi, M.H. Famelart, N. Leconte, B. Garmier, F. Michel and V. Briard, 2003. The size of native milk fat globules affects physico-chemical and sensory properties of Camembert cheese. Lait 83: 131-143.

Mulder, H. and P. Walstra, 1974. The milk fat globule. Emulsion science as applied to milk products and comparable foods. Bucks: Commonwealth Agricultural Bureaux .

Timmen, H. and S. Patton, 1988. Milk fat globules: Fatty acid composition, size and in vivo regulation of fat liquidity. Lipids 23: 685-689.

Walstra, P. 1969. Studies on milk fat dispersion II. The globule-size distribution of cow's milk. Netherland Milk and Dairy Journal 23: 99-110.

INTRODUCTION OF AMS IN ITALIAN DAIRY HERDS: PRELIMINARY OBSERVATIONS ON MILK FAT PARAMETERS FOR LONG RIPENING CHEESE PRODUCTION WITH PRIMIPAROUS DAIRY COWS

F. Abeni[1], L. Degano[2] & F. Calza[1]
[1]Animal Production Research Institute, Cremona, Italy
[2]Institute of Dairy Research, Lodi, Italy

The introduction of automatic milking systems (AMS) in Italy must be considered at the light of the milk quality requirements for cheese-making industry, with special emphasis for typical products. Even if some rules currently do not permit to use milk from AMS for typical long ripened cheeses production, the possible fit of AMS to produce technologically adequate milk must be considered in the research. In special way, two milk fat features must be considered: fat creaming ability and spontaneous milk fat lipolysis.

Thirty-three Friesian heifers, selected for similar expected calving date, were divided into two groups after calving. The first group of nineteen cows was housed in a free stall barn with a traditional 8 + 8 herring-bone milking parlour (MP), and milked twice daily. The second group of fourteen cows was housed in a free stall barn provided with a single box for AMS (De Laval Voluntary Milking System VMS™), with a one-way traffic system. Both groups were fed once daily with total mixed rations, calculated to be similar also taking into account the concentrate supply in AMS during milking, relative to single cow production.

Milk production was automatically recorded at each milking in AMS and for two consecutive milking every week in MP. Samples collection was carried out in five jointed sessions for MP and AMS: the first one in the second decade of April; the second one in the first decade of June; the third one in the third decade of July; the fourth one in the second decade of September; and the fifth one in the first decade of November.

In MP, milk samples were collected from two consecutive milkings, from the measuring bowl. In AMS, milk was automatically sampled at each milking, throughout 48 hours, by the system provided with the robot. Milk was analysed for the following parameters: fat %; natural creaming (NC, % of total fat); free fatty acids (FFA, meq/100 g of fat) as an index of lipolytic activity.

Milk fat percentage in primiparous cows was affected by milking system (3.61 vs. 3.33% for MP and AMS respectively; $P < 0.05$). Natural creaming tends to be lower in AMS (39.74 vs. 36.11% for MP and AMS respectively; $P = 0.06$). Milk from AMS had higher values of FFA content (0.531 vs. 0.700 meq/100 g of fat for MP and AMS respectively; $P < 0.05$), according to the results of other researchers. There was a significant interaction ($P < 0.001$) between milking systems and sampling session for FFA content, with a higher value for AMS in spring, when voluntary milking frequency was higher, and no substantial differences in the warm months, when the milking frequency in AMS tend to decline close to that of MP. These preliminary results on individual milk seem to suggest a role of milking frequency *per se* as a cause of increased lipolytic activity in milk from AMS, as also evidenced in scientific works on the third milking. However, these are only preliminary results from our first controls; further session on individual milk, carried out in other periods of the year, will be necessary to clarify the consistence of actual trend.

TOWARDS SPECIFICATION IN-LINE MONITORING PRIMARY MILK PRODUCTION

P.C.F. Borsboom[1] & J. Dommerholt[2]
[1]Sensortechnology & Consultancy bv, The Netherlands
[2]CR-DELTA, The Netherlands

Udder quarter, its comparison & control are a sine qua non for easy life. Production losses on the dairy farm can be reduced, while maintaining product quality, health & welfare; research shows clearly that the key is the monitoring of udder quarters. Since milk is converted into a range of health products, monitoring at the farm should aim also at processing yield of raw milk. During milk let down, the storage time in the udder quarter is shown and, at the same time, processes like enzyme reaction and other reactions. During lactation we need to look at other natural causes: the calf itself, the reproductive phenomena and most importantly udder maintenance. Crucial management information is to be found in the time relations of both product and the processes within the quarter milking & lactation. In-line quarter monitoring must be functional, non-destructive, reliable, safe and at low cost. Non-destructive and low cost demand is nearly impossible in the culture of chemical analysis. The biophysical approach, that is, using the structure of milk itself, its properties and the phenomena derived from problems of the udder, this is probably the way to go.

The complex colloidal suspension of milk is mainly arranged in dedicated physical structures (casein micelle, average size ~ 0.1 μm, fat particles, ~ 3 μm). It is comparable to blood, where small events change the structural properties; milk is also determined by biological, pathological and biochemical processes of both the cow and udder. The structural behaviour of such translucent materials as milk can be quantified using techniques based on structure-related optical scattering behaviour described by Rayleigh & Mie and applied in bio-medical optics (Tuchin, 2002; Borsboom et al., 1988). Structural changes in milk at nano, micro or macro level are revealed by such optical behaviour as scattering, absorption, fluorescence and polarization. Because milk structures vary by and within species (Blanc, 1981), the information is extremely useful when dealing with genetic properties.

The fast developing electro/optical industry offers a range of high quality low cost components in the wavelength area UV - IR. Using dedicated optical geometries applicable for specific tasks, very adaptable sensors can be produced. Combining milk structural responses with udder quarter flow, volume & conductivity, an appropriate in-line milk production recording is feasible. Calibration and control procedures by embedded software and the use of fibre optics to transfer information to farm management systems complete the set up. A simple design for in-line monitoring applicable in a conventional milking parlour, together with fundamentals of optical metrology of milk and some results, is illustrated.

References

Blanc, B., 1981. Biochemical Aspects of Human Milk - Comparison with Bovine Milk. Wld. Rev. Nutr. Diet. Vol 36, pp 1 - 89 (Karger, Basel 1981).

Borsboom, P.C.F., J.J. ten Bosch, R.P.T. Koeman - An instrument to measure the color-determining properties of bulk translucent materials. SPIE 1988; 1012: 206-211, In-Process Optical Measurements

Tuchin, V.V., 2002. Handbook of Optical Biomedical Diagnostics, Spie Press; ISBN 0-8194-4238-0.

COMPARISON OF SEVERAL METHODS FOR QUANTIFICATION OF FREE FATTY ACID IN BOVINE MILK

T.K. Dalsgaard, & J.H. Nielsen
Danish Institute of Agricultural Sciences, Dept. of Food Quality, Foulum, Denmark

The attention on free fatty acids (FFA) in milk has increased after introduction of automatic milking systems (AMS). Several studies indicate that milk from AMS often contains high concentration of FFA giving rise to off-flavour. Furthermore recent results show that milk fat globules from high-yielding cows seem to be especially affected by mechanical stress. The dairy industry therefore demands reliable and fast analytical methods to evaluate FFA in milk.

Different methods for quantification of fatty acids have been applied in different countries, making comparison of FFA content in milk between countries difficult. In Sweden and Denmark there is a desire to standardize the method applied in the two countries. Fatty acids are quantified as total acidity with the auto-analyzer method and BDI. Both methods have problems with recovery of short chain fatty acids as butanic and hexanic acids. 60% of the butanic acid is quantified using the auto-analyzer method, while the recovery of butanic acid is reduced to 2% when quantified with BDI [IDF, 1991]. Furthermore both methods are time-consuming due to isolation of the free fatty acids prior to analysis. Thus none of the methods applied today makes a reasonable choice for a routine method. Therefore Fourier Transformed infrared spectroscopy (FT-IR) has been proposed for replacement of the two methods. For verification of FT-IR there is a need for a reliable reference method. FT-IR measurements will hence be compared with both BDI and the auto-analyzer method as well as with results obtained with gas chromatography.

INTRODUCTION OF AMS IN ITALIAN DAIRY HERDS: LONG RIPENED CHEESE PRODUCTION FROM ROBOTIC MILKING

R. Giangiacomo[1], T.M.P. Cattaneo[1], G. Contarini[1], F. Abeni[2] & G. Colzani[3]
[1]Institute of Dairy Science, Lodi, Italy
[2]Animal Production Research Institute, Porcellasco, Cremona, Italy
[3]Latteria Soresina, Soresina, Cremona, Italy

Extensive research in robotic milking should include, beside the physiological aspects of animals and their welfare, managing of the plant, quantity and quality of milk, also the suitability of milk to further processing in industrial conditions. Unfortunately not much literature is available under this point of view.

This paper reports the results of cheese making process into long ripened cheese in an industrial plant usually processing Grana Padano cheese. Since the Grana Padano standard does not include milk collection by robotic milking, this name cannot be used for cheese here obtained by this procedure. However, the technological process used in these trials refers to the Grana Padano cheese production.

Two separate batches of milk have been collected, one by automatic milking system (AMS) and one by traditional milking parlour (MP). The two batches, a little over 1 t each, have been delivered to the factory in separate containers. Milk was analysed at arrival for fat, protein, and casein content : the average content was 3.75, 3.34, and 2.72, respectively for AMS, and 3.73, 3.33, 2.71 for MP.

After natural creaming, milk was processed into Grana Padano according to the usual procedure in the factory. Fifteen cheese making processes have been carried out for each batch, giving rise to 30 loafs each. Technological parameters, cheese yield, fat content in cream, total solids in whey indicated no differences between the milk collected by the two milking procedures. Also the expert cheese maker did not notice any difference during the whole process.

Weight loss during the first 5 months of ripening proceeds correctly in both cheeses and no blowing is recorded.

Under the technological point of view these results indicate that robotic milking did not induce modifications or alteration of suitability to processing into a delicate and difficult cheese as Grana Padano.

Starting from the 6th month of ripening, samples of the two cheeses will be analysed sensorially and instrumentally to verify if the ripening process proceeds with the same pattern.

EFFECTS OF AN AUTOMATIC MILKING SYSTEM (VMS®) ON FREE FATTY ACIDS (FFA) IN DIFFERENT MILK FRACTIONS

J. Hamann[1], F. Reinecke[1], H. Stahlhut-Klipp[2] & N.Th. Grabowski[1]
[1]Department of Hygiene and Technology of Milk, School of Veterinary Medicine Hannover, Hannover, Germany
[2]Department of Biotechnology, University of Applied Sciences, Hannover, Germany

Goal of the study

It has been reported that, for reasons not totally cleared yet, the application of automatic milking systems (AMS) evokes a statistically significant increase in the free fatty acids (FFA) content of herd bulk milk. This study was performed to evaluate the influence of an AMS on the FFA content in two milk fractions: the quarter composite (QCM) and the cow composite milk (CCM).

Material and methods

A group of 31 high-yielding German Holstein Frisian cows at different lactation stages and numbers were milked robotically (Voluntary milking system VMS®; DeLaval, vacuum 42 kPa, pulsation rate 60 cycles/min, pulsation ratio 65%). The sampling session lasted over 24 h, and samples included n = 307 QCM and n = 78 CCM. Mean milking frequency was 2.52 milkings/cow/day. The milk samples were analysed for somatic cell count (SCC; Fossomatic®), fat (MilcoScan 4000®) and FFA content (Autoanalyser).

Results

The mean values of SCC, fat and FFA in QCM and CCM in relation to the different milking intervals are summarized in table 1.

Table 1. SCC [lg cells/ml], fat [%] and FFA [mmol/l] in different milk fractions at varying milking intervals.

Milking Interval		< 6 h	6 - 9 h	9 - 12 h	> 12 h
n milkings		14	37	14	13
SCC	QCM	4.81 ± 0.46	4.78 ± 0.58	4.67 ± 0.67	4.55 ± 0.46
	CCM	4.96 ± 0.43	4.95 ± 0.55	4.88 ± 0.65	4.60 ± 0.33
Fat	QCM	4.34 ± 0.85	4.06 ± 0.87	3.45 ± 0.94	3.60 ± 0.90
	CCM	4.54 ± 1.07	4.26 ± 0.85	3.77 ± 1.06	3.75 ± 1.06
FFA	QCM	0.32 ± 0.07	0.29 ± 0.10	0.28 ± 0.09	0.25 ± 0.07
	CCM	0.31 ± 0.09	0.28 ± 0.08	0.28 ± 0.08	0.24 ± 0.07

The FFA values of both milk fractions showed the highest level (0.32/0.31 mmol/l) at an interval of < 6 h. When comparing the extreme intervals (< 6 h and > 12 h), significant differences ($P < 0.01$) were found for FFA, but also for SCC and fat percentage.

Implications

With a mean level of FFA in quarter and cow composite milk samples of approx. 0.28 mmol/l, milk produced by AMS ranges within the same level as herd bulk milk from conventional bucket milking systems does. Therefore, no detrimental influences on the FFA content by applying VMS® could be observed.

SURVEY OF MILK QUALITY ON UNITED STATES DAIRY FARMS UTILIZING AUTOMATIC MILKING SYSTEMS

Jason M. Helgren & Douglas J. Reinemann
Biological Systems Engineering, University of Wisconsin, Wisconsin Madison, USA

Milk quality parameters were recorded for United States dairy farms utilizing automatic milking (AM) from August 2000 to June 2003. The study featured two primary objectives. The first was to assess seasonal variations in milk quality on AM and conventional farms. The second was to assess changes in the quality of milk from AM installations as the amount of time the system had been in operation increased. Farms were admitted to the study over the 3-year period as they began operation with the final data set including 12 AM farms. Bulk Tank Somatic cell count (SCC) and Bulk Tank Total Bacterial Count (TBC) data was analyzed and compared to corresponding data from conventional farms in Wisconsin as well as data from European AM installations.

The geometric mean SCC was 268,000 cells/ml and geometric mean TBC was 13,300 cfu/ml for all farms surveyed. A clear and significant seasonal effect was evident for the SCC data, with higher values observed during the summer months (July, August, and September). There was no significant difference in SCC between AM farms and conventional farms. There was slight evidence of a seasonal effect on TBC. TBC of milk from AM farms was found to be lower than that from conventional farms. There is some evidence that both SCC and TBC decreased as farms experience with AM increased (Figure 1).

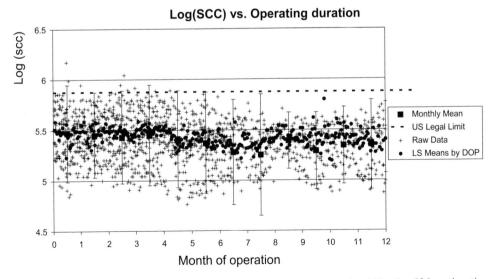

Figure 1. Log(SCC) plotted versus the length of time the farm has been in operation. DOP = Day Of Operation, the number of days the farm has operating.

MILK IODINE RESIDUES: AMS VERSUS CONVENTIONAL

M. McKinzie[1], I. Laquiere[1,] T.C. Hemling[1] & S. DeVliegher[2]
[1]DeLaval, Kansas City, Missouri, USA
[2]Veterinary Epidemiology Unit, Faculty of Veterinary Medicine, Ghent University, Merelbeke, Belgium

Post milking teat dipping has been practiced for over 40 years with iodine containing products being the most widely used. Extensive research has shown the benefit of teat dipping in the reduction of new infections. Numerous studies have also been conducted on iodine as a potential residue in milk in conventional milking systems. That research has been summarized (Hemling "Iodine Residues in Milk", XXI World Buiatrics Congress, Punta del Este Uruguay, December 4-8, 2000.) and shows that iodine is converted to the non-antimicrobial iodide in milk, and that post milking teat dipping may contribute up to about 150 ppb of iodide to milk. The use of an automated milking system could possibly provide new risk factors that affect the level of iodine residues. Under a voluntary milking system the cows will be milked more frequently than under a conventional milking system and the pre-milking teat preparation procedure is automated. We report here data on iodine residues from milk collected during a trial evaluating udder health and teat condition during the transition from conventional to AMS milking while using a high emollient, low iodine teat dip (Proactive Plus™, DeLaval).

Forty lactating cows and heifers from a high yielding dairy herd were randomly allocated to a control and AMS group (VMS™, DeLaval) each group containing both heifers and multiparous cows. In the VMS group, robotic milking was initiated during the study period, while in the control group conventional milking was continued during the whole study period. Milk samples were collected during the pre-trial period and after conversion to AMS. Iodine levels were determined by HPLC method at the (Food and Dairy Chemistry Laboratory, University of Guelph, Guelph, Ontario, Canada).

Results showed low iodine levels in both the Conventional and AMS groups for both the trial and pre-trial period (change in iodine concentration, pre-trial vs. trial, was not statistically different between groups, p=0.50). There were no differences observed between treatments. The iodine levels are well below the recommended 500 ppb level. The low iodine levels are likely the result of the low iodine content (0.15%) of the Proactive Plus teat dip and good teat washing by the VMS during the pre-milking procedure.

	Iodine concentration in milk (ppb)	
Treatment	Conventional milking	AMS
Pre-trial	77.8	72.4
Trial	74.6	75.1

WELFARE

IMPLICATIONS OF AUTOMATIC MILKING ON ANIMAL WELFARE

Hans Wiktorsson[1] & Jan Tind Sørensen[2]
[1]Department of Animal Nutrition and Management, Swedish University of Animal Sciences, Uppsala, Sweden
[2]Department of Animal Health and Welfare, Danish Institute of Agricultural Sciences, Foulum, Denmark

Abstract

The welfare of dairy cows depends on several factors related to social interaction between cows, the management and the environmental conditions. In addition to milking, cow traffic systems and feeding are important issues. While free cow traffic gives most freedom to the cows, long and irregular milking intervals among individual cows counteract the success. Forced traffic results often in unsatisfactory queuing in front of the milking unit (MU) and too few feeding visits. Controlled traffic, which allows access to some basic feeds between milkings without passing the MU, results in better overall welfare conditions. The cows' interest to go to the waiting area in front of the MU depends on the time since last feeding and the number of cows already in the waiting area. Low-ranked cows have to wait longer than the high-ranked when queuing, and generally choose time periods when the high ranked are less active. Although there are small possibilities for them to act voluntarily and in a synchronised way, they cope well with the environment, and studies on stress-related hormones did not revile that they suffer. However, differences in the ability of individual cows, independent of ranking order, to adjust to different degree of guiding calls for management systems with possibilities of individual settings. A concept for assessing animal welfare at herd level has been developed as a tool for the AM-herd, including information on the systems, systems application, animal behaviour and animal health status. The concept has been applied on 8 AM-herds during a year. Each herd received an annual report, which were discussed at the farm. The report included a short overview, a list of indicator values with reference to other results from other farms, and a detailed documentation of raw data from the farm. Major herd differences were found for many indicators emphasising the management impact on animal welfare. The farmers in general appreciated the system as a relevant decision support system.

Introduction

The welfare of dairy cows depends on several factors related to the cows themselves and their social interaction with other cows in the herd, as well as with the manager and management system, feeding and nutritional supply, barn design, cow traffic systems and climatic and other environmental conditions. Compared with conventional dairy cow management, cows in automatic milking (AM) systems are provided more freedom to choose their daily activities and rhythms. In herds with AM there is an interaction between feeding and milking, which may affect animal welfare. In order to optimise the application of the milking unit (MU), feeding is used as a tool, often in relation to regulation of cow traffic. It is assumed that the welfare impact on regulation of feeding and cow traffic in particular may be a welfare problem for low-ranking cows. It is therefore relevant to focus on low- and high-ranking cows when analysing the welfare aspects of feeding and milking management in an AM herd. At herd level, animal welfare is a result of a complex interaction between the production system and management. Thus, the AM herds need also tools that can be used as decision-support tools for improving animal welfare at herd level. Application of such tools, including measurements of welfare indicators at herd levels, can provide us with knowledge on relationships between welfare indicators and between indicators and production, which will increase our knowledge of the benefits of animal welfare in AM herds.

This paper highlights recent results on factors affecting the ability of cows in AM systems to maintain normal feeding behaviour, diurnal rhythm and normal resting and social behaviour, and further describes a system of assessing animal welfare at herd level, followed by experiences obtained by application of the welfare assessment system on 8 AM herds during a year.

Welfare aspects of cow traffic and social interactions in automatic milking systems

Cow traffic systems

In AM systems there is a need to co-ordinate and optimise the milking and feeding systems in such a way that all cows, the subordinate as well as the high-ranked, have access to the feeding area and the MU a sufficient number of times per day. From a cow's point-of-view feeding is given higher priority than being milked (Prescott *et al.*, 1998). Therefore, cows get at least a small amount of concentrate during milking to tempt them to visit the MU. However, this is often insufficient to obtain regular milking intervals among the cows in a herd.

In principle, three types of cow traffic systems have been developed and come into practice; *forced traffic, controlled traffic* and *free traffic*. At *forced cow traffic* the cows have to pass the MU to get access to the feeding area. On their way from the feeding area to the resting area they have to pass a one-way gate, which prevents passage in the opposite direction. *Controlled traffic* is sometimes called free traffic with pre-selection, semi-forced traffic or guided traffic. A commonly used controlled traffic system allows the cows to go straight to the feeding area until they are given milking permission (shortest time between two regular milkings), at which time they are guided back to the common area, or by means of selection gates brought to a closed waiting area in front of the entrance to the MU. *Free cow traffic* allows the individual cow to visit the feeding area as frequently as they want. Free cow traffic gives the most freedom to the cow. The problem with this system is that cows may not pay enough visits to the MU and that many cows have to be fetched. About 15% of all milking occasions were implemented after the cows had been fetched, as shown by Forsberg *et al.* (2002) and Harms *et al.* (2002) (Table 1). In Exp. A over-due cows were fetched after they exceeded 14 hours since their last milking on four occasions per day, and in Exp. B five times during the day if they had "exceeded planned milking time for more than 3-4 hours". Still, on the free cow traffic system 33 percent of all milking intervals were longer than 12 hours. Long and irregular milking intervals may result in udder disturbances, increased somatic cell count in the milk and sometimes mastitis. In addition, the production level will be negatively affected. Harms *et al.* (2002) had also a treatment with completely forced traffic. The milking parameters did not differ from the treatment with forced traffic with pre-selection shown in table 1, but the number of feeding visits was reduced when the level of restrictions in cow traffic increased. Another negative factor with forced cow traffic was that the number of cows queuing in front of the MU was increased. The same was observed by Thune *et al.* (2002).

Table 1. Number of milkings, milking interval and fetched milkings at free cow traffic and controlled cow traffic (forced traffic with pre-selection). Forsberg et al., 2002; (Exp. A), and Harms et al., 2002 (Exp. B).

Parameter	Exp	Free traffic	Forced traffic with pre-selection
No. of cows	A	45	46
	B	49	46
No. of milkings per cow per day	A	2.34[a]	2.63[b]
	B	2.29[a]	2.56[b]
Milking interval	A	10.4[a]	9.3[b]
	B	10.2	9.3
Fetched milkings, %	A	14.5[a]	2.6[b]
of all milkings per day	B	15.2[a]	4.3[b]

[a, b,] Different superscripts within row denote significant differences

Adaptation to control gates

In AM herds, two main types of control gates are used. In a routing gate the cow is identified some distance ahead of the gate, the gate makes a decision and sets her route, and the cow never needs to halt on her passage through the selection unit. In the other type of control gate, the cow is identified and the gate opens if she fulfils the criterion for opening, otherwise she has to back out of the selection unit. This type of control gate often closes rather quickly to stop cows from hitchhiking through the gate. A study on 5 farms (Olofsson *et al.*, 2001) showed that cows need training to understand and adapt to this type of gate. The study was conducted 2 - 3 months after the introduction of the gate and less than 40% of the cows used the gate regularly. Usage of the gate varied between the farms. On one farm that had trained the cows, more than 60% of the cows used the gate daily (Table 2). An additional cause of the low frequency was the location of the gates.

Table 2. Usage of a by-pass gate in 5 herds (Olofsson *et al.*, 2001).

Herd	Regular usage of a by-pass gate (% of cows)
A	33
B	39
C	30
D	61
E	41

cows. That was also found by Ketelaar-de Lauwere et al. (1996) in the daily time budget of high-ranked and low-ranked cows.

Behavioural and physiological stress response for cows with high or low social rank

Cows in an AM system are expected to visit the milking unit voluntarily. However, since a herd consists of both low- and high-ranked cows, it is likely that there will be different milking patterns depending on the ranking order. A diurnal variation between high- and low-ranked cows visiting the milking unit has been observed, where low-ranked cows were milked more frequently during the night (Ketelaar-de Lauwere et al., 1996).

Animal welfare can be measured in several ways. One way is to measure the circulating levels of hormones related to the stress response, such as cortisol and oxytocin. Cortisol is usually related to stress, while oxytocin is related to anti-stress (Johansson et al., 1999). Hopster et al. (2001), who studied the adrenaline and cortisol levels in primiparous cows after queuing in front of the MU in an AM system, found no differences in hormonal levels that indicated stress. They suggested that the gaining of access to the MU by primiparous cows neither provoked overt physiological stress responses nor compromised milk ejection.

Svennersten-Sjaunja and Olofsson (pers. communication; Wiktorsson et al., 2003) measured the cows' hormonal response (cortisol and oxytocin), and performed observations of their behaviour during milking and resting. The six highest-ranked and the six lowest-ranked cows in the AM herd at Kungsangen RC were selected. Eight of the cows were provided with semi-permanent catheters in the jugular vein in order to enable collection of blood samples during resting and milking. From the results it was observed that the low-ranked cows had the highest levels of oxytocin during both milking and resting. The low-ranked cows also had the lowest levels of cortisol in their blood. The results therefore indicate that the low-ranked animals were less stressed, which was not an expected finding. Interestingly, the highest-ranked cow also had the highest level of oxytocin. The behavioural studies showed that the low-ranked cows had a more efficient eating pattern with less time spent in the eating area and fewer visits made to the feed troughs. However, sufficient amounts of feed were consumed. Instead of spending time in the eating area or in the alleys, these cows spent more time standing in the cubicles, suggesting that it was a more relaxing environment. The low-ranked cows were found to be closer to the milking station, especially during resting, indicating a need for them to monitor the milking queue.

Use of welfare assessments as an on-farm decision support tool

A concept for assessing animal welfare at farm level has been developed as a decision support tool for the AM dairy farmer (Sørensen et al., 2002; Rousing et al., 2004). In this concept animal welfare should mirror that positive and negative experiences matter, from the animal's point-of-view. Experiences of the animals cannot be measured directly. They have to be assessed indirectly. Two kinds of information may be relevant: 1) Information about the system and how it is managed, and 2) information on how animals respond to the way they are kept and the way they are treated. Each source of information can be subdivided into four sources of information: the system, systems application, animal behaviour and animal diseases as shown in figure 1. The system and systems application provide information on risk factors for welfare problems. Direct measures on the animals provide information on the animals response to the environment. In an on-farm situation

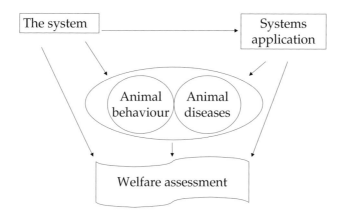

Figure 1. Sources for assessing animal welfare on an animal farm (Sørensen et al., 2001).

all four data sources provide valuable information on the assessment of animal welfare at farm level.

The welfare indicator

Aggregating welfare indicators into a welfare assessment protocol calls for a systematic procedure, as described by Rousing *et al.*, (2001). Three steps are suggested:
1. All suggested indicators should be thoroughly evaluated for their independent relevance to animal welfare.
2. Considering that we have information on all the indicators on the list except for the indicator in question, we can evaluate the marginal increase in information adding this indicator to the list. For example; observation of abnormal getting up behaviour in cows may be supplemented or replaced by clinical registration of pressure sores and hock lesions. However, behavioural observations might provide important information when investigating the probable cause of pressure sores and/or hock lesions.
3. The feasibility of the suggested indicators for on-farm studies (investigate each of the indicators for suitability for on-farm studies) are important. This evaluation relates to time and costs when carrying out registrations or tests. Selection of an indicator depends on whether information is already routinely available or the information can be obtained as a supplement to ordinary consultations by, e.g. veterinarians or husbandry advisors. This third step regards developing methods and tests for use on farms.

A protocol describing a full set of indicators for assessing animal welfare on an AM-dairy farm has been developed (Hindhede *et al.*, 2002). Each indicator is described in terms of independent value, marginal value and suitability for on-farm use. The protocol documents the current measures included in a welfare assessment system. It is also a research tool for developing operational assessment systems with different resource demands. Behaviour studies are conducted 6 times during a year, focusing on: man-animal relationship, behaviour at/in AMU, getting up behaviour, resting behaviour, social behaviour, diurnal behaviour pattern and usage of the stable (Rousing *et al.*, 2004).

Disease can be regarded as an important welfare indicator, because it is in many cases associated with negative experiences such as pain, discomfort or distress. The disorders that have the greatest impact on welfare are either acute disease processes, causing suffering, or long-term progressive conditions involving chronic pain. One indicator in a welfare assessment at farm level may be the prevalence and intensity of certain health problems in the herd. A protocol for systematic clinical examination is developed, focussing on important welfare aspects. As a supplement to these important indicators for disease, incidence and death are included. Welfare indicators based on regular clinical examinations are measured 6 times a year focussing on: hoof and leg disorders, lameness, skin lesions, udder infections, body condition, and clinical diseases.

The welfare of farm animals is affected by the production system itself as well as the way the individual farmer applies the system. Knowledge of how system and management might affect the animals can be included in a welfare assessment system and provide information of risk of welfare problems, as well as causal factors. Any strategy requiring system and management routines to be recorded will have certain limits and pitfalls. Although different aspects of these indicators have been studied under experimental conditions, there is still considerable ignorance of the effect on welfare of a number of minor features in different housing systems. Furthermore, interactions between different factors are currently poorly understood. The marginal welfare information value is typically low, so there is still a need for a strategy that focuses directly on the livestock response. Most system indicators and some management indicators are reasonably easy to define and measure, whereas several management indicators are difficult to assess but nevertheless have a serious impact on animal welfare. Surveying housing system and housing equipment as well as interviews with the farmer seems to be relevant methods of measurement.

An expert panel evaluation on the relevance of the indicators concerning behaviour, health, system and management showed that the independent relevance of the selected indicators were found to be high (Rousing et al., 2004).

The welfare assessment report

The results of welfare assessment are presented to the farmer as a welfare assessment report. Details on how to construct such a report are described and discussed by Bonde et al. (2001) in general and for an AMS-dairy herd in particular (Hindhede et al., 2004). The welfare assessment system provides three types of information to the farmer depending on the aim of the assessment, i.e., to give an overview, to give an evaluation, and to give a full documentation. The overview should give the farmer a clear picture of the actual welfare status of the farm. This is a prerequisite when determining the priority of animal welfare considerations in a whole farm framework. It consists of a short summary of the finding of the welfare assessment (Hindhede et al., 2004). The farmer often requests the evaluation and interpretation of welfare on the farm. This evaluation may rely on a comparison with results from similar production systems, with previous results from the herd, or predefined goals set by the farmer. Further, recommended guidelines may be available for some indicators (for example the length and width of cow cubicles), and therefore may be relevant as a reference for evaluation. Documentation is important for linking the conclusion and evaluation to the exact measurements on the farm. In order to accept that the conclusions are based on farm specific circumstances, the farmer needs documentation in terms of familiar recordings from the farm. All results may be printed in the report or at least the results indicating welfare problems in the herd.

A full set of welfare indicators from the protocol has been measured during a year on 8 AMS dairy herds. Major differences were found between herds in many welfare indicators. To illustrate the variation result from a fear test is given in figure 2. Six times during a year 30-40 cows on each farm were tested by a person approaching each cow in the cubicle system and the avoidance distance measured in 5 categories as described by (Rousing, 2003). As appears from figure 2, the percentages of cows avoiding man differed from 5 to 62%.

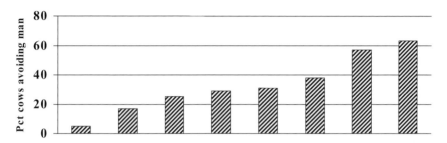

Figure 2. Average percent of cows avoiding man in an avoidance test in the walking areas (mod a. Hindhede et al., 2003).

The role of welfare assessments in a decision support framework

It is useful to distinguish between decisions made at the operational level (day-to-day management), tactical planning during the coming year or season, and strategic planning which may involve major investments. Operational management in terms of daily routines often has a considerable effect on animal welfare. Operational management might often be changed or modified with relatively little change in input and with relatively small impact on other aspects on the farm. Experience from discussions with the AM-farmers in relation to the welfare assessment report revealed that they found the reports relevant as a decision support tool (Hindhede *et al.*, 2004). The farmers gave, e.g., the following interesting responses:

- Appreciated the possibilities for comparisons with other herds with AM, prior observations on the herd, as well as a defined goal on the farm or the adviser's judgement as frame of reference for welfare assessment.
- Agreement on welfare assessment (determining causal relations and exclusion of false supposed reasons) being strengthened by combining/linking different results concerning behaviour, clinical examination, housing system and management.
- Appreciation of conclusions with suggestions for optimising animal welfare.
- Suggestions of new indicators concerning habituation, etc., as replacement for others.

Tactical management in terms of plans for replacement and reproduction management may influence animal welfare, but will typically also have major impacts on effects on productivity and production level in an AMS dairy herd. An example of a tactical change that influences animal welfare and production is a change from zero grazing to a pasture period in the summer. It may be necessary to collect cows for milking several times a day from pasture, which will increase the time queuing for milking.

Strategic management, e.g. to increase the herd size and invest in a second or third AMU, may have a major impact on animal welfare. Since such decisions will influence the whole farm's economy as well as having environmental impacts, it is important to analyse the consequences of these types of actions for all stakeholders on an animal farm. Suggestions for a procedure for this activity are given by Sørensen *et al.* (2001).

Following the data-recording period, a full welfare assessment report was presented to the farmer. Based on this report, operational management changes were discussed. It is suggested that the farmer uses the information from annual reports as input for a strategy planning process when having received information from a 2-3 year period as input for a strategic planning process (Hindhede *et al.*, 2004).

Acknowledgement

Special thanks to co-workers and co-authors of papers upon which most of the present paper is based.

References

Bonde, M., T. Rousing and J.T. Sørensen, 2001. The welfare assessment report. Overview, documentation and evaluation. Acta Agriculturae Scandinavica, Section A, Animal Science Supplementum 30, 58-61.

Forsberg, A-M., G. Pettersson and H. Wiktorsson, 2002. Comparison between free and forced cow traffic in an automatic milking system. Proc. Technology for milking and housing of dairy cows. NJF seminar No 337: 11 - 13 Febr. 2002, Hamar, Norway, 25.1 - 25.6.

Harms, J., G. Wendl and H. Schön, 2002. Influence of cow traffic on milking and animal behaviour in a robotic milking system. The First North American Conference on Robotic Milking. March 20-22,2002, Toronto, Canada, II: 8 - 14.

Hermans, G.G.N., A.H. Ipema, J. Stefanowska and J.H.M. Metz, 2003. The Effect of Two Traffic Situations on Behavior and Performance of Cows in an Automatic Milking System. Journal of Dairy Science.Vol 86, 6: 1997-2004.

Hermans, G.G.N., M. Melin, G. Pettersson and H. Wiktorsson, 2004. Behavior of high-and low-ranked dairy cows after redirection in selection gates in an automatic milking system. Proceedings from the International Symposium Automatic Milking March 24-26 2004 Lelystad.

Hindhede, J., T. Rousing, C. Fossing and J.T. Sørensen, 2002. A protocol for assessing animal welfare in an automatic milking system. Report produced within the EU project "Implication of the introduction of automatic milking on dairy farms" (QLK5-2000-31006) as part of the EU-program "Quality of Life and Management of Living Resources". Deliverable D23, February 2002, 18 pp. Available at http://www.automaticmilking.nl

Hindhede, J., T. Rousing, I. Klaas and J.T. Sørensen, 2003. Velfærdsvurderingsrapport for en malkekvægsbesætning med autiomatiks malkning. Afd. For Husdyrsundhed og Velfærd DJF 2003, 43 pp.

Hindhede, J., T. Rousing, I. Klaas and J.T. Sørensen, 2004. A welfare assessment report as a decision support tool in an AMS herd: Practical experience. Proceedings from the International Symposium Automatic Milking March 24-26 2004 Lelystad

Hopster, H., J.T.N. van der Werf and G. van Reenen, 2002. Impact of Queuing for Milking on Heifers in Robotic Milking Systems. The First North American Conference on Robotic Milking. March 20-22,2002, Toronto, Canada, VI: 24-31.

Johansson, B., I. Redbo and K. Svennersten-Sjaunja, 1998. Effect of feeding before, during and after milking on dairy cow behaviour and the hormone cortisol. Animal Science 68: 597 - 604.

Ketelaar-de Lauwere, C.C., S. Devir and J.H.M. Metz, 1996. The influence of social hierarchy on the time budget of cows and their visits to an automatic milking system. Applied Animal Behaviour Science, 49: 199 - 211.

Olofsson, J., G. Pettersson, H. Wiktorsson, T. Ekman and M. Sundberg, 2001. Teknisk provning av DeLavals automatiska mjolkningssystem (VMS™). Report 249. Department of Animal Nutrition and Management, Swedish University of Agricultural Sciences, Uppsala, Sweden.

Prescott, N.B., T.T. Mottram and A.J. Webster, 1998. Relative motivations of dairy cows to be milked or fed in a Y-maze and an automatic milking system. Applied Animal Behaviour Science, 57: 23 -33.

Rousing, T. 2003. Welfare assessment in dairy cattle herds with loose-housing cubicle systems. Development and evaluation of welfare indicators. Phd thesis DIAS Report 45, 101 pp.

Rousing, T., J. Hindhede, I. Klaas and J.T. Sørensen, 2004. Assessing animal welfare for decision support in an AMS herd: An expert evaluation Proceedings from the International Symposium Automatic Milking March 24-26 2004 Lelystad

Rousing, T., M. Bonde and J.T. Sørensen, 2001.How to aggregate welfare indicators into an operational welfare assessment system: a bottom up approach. Acta Agriculturae Scandinavica, Section A, Animal Science Supplementum 30, 53-57.

Sørensen, J.T., J. Hindhede, T. Rousing and C. Fossing, 2002. Assessing animal welfare in a dairy cattle herd with an automatic milking system. In: Proc. First North American Conference on Robotic Milking, March 20-22, Toronto, Canada, VI-54- 59.

Sørensen, J.T, P. Sandøe and H. Halberg 2001. Animal welfare as one among several values to be considered at farm level: The idea of an ethical account for livestock farming. Acta Agriculturae Scandinavica, Section A, Animal Science Supplementum 30, 11-17.

Thune, R.O, A.M. Berggren, L. Gravås and H. Wiktorsson, 2002. Barn layout and cow traffic to optimise the capacity of an automatic milking system. The First North American Conference on Automatic Milking, March 20-22, 2002, Toronto, II: 45 - 50.

Wiktorsson, H., G. Pettersson, J. Olofsson, K. Svennersten-Sjaunja, and M. Melin, 2003. Welfare status of dairy cows in barns with automatic milking. Relations between the environment and cow behaviour, physiologic, metabolic and performance parameters. Deliverable D24. Report within EU project QLK5-2000-31006. Available at http://www.automaticmilking.nl

A WELFARE ASSESSMENT REPORT AS A DECISION SUPPORT TOOL IN AN AMS HERD: PRACTICAL EXPERIENCES

Jens Hindhede, Tine Rousing, Ilka Klaas & Jan Tind Sørensen
Danish Institute of Agricultural Sciences, Department of Animal Health and Welfare, Foulum, Denmark.

Abstract

A welfare assessment report has been developed and applied during a year in eight dairy herds with automatic milking systems. The welfare assessment seems to be accepted as a "production goal" for the dairy farm similar to productivity, product quality, environmental impact etc. The welfare assessment system seems to be a convenient decision support system for optimising animal welfare in herds with AMS. The system may have importance as input for a strategy planning process when having received information from annual reports from a 2 - 3 year period.

Background

During the last few year farmers, consumers, politicians and scientists have increased their interest in animal welfare in agricultural systems with animal production. Animal welfare (AW) can vary substantially between similar production systems indicating the major influence of management. In order to monitor and improve AW the farmers need a decision support tool providing her/him with information on AW on the farm. Animal welfare cannot be measured directly but needs to be assessed through indirect indicators. Methods for assessment of AW have been developed at the Danish Institute of Agricultural Sciences for different animal production systems (Sørensen et al., 2003, Rousing, 2003). In the Danish systems the farmer received information on animal welfare in an annual report, which immediately after finalisation was presented to and discussed with the farmer.

Recently, a system for assessment of cow welfare in herds with Automatic Milking System (AMS) has been developed (Sørensen et al., 2002, Rousing et al., 2004, Wiktorsson & Sørensen, 2004). The proportion of welfare indicators, referring to information on housing system, management routines, dairy cow behaviour and health, is described by Rousing et al. (2004). This system has been applied/tested in 8 Danish dairy herds with AMS. This paper aims briefly to present the concept of the annual report in the welfare assessment of an AMS-herd and further to discuss experiences gained by applying it in the 8 herds emphasising the farmers perception of the report as a Decision Support Tool (DST).

The welfare assessment report

An important and integrated part of the welfare assessment (WA) is the WA-report for communication with the farmer. Bonde et al. (2001) discussed the structure of such a report, intended as a decision support tool for the farmer in animal welfare considerations on the farm by providing:

- An overview, which allow priorities to be set among welfare-motivated interventions on the farm.
- An evaluation of the present welfare situation in the dairy herd
- Relevant data from exact observations in the herd as a documentation of the welfare situation

The developed WA-report for an AMS-herd contains the following seven chapters:

1. Background and purpose of the welfare assessment on herd level. A short guide for studying the welfare assessment report
2. Summary and conclusions for the farm. Results from behavioural observations and tests, clinical health examinations and descriptions of the system and management are summarised and interpreted separately as well in combination
3. General description of the indicators: Their expected influence on animal welfare and the observation method
4&5. Two chapters contain results concerning behaviour respectively clinical diseases for all 8 farms. The averages for all 6 observations during the year was presented to facilitate a comparison to other farms with AMS
6. Main results concerning system and management on-farm compared to available standards or recommended guidelines
7. Farm specific recordings concerning behaviour and health of the 6 repeated sessions showing the development through the year and seasonal changes in animal welfare. Detailed and specific results concerning system and management on-farm compared to available standards or recommended guidelines

The WA-report from each herd was made anonymous prior to publishing.

Application of the welfare assessment report in eight private herds with AMS

The chronology for applying the welfare assessment as a decision support tool in AMS-herds is shown in figure 1, beginning with measurements and test on farm, preparing an annual WA-report, discussing the report results with the farmer and his staff and - if relevant - changing operational management. More radical changes may be relevant on a strategic level.

The farms
Eight herds participated in a one-year study and all operated with Lely Astronaut milking units, hereof three herds with two units and five herds with a single unit. Herd sizes ranged from 60 - 120 Holstein Friesian cows. All herds practised free cow traffic. In 5 of the herds cows were grazing moderate amounts in the summer. All herds had at least two years experience with AMS.

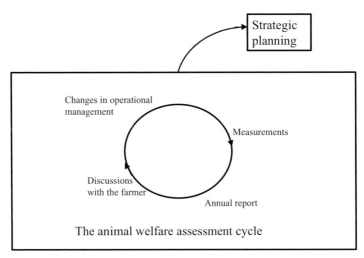

Figure 1. The annual welfare assessment report as a decision support tool.

Recording frequency and data collection

The observation period was one year and the registrations on each farm were carried out every second month or 6 times a year.

Behaviour measurements and behaviour tests were performed to show whether the animals were adapting to the production system or whether the animals showed any signs of strain. Animal behaviour was recorded through 6 two-days visits by a technician and included avoidance test of about 40 cows randomly assigned.

Clinical examination was performed in 6 half-day visits by a vet. During each visit 40 - 60 cows were randomly assigned for clinical examination (udder and teat conditions, body condition scores, lameness, pressure lesions, hygiene etc.). Cows tested in the avoidance test were also selected for clinical examinations.

Measurements concerning system and management were regularly updated by the technicians and supplied with information from farmer interviews.

All the data were collected, analysed, evaluated and presented to the farmer in a readable and problem-oriented WA-report containing 40 - 50 pages.

Presentation of the WA-report to the farmer

The WA-report was sent to the farmer about one week in advance of a visit lasting 2 - 3 hours, where the report was presented to the farmer and the staff involved in the milk production by the scientists and technicians involved in data recording and preparing the report.

Three specialists presented the WA-report, each focusing on her/his expertise: in animal behaviour, clinical diseases and animal husbandry in general (housing, feeding, technology etc.). Comments from the farmer and the staff in order to remove any uncertainty had high priority. All comments were discussed and the final report adjusted after the visit.

Results from the WA-report

In the following some results from the report (extracts from tables) will be presented.

Results from the avoidance test, where a person approached individual cows revealed a significant difference between the 8 herds in the proportions of cows avoiding man, namely from 5 to 63% (Table 1).

Table 1. Avoidance test in 8 AMS-herds. Cows avoiding man in the walking area, %.

Herd number	1	2	3	4	5	6	7	8
Cows avoiding man in the walking area, %	57	17	38	21	25	5	29	63

The prevalence of lameness in the eight herds differed from 10 to 45% (Table 2).

Table 2. Prevalence of lameness in 8 AMS-herds.

Herd number	1	2	3	4	5	6	7	8
Cows with lameness, %	13	17	14	18	21	11	35	10

Table 3 show the prevalence of cows with severe pressure lesions on the hock for each of the eight herds. The prevalence ranged from 10 to 35% of the cows.

Table 3. Prevalence of cows with severe pressure lesions on the hock in 8 AMS-herds.

Herd number	1	2	3	4	5	6	7	8
Cows with severe pressure lesions on the hock, %	7	26	0	3	4	3	9	2

Herd 2 had the highest prevalence of cows with severe pressure lesions, 26%. The seasonal variation is illustrated in table 4. The level is particularly high in March and May. Discussion with the farmer revealed, that the pressure lesions in the herd developed already for heifers, when kept on slatted floor without a bedded/soft resting area.

Table 4. Prevalence of cows with severe pressure lesions on the hock during season in an AMS-herd.

Season	Jan.	March	May	July	Sept.	Nov.
Cows with severe pressure lesions on the hock, %	12	32	46	25	20	15

The hygiene level in the herds was described as the percent of cows seriously dirty (manure). Some herds had clean cows whereas in others herds 22% of the cows were thoroughly dirty (Table 5).

Table 5. Percent cows seriously dirty (manure) in 8 AMS-herds.

Herd number	1	2	3	4	5	6	7	8
Cows seriously dirty, %	4	10	1	7	15	3	7	22

An extract from the system description is seen in table 6. Welfare indicators, referring to detailed information on housing system revealed often inappropriate dimensions of cubicles/stalls as well as corridors/passage areas. In combination with inconvenient floor type, -material, -surface and -condition, these too narrow passages with dirty and slippery floor loaded especially low ranking cows and the cow traffic in general.

Table 6. System description. Cubicles dimension and floor.

	Measured	Recommended
Number of cubicles	2 x 69	1 cubicle per cow
Floor type	Madras	Deformable
Length, cm	246	260 towards wall
Width, cm	112	120

Farmers response on the WA-report

Comments and questions as well as the farmer's perceptions of the WA-report as a relevant decision support tool were collected and will be presented in the following.

- Agreement on welfare assessment (determining causal relations and exclusion of false supposed reasons) as being strengthened by combining results concerning behaviour, clinical examination, housing system and management.
- Management is crucial for animal welfare in the milk production, but it is very difficult to assess, which means WA-AMS is important as a decision support tool
- Some farmers like to have behavioural as well as health information of the same cows
- Information on "Getting up behaviour" seems important when prevalence of pressure lesions and cubicles dimension are in focus
- Suggestion of new indicators such as queuing at the AMU, habituation to the AMS etc. as replacement for others considered as less important
- Further ethical aspects could be emphasised in the WA. For example, it will imply unnecessary killing of a foetus, when breeding cows - already determined to be slaughtered - to avoid disturbances from the cow in heat.
- Farmers appreciated the possibilities to compare their own results with those of the other herds with AMS
- Farmers appreciated a defined goal on the farm or the adviser's judgement as frame of reference for welfare assessment. However, the farmers asked often for 'acceptable' standards

- Farmers appreciated the presented conclusions and the suggestions to optimise animal welfare
- Acknowledgement of the attribution of the WA to reveal well-known problems and as such support the motivation and decision for changes in system or management
- Farmers were often surprised by the results from the avoidance test. The level as well as the variation between herds in cows avoiding man seems not to differ serious from the findings of Rousing (2003) in conventional loose housing systems for dairy herds with manual milking systems. But there were often significant deviations in farmer perception of the importance of these results. Some mentioned, that "cow-temper" seems important for the voluntary cow traffic in the AMS. One farmer was convinced that the herdsman always has to give way for the cows.
- Experiences from WA-AMS may be appropriate to include in building a new milk production system or when reconstructing an existing production system
- The WA-AMS should be reduced and more concentrated - fewer indicators and/or lower test frequency.
- The farmers had preference for unbiased advisers (independent of/not related to the farm) in the WA measurement, analysis, preparing the report and presentation on farm
- The systematic clinical examination was regarded as quick and suitable, because animal health data on farm is often incomplete
- Even though the farmer could perform most of the WA-AMS him-/herself, he/she would prefer external persons to ensure, that the WA will be performed, and because it is difficult to control oneself

The conclusion from our experience with applying the welfare assessment report in eight dairy herds with AMS is:
- The WA-AMS seems to be accepted as a "production goal" for the dairy farm similar to productivity, product quality, environmental impact etc.
- The AW-AMS seems to be a convenient decision support system for optimising animal welfare in herds with AMS
- The AW-AMS may have importance as input for a strategy planning process when having received information from annual reports from a 2 - 3 year period

References

Bonde, M., T. Rousing and J.T. Sørensen, 2001. The welfare assessment report. Overview, documentation and evaluation. Acta Agriculturae Scandinavica, Section A, Animal Science Supplementum 30, 58-61.

Hindhede, J., Rousing, T., Fossing. C. & Sørensen, J.T., 2002. A protocol for assessing animal welfare in an automatic milking system. Report produced within the EU project "Implication of the introduction of automatic milking on dairy farms" (QLK5-2000-31006) as part of the EU-program "Quality of Life and Management of Living Resources". Deliverable D23, February 2002, 18 pp. Available at http://www.automaticmilking.nl

Hindhede, J., T. Rousing, I. Klaas and J. T. Sørensen, 2003. Velfærdsvurderingsrapport for en malkekvægsbesætning med automatisk malkning (Anonym). Afd. For Husdyrsundhed og Velfærd DJF 2003, 43 pp.

Rousing, T. 2003. Welfare assessment in dairy cattle herds with loose-housing cubicle systems. Development and evaluation of welfare indicators. Phd thesis DIAS Report 45, 101 pp.

Rousing, T. & Waiblinger, S. 2003. Fear testing in commercial dairy herds: test reliability and effect of test person. Applied Animal Behaviour Science (In press).

Rousing, T., J. Hindhede, I. Klaas and J.T. Sørensen, 2004. Assessing animal welfare for decision support in an AMS herd: An expert evaluation. In: Proceedings from the International Symposium Automatic Milking March 24-26 2004 Lelystad

Sørensen, J.T., J. Hindhede, T. Rousing and C. Fossing, 2002. Assessing animal welfare in a dairy cattle herd with an automatic milking system. In: Proc. First North American Conference on Robotic Milking, March 20-22, Toronto, Canada, VI-54- 59.

Sørensen, J.T., Bonde, M.K., Møller, S.H. & Rousing, T., 2003. A concept for assessing animal welfare at farm level as a farm management tool. In: Proc. 10th International Symposium of Veterinary Epidemiology and Economics (on CD-rom), 17-21 November, Vina del Mar, Chile, Abstract. No. 728, 3 pp.

Wiktorsson, H. & J.T. Sørensen, 2004. Implication of automatic milking on animal welfare. In: Proceedings from the International Symposium Automatic Milking March 24-26 2004 Lelystad

RELATIONSHIPS BETWEEN TIME BUDGETS, CORTISOL METABOLITE CONCENTRATIONS AND DOMINANCE VALUES OF COWS MILKED IN A ROBOTIC SYSTEM AND A HERRINGBONE PARLOUR

D. Lexer[1], K. Hagen[1], R. Palme[2], J. Troxler[1] & S. Waiblinger[1]
[1]Institute of Animal Husbandry and Animal Welfare, University of Veterinary Medicine Vienna, Vienna, Austria
[2]Institute of Biochemistry, University of Veterinary Medicine Vienna, Vienna, Austria

Abstract

In the present study we investigated the relationships between time budgets, concentrations of cortisol metabolites in the faeces, and dominance values in two groups of 30 dairy cows. One group was milked in a single-box robotic milking system with partially forced cow traffic (R-group). The other group was milked twice daily in a 2x6 herringbone-parlour (HP-group). Both groups consisted of Brown Swiss and Austrian Simmental cows and were housed in identical conditions.

Direct observations of social behaviour and 24-hour video recordings were conducted during six two-day blocks. The dominance value (DV) of each cow was calculated from the antagonistic interactions observed directly. The time budget (*lying, feeding, other activity*) of each cow was calculated from scan samples of the video tapes. On the day after each observation block, faeces of all individual cows were collected to determine the concentration of cortisol metabolites as an indicator of baseline adrenocortical activity and possible chronic stress (CC).

The groups did not differ significantly (MWU tests, alpha=0.05) in *lying* or *other activity*. Cows in the R-group spent significantly less time *feeding* than cows in the HP-group (9.9%±0.4% vs 10.6%±0.3%). The Pearson correlation matrix for the time budget parameters and DV within groups indicated negligible relationships among the parameters. In a between-group comparison for the subset of lactating cows only, the groups did not differ significantly in feeding behaviour. The results were otherwise similar to those for the full data set. There were no group differences with regard to CC.

Within the R-group, CC of lactating cows was significantly correlated with lying (rp=-0.56, p=0.007) and other activity (rp=0.57, p=0.005). Relationships between DV, time budget and CC were otherwise negligible. Social rank did not influence the relationship between CC and time budget in either group.

Introduction

Robotic milking is increasingly common on dairy farms and needs to be evaluated for its effect on cow welfare. Previous studies on dairy cows' stress responses in automatic milking systems have focused almost exclusively on stress experienced through the milking procedure per se (Hagen et al., 2003, Hopster et al., 2002). However, for a comprehensive evaluation of animal welfare consequences of robotic milking, effects of the overall system on chronic stress responses of dairy cows should also be addressed.

Therefore, we investigated the relationships between time budgets, adrenocortical activity and social rank of dairy cows. Emphasis was on Austrian conditions with regard to cow breeds, and on the comparison between a robotic system and a representative conventional milking system.

Animals, material and methods

Animals and housing

The study was carried out in two groups of dairy cows at the research farm of the Landwirtschafliche Bundesversuchswirtschaften GmbH in Wieselburg, Austria, between July and October 2001. Each group consisted of 15 Austrian Simmental and 15 Brown Swiss, in total 30 cows per group. One group was milked in a single box robot (Astronaut®, Lely Industries NV, Netherlands) with partially forced cow traffic (R-group). The other group was milked in a 2x6 herringbone parlour (Happel Ltd., Germany) twice a day (HP-group). Except for milking, the two groups were kept in similar management conditions. They were housed in the same uninsulated building in separate loose-housing pens with slatted floors. Each pen had 30 cubicles with soft rubber mats and small amounts of straw. Each cow had its own specific feeding place (American Calan Inc., USA). All animals were fed roughage ad libitum once a day: R-group in the morning, HP-group in the evening. Concentrates were delivered in the milking robot and in concentrate feeders (one in the R-group, two in the HP-group). Water was supplied ad libitum. In both groups, some cows were dry during parts of the study period.

Data collection
Time budgets and dominance values

Direct observations (7.5 hours per day and group) and 24-hour video recordings were carried out during six blocks of two consecutive days each (amounting to a total of 12 days per group). Consecutive observation blocks were at least one week apart. Direct observations involved continuous behaviour sampling of the cows' social interactions. All agonistic interactions leading to displacement (pushing away, chasing away, chasing up, threatening) were grouped for analysis. The dominance value (DV) of each cow was calculated following Sambraus (1975). The video tapes were scan-sampled every 5 minutes for *lying, feeding* and *other activity*, allowing calculation of the time budget for each cow.

Adrenocortical activity (cortisol metabolite concentration)

A non-invasive method was chosen to determine the adrenocortical activity: the concentration of cortisol metabolites was used as an indicator of baseline cortisol levels and chronic stress (CC). In ruminants, faecal metabolites are excreted with a delay of about 10-12 hours (Palme et al., 1999). Measured concentrations of cortisol metabolites in the faeces thus reflect glucocorticoid production 10-12 hours earlier. We collected faecal samples of each cow between 10:00 and 12:30 on the day after each observation block. The metabolite concentrations thus reflected the cortisol production of the night before.

The faecal samples were analysed for concentrations of 11,17-dioxoandrostanes (11,17-DOA). Procedures for preparation of biochemical analysis were as described by Palme and Möstl (1997). Samples from cows that were more than 23 weeks into gestation were excluded from the analysis, because the placenta produces large amounts of steroids and the steroid metabolites may cross-react with the antibody used in the analysis (Möstl *et al.*, 2002). As most non-lactating cows were more than 23 weeks into gestation, the statistical analysis of CC was conducted within the subgroup of lactating cows only.

Statistical analysis

For group comparisons, the data of each 2-day observation block were grouped, and the proportions of *lying, feeding* and *other activity* were calculated per group (mean, N=6). For the concentration of cortisol metabolites (CC) the median of the samples (N=2 to 6 per animal) was used for each cow. Between-group differences were tested with Mann-Whitney-U-tests. All tests were performed with alpha=0.05 (two-tailed).

Relationships between time budgets, DV and CC were evaluated with Pearson correlation coefficients. To identify a possible effect of social rank on the relationships among time budgets and CC, we tested Pearson correlation coefficients for categories of DV against the corresponding correlation coefficient for the whole (lactating) group after z-transformation according to Hotelling (Sachs 1997). Following Ketelaar-de Lauwere *et al.* (1996), cows with DV>0.60 were classified as high ranking, cows with DV<0.40 as low ranking, and the rest as middle ranking.

Results

The R-group did not differ significantly from the HP-group in the proportion of time spent *lying* (48.6%±3.0% vs 47.6%±2.5%) or in *other activity* (41.5%±3.0% vs 41.9%±2.5%). Cows in the R-group spent significantly less time *feeding* compared to cows in the HP-group (9.9%±0.4% vs 10.6%±0.3%). However, this effect was not sustained when the analysis was carried out for the subgroup of lactating cows only (11.0%±0.9% vs 10.8%±0.5%). Lactating cows did not differ between groups with regard to CC (66 nmol/kg [range: 38 - 148] vs 71 nmol/kg [range: 35 -125]).

Correlation coefficients for DV with the time budget parameters within groups indicated negligible relationships (r_p from 0.06 to 0.19, non significant). In the R-group (for lactating cows only, N=22), there was a negative correlation between CC and *lying* (r_p=-0.56, p=0.007), and a positive correlation between CC and *other activity* (r_p=0.57, p=0.005), but no correlation between CC and *feeding* (r_p=0.03, p=0.90). In contrast, correlation analysis revealed no significant relationships between CC and time budgets in the HP-group, but correlation coefficients were in the same direction (lactating cows only, N=26; *lying*: r_p=-0.20, p=0.24; *other activity*: r_p=0.24, p=0.23; *feeding*: r_p=-0.08, p=0.70).

To test for a possible effect of social rank on the relationship between CC and time budget variables, the correlation coefficients reported above for whole groups were compared with correlation coefficients within the DV-categories. Table 1 shows the results for the relationship between CC and *lying* within the R-group. All z-values were non-significant, indicating no major influence of social rank on the relationship between CC and time spent *lying*. Similar results were obtained for the relationships between CC and *feeding* or *other activity*. Equally, results for the HP-group did not reveal interdependencies among DV, CC and time budget variables.

Table 1. Pairwise comparison of Pearson correlation coefficients for the relationship of CC and lying within DV categories with overall correlation (rp=-0.56, p=0.007).

DV category	r_p	N	z
High	-0.62[ns]	7	0.04[ns]
Middle	-0.13[ns]	6	0.79[ns]
Low	-0.72[*]	9	0.45[ns]

z = 1.96 (2-tailed), for n<50 z-transformation according to Hotelling.

Discussion

Time spent *feeding* was significantly lower in the R-group with partially forced cow traffic than in the HP-group. This corresponds with findings by Ketelaar-de Lauwere *et al.* (1998). However, separate tests for lactating cows only revealed no significant differences between the groups. Thus, the lower time budget for *feeding* in the R-group might have been exclusively due to non-lactating cows. Non-lactating cows are fed on a less attractive diet compared to lactating cows and thus spend less time feeding.

It was surprising that there were no statistically significant relationships between DV and time budget variables. Cows of low social rank usually spend more time standing, because they have to wait in front of the milking robot (Wiktorsson *et al.*, 2003). A possible explanation for our finding might be that the size of the R-group was relatively small compared to producer recommendations and other studies, and waiting time in front of the robot may thus have been brief even for cows of low social rank. The low-ranking cows may also employ a strategy of avoiding the milking times preferred by higher ranking cows.

A novel approach in our study was the use of faecal cortisol metabolites to measure baseline levels of adrenocortical activity as a possible indicator of chronic stress as potentially caused by the overall management system. There were no significant differences in CC between the groups and the median values are comparable to other herds (Mülleder *et al.*, 2003). Within the R-group, there was a negative relationship between CC and *lying*, and a positive relationship between CC and *other activity*. The partially forced cow traffic may have caused this effect by increasing stress for cows that lay down less. As the groups did not differ overall, it is likely that some other factor was more stressful in the HP-group. One explanation might be that driving the cows to the milking parlour twice a day as well as standing in the narrow waiting area in front of the HP-system might be sources of stress (Hemsworth and Coleman 1998).

Conclusion

In summary, we find that absolute mean values of analyzed parameters in this study are comparable to findings in the literature. Groups did not differ significantly with regard to time budget variables or cortisol metabolite concentrations in their faeces. While these results indicate no adverse effects of the robotic milking system, in-depth analysis of within-group interrelationships among parameters might be a valuable extension of animal welfare assessments comparing alternative milking systems.

Acknowledgements

This study was part of a joint research project of the *Landwirtschafliche Bundesversuchswirtschaften GmbH* in Wieselburg, the *Österreichische Agentur für Gesundheit und Ernährungssicherheit GmbH Milchwirtschaft Wolfpassing*, and the *Institute of Animal Husbandry and Animal Welfare of the University of veterinary Medicine Vienna*. Financial support for work reported in this contribution was provided by the Austrian Federal Ministry of Agriculture, Forestry, Environment and Water Management under grant no. 1206 sub.

References

Hagen, K., D. Lexer, R. Palme, J. Troxler, S. Waiblinger, 2003. Behaviour, heart rate and milk cortisol of Simmental and Brown Swiss cows during milking in a robotic system compared to a herringbone parlour. In: Proceedings of the 37th International Congress of the ISAE, 24.-28.6.2003, Abano Therme, Italy, pp. 97.

Hemsworth, P.H., G.J. Coleman, 1998. Human-livestock interactions: The stockperson and the productivity and welfare of intensively farmed animals. New York, CAB International.

Hopster, H., R.M. Bruckmaier, J.T.N. Van der Werf, S.M. Korte, J. Macuhova, G. Korte-Bouws, C.G. Van Reenen, 2002. Stress responses during milking; comparing conventional and automatic milking in primiparous dairy cows. Journal of Dairy Science 85: 3206-3216.

Ketelaar-de Lauwere, C. C., S. Devir, J.H.H. Metz, 1996. The influence of social hierarchy on the time budget of cows and their visits to an automatic milking system. Applied Animal Behaviour Science 49 (2): 199-211.

Ketelaar-de Lauwere, C.C., M.M.W.B. Hendriks, J.H.M. Metz, W.G.P. Schouten, 1998. Behaviour of dairy cows under free or forced cow traffic in a simulated automatic milking system environment. Applied Animal Behaviour Science 56: 13-28.

Möstl, E., J.L. Maggs, G. Schrötter, U. Besenfelder and R. Palme, 2002. Measurement of cortisol meatbolites in faeces of ruminants. Veterinary Research Communications 26 : 127-258.

Mülleder, C., R. Palme, C. Menke, S. Waiblinger, 2003. Individual differences in behaviour and in adrenocortial activity in beef-suckler cows. Applied Animal Behaviour Science 84: 167-183.

Palme, R. and E. Möstl, 1997. Measurement of cortisol metabolites in faeces of sheep as a parameter of cortisol concentration in blood. International Journal of Mammalian Biology 62 (Suppl.II): 192-197.

Palme, R., C. Robia, S. Messmann, J. Hofer and E. Möstl, 1999. Measurement of faecal cortisol metabolites in ruminants: a non-invasive parameter of adrenocortical function. Wiener Tierärztliche Monatschrift 86: 237-241.

Sachs, L., 1997. Angewandte Statistik. Springer Verlag, Berlin, 881 pp.

Sambraus, H.H., 1975. Beobachtungen und Überlegungen zur Sozialordnung von Rindern. Züchtungskunde 47: 8-14.

Wiktorsson, H, G. Pettersson, J. Olofsson, K. Svennersten-Sjaunja, M. Melin, 2003. Welfare status of dairy cows in barns with automatic milking. Report of the EU project: Implications of the introduction of automatic milking on dairy farms. Http://www.automaticmilking.nl

ASSESSMENT OF WELFARE OF DAIRY COWS MILKED IN DIFFERENT AUTOMATIC MILKING SYSTEMS (AMS)

I. Neuffer, R. Hauser, L. Gygax, C. Kaufmann & B. Wechsler
Swiss Federal Veterinary Office, Centre for proper housing of ruminants and pigs, Taenikon, Switzerland

Abstract

The question whether the milking of dairy cows in automatic milking systems (AMS) is more stressful for the animals than milking in conventional parlours could not be answered conclusively in previous studies. These studies were typically conducted on single farms and provided only information for a specific AMS model. Consequently, comparative studies with different AMS models need to be conducted. In the present study, the welfare of dairy cows milked in either one of two AMS models or an auto-tandem parlour was assessed on four farms each.

On each farm, the behaviour during milking of 20 focal animals was videotaped. Additionally, 10 of the focal cows on each farm were equipped with a belt that allowed constant heart rate variability (HRV) recordings. Milk samples of all cows were analysed for somatic cell count and milk cortisol.

In the analysis, data on cow behaviour and HRV were combined with data originating from the AMS management software (e.g., milk yield, cow age, state of lactation, breed). Altogether, more than 1600 milkings of 234 cows were included in the study.

Differences between the milking systems were found in behavioural (i.e. stepping, leg lifting and kicking) and HRV parameters. Milk cortisol was not influenced by the milking system. These results allow the preliminary conclusion that a difference between the milking systems exists.

Introduction

In Switzerland, mass-produced housing systems and equipment for farm animals are subjected to an authorisation procedure regarding animal welfare since 1981. Conventional milking systems are not included in this test procedure. However, the procedure is applied to AMS because of the lack of constant human supervision during the milking process. Thus, the Swiss Federal Veterinary Office will use data of the research project presented here as a basis for a decision whether and with which specific requirements (e.g. number of cows per milking unit) AMS may be used in Switzerland. For the duration of the test procedure, AMS are authorized on a temporary basis.

The influence of AMS on the welfare of cows has already been investigated in several studies, often including only one AMS model in use on a single experimental farm (Hopster *et al.* (2002), Wenzel *et al.* (2003)). Results of studies carried out on several farms and comparing the effects of different AMS models on animal welfare have not been published yet. As a consequence, it is not known yet whether the welfare of cows can be considered as equal when different AMS models are in use.

The conclusions of previous studies focusing on welfare in cows milked in AMS vary considerably. Hopster *et al.* (2002) observed only marginal differences in functional, behavioural and physiological stress responses between cows milked in an AMS or in a tandem parlour and concluded that cows are not stressed during milking in an AMS. In this study, the responses in oxytocin, cortisol, plasma adrenaline and noradrenaline concentrations, heart rate, milk flow, residual milk and number of steps were low and typical for cows being milked. Contrary to this, Wenzel *et al.* (2003) found that cows milked by AMS showed differences in their behaviour and physiology compared to cows milked in a parlour. In the AMS, step behaviour occurred more often during milking, heart rates rose significantly prior to milking and milk cortisol levels were higher compared to controls.

The investigation presented here was carried out on commercial farms. Non-invasive methods were chosen due to the large number of animals and farms and due to the fact that the animals used were privately owned. In line with the experimental approaches of Wenzel *et al.* (2003) and Hopster *et al.* (2002), a combination of behavioural and physiological parameters was employed, i.e. video observations of behaviour, milk cortisol analysis and heart rate measurements. In addition to heart rate, we also measured heart rate variability. The physiological variability in heart rate is considered a better indicator of an individuals' capacity to cope with its environment than simple heart rate (Porges, 1995).

The aim of the present study was to overcome the disadvantages of restricted studies carried out on experimental farms with small numbers of animals and to answer the question whether automatic milking compromises the welfare of dairy cows under practical conditions.

Animals, materials and methods

Farms

Twelve Swiss farms were included in the present study. Four farms each were selected on which either one of two AMS models or conventional milking systems were in use. The investigation focused exclusively on farms with one-box-systems. Modern auto-tandem milking parlours were chosen to represent the conventional milking systems because of their similarity to AMS, i.e. automatic entry and exit gates, individual entry of the cows and individual milking stalls. To avoid possible effects of the transition from conventional to automatic milking, investigations were only conducted on farms with AMS installed for a minimum of six months prior to the start of the study.

Animals

Twenty lactating animals of each herd were chosen for video observations. Cows were selected according to the criteria good general health, no lameness, and calm behaviour during milking based on the advice of the farmer. The aim was to choose a cross-section of the herd with animals of different ages and states of lactation that were considered trouble-free during milking. Ten of the 20 focal cows were equipped with a belt for telemetric heart rate recording (see below) on the day before the first observation. The other focal animals could be identified by their collar numbers.

Data recording

To avoid any influence of high temperature and insects on animal behaviour, the investigations were carried out in autumn/winter 2001/2002, spring 2002 and autumn/winter 2002/2003. Each farm was examined for five consecutive days with the first day used for preparation of the equipment and the last for dismantling the material. The goal was to collect data of at least six milkings per cow during the three days of observation, with the expectation of a minimum milking frequency of two milkings per day.

Video observations of behaviour and performance of milking system

Two cameras were installed around the AMS: one camera showing udder and hind legs of cows during milking and one camera recording general activity of cows around the AMS and in the waiting area.

Movements of hind legs (stepping, leg lifting and kicking) were analysed with stepping defined as shifting weight from one hind foot to the other and the foot lifted less than 10 cm. If the foot was lifted more than 10 cm, the movement was recorded as leg lifting. Kicking was defined as a faster, more directional movement than leg lifting.

Simultaneously, the duration of the attachment of the milking cluster, the duration of milking and the number of attempts to attach each teat cup were registered by analyzing the videotapes.

A program combining the time signal from the video recorder with entered codes was used for analysis of video observations (ETHO, R. Weber, FAT, Taenikon).

Heart rate variability measurement

For continuous telemetric assessment of heart rate variability (HRV), 10 cows were equipped with a custom-built elastic belt mounted behind the forelegs of the cows, which carried three electrodes, an amplifier and a radio transmitter. Continuous transmission was possible for three to four consecutive days. The corresponding receiver unit was connected to a demodulator, which transformed the analog into a digital signal. Finally, the peaks in the R-R interval curve were detected and saved as beat-to-beat intervals for later analysis. Two parameters of the Time-Domain-Analysis, namely standard deviation of the beat-to-beat interval (SDRR) and square root of the mean squared differences of successive beat-to-beat intervals (rMSSD), were examined. Time-domain parameters can be used to assess the autonomic effects of drugs and other interventions including exercise and psychological and physical stress (Kleiger *et al.*, 1995). Due to the great interindividual variability in the absolute level of HRV, data of a given cow during visits to the AMS or the milking parlour were compared to data of the same cow lying quietly.

Lying behaviour of the 10 focal animals used for HRV measurement was recorded during daytime using the time-sampling-method over the 3 days of data collection on a given farm.

Milk sampling

In the AMS, automatic milk sampling units were installed for 24 hours during the investigation, thus collecting at least two milk samples per cow. In the tandem parlours, a second person needed to be present for the milk sampling. As this was considered to be a potential source of disturbance, milk samples from tandem parlours were collected during the days directly before or after the investigation. Somatic cell counts and milk cortisol levels were analysed. Cortisol measurements were available for 1457 milkings of 468 individuals.

Automatic milking – A better understanding

Management data

General data concerning the farm, i.e. herd size, feeding regime, farm area as well as specific data concerning the animals, i.e. age, number of lactations, lactational state, and milk yield were collected by means of a questionnaire and by gathering information from the management programs of the AMS and the milking parlours.

Analysis and statistics

Data was analysed using linear mixed effects models in Splus2000 and R 1.8.1. Assumptions of the models were checked using graphical analysis of the residuals. For some response variables transformations (log, square-root) or the use of alternative distribution families (binomial, poisson) was necessary.

Stepping, leg lifting and kicking during milking were investigated separately. The behavioural parameters were compared between the milking systems, and correlations with somatic cell count, duration of milking, day of lactation, age, milk yield and breed were calculated. For farms equipped with AMS, only milkings that were completed without technical problems (i.e. without unsuccessful attachment of teat cups) or human intervention were included in the analysis. Data of 226 cows was analysed, with an average of 6.1 milkings per cow.

Following the guidelines of the Task Force of the European Society of Cardiology and the North American Society of Pacing and Electrophysiology (Task Force, 1996), data periods of 5 min were selected out of the continuous HRV recordings. Only cows for which data of at least three milkings and 20 periods of quiet lying were of good quality (less than 5% defective values) were included in the analysis. Data of 58 cows from 11 farms fulfilled these requirements. The relatively high data loss was mainly due to our custom-built heart rate recording system.

Results

Behaviour during milking

The three investigated milking systems differed significantly in the frequency of stepping, leg lifting and kicking with p-values of <0.05, <0.01, and <0.05, respectively (n=1359 milkings). Stepping was observed during most milkings (n=1309), whereas leg lifting (n=527) and kicking (n=146) were less frequent. In AMS1, stepping was observed more often than in AMS2, whereas leg lifting occurred more often in AMS2. Kicking happened more frequently in the milking parlours than in the farms equipped with AMS. Apart from the milking system, several factors have been found to influence the behaviour of the cows during milking. The frequency of stepping was influenced by somatic cell count (p<0.05), day of lactation (p<0.05) and total duration of cleaning the udder, attaching the milking cluster and milking (p<0.001). No influence of the time since the previous milking, daily milk yield or breed was found.

Leg lifting was influenced by the interval between milkings (p<0.05), the total duration of cleaning the udder, attaching the milking cluster and milking (p<0.0001), daily milk yield (p<0.01) and breed (p<0.001).

Heart rate variability

RMSSD and SDRR showed similar patterns. Changes in these measures between milkings and at rest differed significantly among the systems with the interaction of system and effect of milking showing a p-value of <0.0001. The changes between milking and lying quietly are smaller in AMS1 compared to the milking parlours, in contrast to AMS2, where the differences are larger than in the milking parlours.

Milk cortisol

There was no obvious relation between the AMS model and level of cortisol secretion though there were indications that cows on AMS farms had disturbed circadian patterns of cortisol.

Discussion

Behaviour

Stepping during milking could be observed more often on farms equipped with AMS1 than on farms with a tandem parlour. In AMS2, this difference was not significant. Wenzel *et al.* (2003) observed step behaviour more often in the AMS than in the milking parlour. This is in contrast to the results of Hopster *et al.* (2002) who did not find a difference in the total number of steps between cows milked in an AMS or a milking parlour.

In agreement with Wenzel *et al.* (2003), kick behaviour was rare with no significant difference between the milking systems. No kick behaviour was observed in the investigation of Hopster *et al.* (2002). It must be taken into account that the two mentioned studies concentrated on the milking system only without calculating effects of measured factors such as age, lactation period or breed.

HRV

Minero *et al.* (2001) stated that HRV levels in cows show a great individual variability. Accordingly, we performed intraindividual comparisons of indices during the different challenges resting and milking. Significant differences in HRV parameters between resting and milking, and between the milking systems were found in our investigation. Our results show a decrease in short-term variability (rMSSD), as well as long-term variability (SDRR) with increasing levels of stress load. This correspond to similar results in Mohr *et al.* (2002), whose results in the time-domain indicate a clear reduction tendency of vagal tone as stress level increases. Calves exposed to external (heat / insects) or internal (sickness) stress showed a significant reduction in several HRV parameters compared to healthy individuals (Mohr *et al.*, 2002).

Conclusion

In contrast to the results of Wenzel *et al.* (2003) but in accordance with the study of Hopster *et al.* (2002), our data does not show clear tendencies in the differences between the milking systems.

In our investigation, the behaviour parameters frequency of stepping and frequency of leg lifting during milking were influenced by the milking system, as well as the difference between milking and resting in both HRV parameters. No effect of the system was found to influence milk cortisol. These results allow the preliminary conclusion that a difference between the milking systems exists, but that the results show no consistent tendency.

References

Hopster, H., R.M. Bruckmaier, J.T.N. Van der Werf, S.M. Korte, J. Macuhova, G. Korte-Bouws and C.G. Van Reenen, 2002. Stress responses during milking: Comparing conventional and automatic milking in primiparous dairy cows. Journal of Dairy Science 85: 3206-3216.

Kleiger, R.E., P.K. Stein, M.S. Bosner and J.N. Rottman, 1995. Time-Domain measurements of heart rate variability. In: Malik, M. and A.J. Cramm (editors), Heart Rate Variability. Armonk, NY. Futura Publishing Company, Inc., pp. 33-45.

Minero, M., E. Canali, V. Ferrante and C. Carenzi, 2001. Measurement and time domain analysis of heart rate variability in dairy cattle. The Veterinary Record 149: 772-774

Mohr, E., J. Langbein and G. Nuernberg, 2002. Heart rate variability: A noninvasive approach to measure stress in calves and cows. Physiology and Behaviour 75: 251-259

Porges, S.W., 1995. Cardiac vagal tone: a physiological index of stress. Neurosciences and Biobehavioral Reviews 19: 225-233.

Task Force of the European Society of Cardiology and the North American Society of Pacing and Electrophysiology, 1996. Heart rate variability: Standards of measurement, physiological interpretation, and clinical use. European Heart Journal 17: 354-381.

Wenzel, C., S. Schönreiter-Fischer and J. Unshelm, 2003. Studies on step-kick behavior and stress of cows during milking in an automatic milking system. Livestock Production Science 83: 237-246.

ASSESSMENT OF ANIMAL WELFARE IN AN AMS HERD: EXPERT EVALUATION OF A DECISION SUPPORT SYSTEM

Tine Rousing, Jens Hindhede, Ilka Christine Klaas, Marianne Bonde & Jan Tind Sørensen
Danish Institute of Agricultural Sciences, Department of Animal Health and Welfare, Foulum, Denmark

Abstract

An operational welfare assessment protocol integrating behavioural measures and clinical observations with information on the housing system and management routines has been developed for decision support in AMS herds. As part of an evaluation of the extent to which the welfare assessment system fulfils is intention of being a decision support tool for the dairy farmer - an expert panel evaluation of the welfare protocol has been carried out. The expert panel consisted of 18 researchers and 3 farm advisers participating in an AMS Conference held by the Danish Livestock Association and the Swedish Dairy Association in Lund, Sweden, in June 2003. The conference delegates were asked to define the term animal welfare, and to score the individual welfare relevance of each of 38 listed measures in a welfare assessment protocol. Furthermore, they were asked to nominate the 10 most important measures in the protocol for inclusion in an on-farm welfare assessment system aiming at decision support in an AMS herd. The expert panel generally appreciated the listed protocol measures as highly relevant. When nominating the 10 most relevant measures the prevalence of clinical lameness was considered the most important measure to include in a welfare assessment protocol. Furthermore, 15 out of 19 experts nominated measures from all 4 information sources: housing system, management, behaviour and health. Nomination of the most relevant measures for inclusion in an operational welfare assessment system seems to be related to the experts' perception of the term animal welfare. Experts supportive of a welfare definition based on animal's mental state attached great importance to measures related to 'cow comfort' when nominating relevant welfare measures, while experts supportive of a welfare definition based on 'physiological fitness' stressed the importance of health measures and the fulfilment of physiological needs. The experts did not focus on measures relating specifically to management routines characteristic of AMS herds as being of major relevance for animal welfare assessment.

Background

Ultimo 2003 308 dairy farms in Denmark were practising automatic milking according to the Danish Cattle Data Base. Automatic Milking Systems (AMS) change the production system in itself as well as the daily management routines including surveillance of the individual animals in comparison with manual milking in loose housing systems. AMS offers not only fully automated individual milking but also possibilities for individual feeding of concentrates in automatic milking units (AMU), as well as automatic and individual separation of animals for milking, grazing, veterinary treatment, insemination etc.

In recent years consumers, and politicians as well as the agricultural business are showing growing concern about animal welfare in livestock production. The interactions between the production system and the management routines are known to affect animal welfare. Studies by for example Thomsen and Kjeldsen (2003), Winckler and Willen (2001) and Fregosi and Leaver (2001) report considerable herd differences within comparable housing systems in e.g. mortality, lameness and lying and feeding behaviour, respectively. These findings indicate that herd specific welfare problems exist in today's dairy cattle production systems - and further, that it is often possible for the farmer to improve animal welfare through management. However, the knowledge of how AMS effects the welfare of the cows is limited. Consequently, if the individual farmer wants to improve animal welfare in his/her herd there is a need for developing methods for assessment of animal welfare at herd level aiming at identifying problems as well as their causes.

Agreement on the definition of animal welfare and the aim of the welfare assessment is imperative for any assessment of animal welfare. Animal welfare assessment systems based on different welfare definitions have been developed in Europe (Johnsen et al., 2001). These welfare assessment systems are serving different purposes such as certification of the level of animal welfare in individual farms, evaluation of different production systems, and assistance to individual farmers of identifying and rectifying welfare problems on the farm, respectively. At the Danish Institute of Agricultural Sciences (DIAS) operational welfare assessment systems of the latter type have been developed. The first prototype of animal welfare assessment was regarding conventional dairy and pig production. These welfare assessment systems were included in the project 'Ethical Account for Livestock Farming' (Sørensen et al., 2001). Since then, welfare assessment systems have been developed focusing specifically on pig production systems with loose housed pregnant sows (Bonde 2003), mink production (Møller et al., 2003), organic egg production (Hegelund et al., 2003), dairy cattle herds with loose housed cows (Rousing 2003) and dairy herds with AMS (Sørensen et al., 2002).

These activities are based on the animal welfare definition by Simonsen (1996) reflecting 'the positive and negative experiences of the individual animal throughout a lifetime'. The mental experiences of an animal cannot be measured directly, but has to be assessed indirectly through welfare indicators. Therefore, the welfare assessment systems mentioned all integrate animal behaviour and health with information on housing and management routines. Information on the housing system and current management is expected to indicate risks of welfare problems or potentials for good welfare. Furthermore, these informations may assist in identifying causal relations needed for decision support. Relevant behaviour measures and measures of clinical health are included because they reflect the animals' response to the existing housing system and the current management. Selection of relevant welfare indicators has been approached by evaluating indicator candidates step by step concerning their independent welfare relevance, their marginal welfare value when included in a full welfare assessment system, and their applicability for on-farm welfare assessment (Rousing et al., 2001).

A prototype of a welfare assessment system intended as an advisory tool for use in dairy herds applying AMS has been developed at the Danish Institute of Agricultural Sciences

(Hindhede *et al.*, 2002). The selected welfare indicators is described by Hindhede *et al.*, 2002, and an overview is presented in table 1.

An expert panel evaluation has been carried out as part of an evaluation procedure focusing on the success of the welfare assessment system in fulfilling its intention of identifying welfare problems, assess the development over time and allow the farmer to respond appropriately. This paper aims at reporting results of the expert panel evaluation of the suggested welfare assessment system as a decision support tool for the AMS farmer.

Materials and methods

A questionnaire was designed based on the developed welfare indicator protocol, and a prototype was tested by a Danish researcher in cow welfare, who was unfamiliar with the project. Based on his comments the final version of the questionnaire was produced. The questionnaire study was carried out during a 11/2-hour session as part of an AMS Conference held by the Danish Livestock Association and the Swedish Dairy Association in Lund, Sweden, in June 2003. 21 out of 28 conference participants that were not involved in the project consented to take part in the investigation. The 21 panel members stated the following expertise: Health (mainly udder health, mastitis) (5), milking and milk quality (5), housing (3), behaviour and welfare (3) and feeding (2). 3 panel members did not state any overall expertise.

After an oral introduction to the welfare assessment system the conference participants were asked to fill in the questionnaire. They were asked to state their opinion on the term animal welfare by indicating if their understanding of the term animal welfare were 'mostly related to physiological fitness and health, the animals' feelings and experiences, or having naturalness as the point of reference'. Further, the panel members were asked to imagine that they were to carry out a welfare assessment in a dairy cow herd practising automatic milking. They were presented with the list of 38 measures shown in table 1, and were asked the question 'How important is it to be informed about each of the 38 suggested concerns if you should assess the animal welfare in a herd'. The panel members were asked to answer this question without previous discussion and by scoring each measure on a scale from 1-5; where 1 = not relevant and 5 = most relevant. They were instructed not to include any concerns about the costs related to carrying out the welfare assessment when scoring the welfare relevance of the listed indicators. Finally, the panel participants each nominated the 10 most important measures in the protocol for inclusion in an on-farm welfare assessment system aiming at operational decision support in an AMS herd.

Results

17 panel members indicated their opinion on welfare definition. 6 supported a welfare definition relating to physiological fitness/health, 11 supported a welfare definition based the animal's mental state. All 21 panel members accomplished the individual scoring of all 38 measures, and 19 panel members nominated the ten most relevant measures.

The expert panel generally appreciated the listed protocol measures as highly relevant. Table 1 summarises the results of the expert panel evaluation, presenting the median scores for the welfare relevance of all 38 measures. The means of the median scores were 3,8 for measures relating to the housing system, 4,1 for management information, 3,9 for the behavioural measures, and 4,3 for the health measures.

When nominating the 10 most relevant measures 15 out of 19 experts nominated measures from all 4 information sources: housing system, management, behaviour and clinical health. Table 2 illustrates the 10 most frequently nominated measures. The table summarises the overall nominations and present the results relating to the two stated welfare definitions. Lameness was nominated most frequently as relevant for inclusion in an operational welfare assessment system. The panel nominated three other measures relating to animal health most relevant: body condition, teat and udder lesions, and incidence of

Table 1. Welfare indicators referring to information on housing system, management routines, dairy cow behaviour and health. Median scores of individual welfare relevance (1: not relevant - 5: highly relevant).

Housing system	(Median score)
Housing design (Localisation of cubicles, feeding area, AMU's, drinking facilities, and calving boxes etc.)	4
Surfaces and materials (Quality of flooring, prominent spigots and unprotected bolts in exposed positions etc.)	4
Cubicles (Dimensions, type of cubicle partition, placing of neck and chest bars, usage of rubber mats, straw bedding or mattresses)	5
Passages (Widths, blind passages, gates for guidance of 'cow traffic', usage of and quality of scraping system)	4
'Collecting area' (Possibility for occasionally establishment, flexibility regarding capacity, dimensions)	4
AMU (Dimensions, widths of entrance and exit passages, possibility for separation before and after milking, usage of cow trainer etc.)	4
Feeding table and forage fence (Overall capacity, type of forage fence, dimensions, possibility for fixation)	4
Concentrate feeder outside AMU (Usage of and criteria for feeding additional concentrates)	3
Drinking facilities (Number and capacity; water pressure of bowl drinkers, length and widths of water troughs)	4
Brushes (Usage of, number, quality)	3
Food bath (Usage of, dimensions)	3
Indoor climate (Natural ventilation ('open' /'half open' walls), usage of and type of ventilating plant)	4
Indoor lightning (Hours pr. day with 'artificial' lighting, lighting at resources such as at and within AMU's and at feeding table)	3

Animal behaviour	
Man-animal relationship (Fear test; behavioural response at human-approach)	4
Social behaviour (Social grooming and antagonistic behaviour; diurnal rhythm, location in shed)	4
Behaviour during milking (Hesitations to enter AMU, being driven into AMU, stepping, kicking, defecating and urinating during milking)	4
Getting-up-behaviour (Uneasy, abrupted or abnormal getting-up behaviour)	4
Usage of resources (Feeding, resting and queuing; diurnal rhythm)	4
AMU visits (Number and time of day)	4
Refusals at/in AMU (Number and time of day)	3

Stocking rate (Number of animals per cubicle, AMU, drinking facility, and capacity at feeding table)   5

Cleaning and littering (Frequency, quantity and quality of litter, quality and cleanness of mats and
mattresses)   4

Feeding (Policy, quantity and quality; ad lib. / restrictive, TMR, number of feedings, feeding
concentrates in AMU and in additional dispensers etc.)   4

Watering (Quality of water, not functioning water bowls, low water pressure of water bowls, and low
water surface of water troughs)   4.5

Milking criteria (Policy of milking frequency, minimum interval between milkings, and refusals)   3

Introduction policy to milking and milking routines (Introduction procedure, policy of fetching,
assistance of 'alarm' cows etc.)   4

Grazing (Practising grazing, grazing hours per cow per day, stocking rate on pasture, length and quality
of passages to pasture, watering on pasture, possibility for cows' to seek shade and shelter etc.)   4

Animal health

Clinical examination (Scoring)

  Body condition ('poor' - 'fat')   4

  Dirtiness (one or more dirty lower legs - general dirtiness with inflammation of the skin)   4

  Skin lesions (no lesions - infected ulcers)   4

  Ecto parasites (no or few parasites or considerable occurrence of skin parasites)   4

  Pressure sores (no swellings - swellings with infection)   5

  Lameness (normal gait - severe lameness/reluctance to bear weight on a limb)   5

  Hoof length (normal hoof length - severely overgrown hooves)   4

  Teat and udder lesions (no lesions - infected lesions, and dry quarters)   5

  General condition (assessment of overall appearance, appearing 'poorly' or 'fit')   4

Incidence of clinical disease (incidence of mastitis, paratuberculosis, puerperal paresis, and
metabolic diseases)   4

Mortality   4

---

clinical disease. Other frequently nominated indicators were dirtiness, information on cubicles (dimensions, cubicle partition, placing of neck and chest bars and flooring) and housing design, stocking rate and watering, and characterisation of getting-up-behaviour and behaviour during milking.

Measures relating to management routines characteristic of herds practising automatic milking were found of minor relevance when assessing animal welfare. Thus, introduction policy to milking and milking routines were each nominated by 3 panel members, information on visits to the AMU were nominated twice and information on the AMU itself, milking criteria including policy of milking frequency, minimum interval between milkings, as well as refusals at AMU were only nominated once.

For both welfare definitions (physiological fitness/health and mental state) the experts nominated measures from all 4 information sources. However, a pattern in the nominations relating to welfare definition seemed to appear. Experts supporting a welfare definition based on the animal's mental state emphasised the importance of measures relating to *cow comfort*

Table 2. The ten most frequently nominated welfare indicators for inclusion in an operational decision support system in an AMS herd. Results presented as a total among the panel members and relating to welfare definitions.

10 most frequently nominated welfare indicators (number of nominations)

In total (19 experts)	Based on welfare definition: 'Feelings'/ experiences (11 experts)	Based on welfare definition: Physiological fitness (6 experts)
Lameness (13)	Cubicles (4)	Lameness (8)
Cubicles (11)	Getting-up-behaviour (4)	Housing design, Watering (6)
Stocking rate, Watering, Dirtiness, Incidence of clinical disease (10)	Surfaces and materials, Passages, Behaviour during milking, Stocking	Feeding, Dirtiness, Teat and udder lesions (5)
Getting-up-behaviour, Teat and udder lesions (9)	rate, Dirtiness, Ecto-parasites, Lameness, Incidence of clinical	Surfaces and materials, Getting-up-behaviour, Stocking rate, Body
Housing design, Behaviour during milking, Body condition (8)	disease (3)	condition, Incidence of clinical disease (3)

(information on the cubicle, uneasy / abnormal getting-up behaviour, as well as surfaces and materials were most frequently nominated). These experts also regarded dislike or fear behaviour during milking as relevant to include in a welfare assessment in AMS herds. Among the clinical measures lameness, dirtiness, incidence of clinical disease and ecto-parasites were nominated equally frequently.

Whereas non-clinical measures were of highest priority in the group of experts supporting a welfare definition based on the animal's mental state, the experts supporting a welfare definition based on physiological fitness most frequently nominated measures relating to first of all *disease*. 8 out of the 9 experts nominated lameness. Further, teat lesions, body condition and incidence of clinical disease were emphasised. Moreover, *logistics* (housing design) and fulfilment of *physiological needs* (water and feed) were also frequently nominated.

Conclusion

Evaluation of welfare assessment systems is a difficult task, as the welfare state in a herd is not directly measurable. Even though the number of panel members were limited and their expertise in relation to welfare assessment were varying - the results of the present expert panel has strengthened the welfare assessment system. The panel has supported the suggested measures for inclusion in a welfare assessment protocol and has demonstrated that integration of different information is decisive for operational welfare assessment at herd level.

References

Bonde, M.K.. 2003. Welfare assessment in a commercial sow herd- development and evaluation and report of the method. PhD-thesis, DIAS report No 46, 98 pp.

Fregonesi, J, Leaver, J.D., 2001. Behaviour, performance and health indicators of welfare for dairy cows housed in strawyard or cubicle systems. Livest. Prod. Sci., 68, 205-216.

Hegelund, L. Sørensen, JT. & Johansen, N.F. 2003. Developing a welfare assessment system for commercial organic egg production system. Animal welfare (in press)

Hindhede, J., Rousing, T., Fossing. C. & Sørensen, J.T., 2002. A protocol for assessing animal welfare in an automatic milking system. Report produced within the EU project "Implication of the introduction of automatic milking on dairy farms" (QLK5-2000-31006) as part of the EU-program "Quality of Life and Management of Living Resources". Deliverable D23, February 2002, 18 pp. Available at http://www.automaticmilking.nl

Johnsen, P.F., Johannesson, T., Sandøe, P., 2001. Assessemnt of fam animal welfare at herd level: many goals, many methods. Acta Agric. Scand., Sect. A, Anim. Sci. Suppl. 30, 26-22.

Møller, S.H., Hansen, S.W., Sørensen, J.T., 2003. Assessing animal welfare in a strictly synchronous production: The mink case. Animal welfare (in press).

Rousing, T, Bonde, M.& Sørensen, J.T. 2001.How to aggregate welfare indicators into an operational welfare assessment system: a bottom up approach. Acta Agric. Scand., Sect. A, Anim. Sci. Suppl. 30, 53-57.

Rousing, T. 2003. Welfare assessment in dairy cattle herds with loose-housing cubicle systems- development and evaluation of welfare indicators. PhD-thesis, DIAS report No 45, 101 pp.

Simonsen, H. B., 1996. Assessment of animal welfare by a holistic approach: Behaviour, health and measured opinion. Acta Agric. Scand., Sect. A, Anim. Sci. Suppl. 27, 91-96.

Sørensen, J.T., Sandøe, P. & Halberg, N. 2001. Animal welfare as one among several values to be considered at farm level: The idea of an ethical account for livestock farming. Acta Agric. Scand., Sect. A, Anim. Sci. Suppl. 30, 11-16.

Sørensen, J.T., Hindhede, J., Rousing, T. & Fossing, C., 2002. Assessing animal welfare in a dairy cattle herd with an automatic milking system. In: Proc. First North American Conference on Robotic Milking, March 20-22, Toronto, Canada, VI-54-VI-59.

Thomsen, P. T., Kjeldsen, A. M., 2002. Dødelighed blandt danske malkekøer - 2 (in Danish), http://www.lr.dk, Kvæginfo nr. 1039.

Winckler, C., Willen, S., 2001. The reliability and repeatability of a lameness scoring system for use as an indicator of welfare in dairy cattle. Acta Agric. Scand., Sect. A, Anim. Sci. Suppl. 30, 103-107.

AUTOMATIC MILKING AND GRAZING IN DAIRY CATTLE: EFFECTS ON BEHAVIOUR

L.F.M. Heutinck, H.J.C. van Dooren & G. Biewenga
Applied Research, Animal Sciences Group, Wageningen UR, Lelystad, The Netherlands.

Abstract

Milking dairy cattle with an automatic milking system (AMS) can be a reason for a farmer to minimize pasture time during the grazing season, since it is assumed that it may lead to a decrease in use of AMS capacity and an increase in labour required for fetching cows with long milking intervals from pasture. Previous research in The Netherlands showed that grazing in combination with automatic milking can be successful under various management conditions. Knowledge about cow behaviour may reveal useful information in this respect. The aim of this research was to study the effect of grazing combined with automatic milking on the potentially relevant behavioural aspects of a dairy herd.

A commercial herd of about 60 cows was observed before and during the grazing seasons of 2001, 2002 and 2003. The cows were milked by an AMS which was introduced in September 2000. Access to pasture was restricted and regulated by a selection gate. Locations in the barn and at pasture and standing/lying positions were observed during 48 h periods using scan sampling with 15 min intervals.

Herd behaviour as well as behaviour of 10 selected individual cows showed comparable results. The mean percentage of cows in the AMS was lower during the grazing season than at the end of the winter season, although it was on an acceptable level. No significant differences were found for mean presence in the waiting area in front of the AMS or for the total lying behaviour. These results apply to the management practices used at the farm. Further research on the effect of grazing systems, pasture distances, fetching regimes, cow traffic systems or occupancies of the AMS and their interaction on herd and individual level is needed to optimise the management of dairy cattle offered grazing combined with automatic milking.

Introduction

In The Netherlands as well as in other European countries (for example Denmark, Sweden) the number of automatic milking systems (AMS) replacing a conventional milking parlour on dairy farms is increasing substantially. It is often assumed that grazing does not combine well with automatic milking, since it may lead to a decrease in use of capacity of the AMS and an increase in labour required for fetching cows with long milking intervals from the pasture site. However, cow health and welfare as well as the ´green´ public image of the dairy industry benefit from giving dairy cattle the opportunity for grazing (see references in Ketelaar-de Lauwere et al., 1999). Therefore, more information on how to combine the use of an AMS with grazing is needed. On experimental scale (24 cows) Ketelaar-de Lauwere et al. (1999) found that grazing combined with use of an AMS on a voluntary basis was possible for restricted -, zero -, and unrestricted grazing. In an observational study involving

25 Dutch dairy farms it was found that grazing in combination with automatic milking is possible under various management conditions while other farmers are less successful (Ruis-Heutinck *et al.*, 2001). Information about herd and individual cow behaviour on commercial dairy farms combining automatic milking with grazing is lacking, however, although this knowledge may reveal which management practices may be successful and which not. The aim of this research was to study the effect of grazing combined with automatic milking on the potentially relevant behavioural aspects of a dairy herd.

Material and methods

Animals, housing and management

The animals were part of a commercial herd of about 60 dairy Holstein Friesian cows at the experimental dairy farm Nij Bosma Zathe in The Netherlands. The lay-out of the housing facility is shown in figure 1. At Nij Bosma Zathe an AMS was introduced in September 2000. In the following 3 grazing seasons (2001, 2002 and 2003) a study of the behaviour of the herd and 10 selected cows within this herd was carried out. The selection of individual cows was semi-random, based on lactation number (1, 2 and 3-or-more) and lactation stage. The latter varied from zero up to about 5 months at the start of the first observation each year.

The cows were milked with a two-units-in-row Galaxy AMS. The minimum milking interval was set at 6, 8 or 10 h depending on production level and days in lactation.

In 2001 and 2002 (except for the last observation period) a one-way cow routing system was used, i.e. cows had access to the feeding gate only by passing the AMS. In 2003 free cow traffic was used. Concentrates were supplied in the AMS and in a concentrate feeder (with entrance in the feeding compartment). Maize silage was supplied in the morning and in the late afternoon in 2001 and 2002. In 2003 the supplied roughage was a mixture of grass and maize silage. On average an amount of 7.5 kg dry matter roughage per cow per day was available at the feeding gate. Water was available in two water troughs located at opposite sides of the barn, and at the entrance of the pasture site.

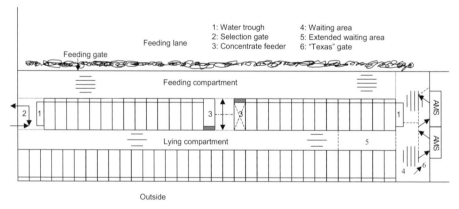

Figure 1. Barn lay-out at Nij Bosma Zathe.

Pasture

Pasture sites (about 4 ha each) were located with distances ranging from 50 to 800 m from barn to pasture entrance. Access to the pasture site was regulated by a selection gate at the entrance of the barn. When milking intervals were within two hours before the end of the minimum milking interval, access to pasture was denied. A restricted daytime grazing system was offered (the selection gate allowed cows to go to pasture between 6 am and 4.30 pm) with fetching of the cows at about 8 pm except for the last observation period of 2003 when cows had 24 hour access to pasture. Cows were fetched at 6 am and 4 pm in this system.

Sward height varied between the observation periods.

Behavioural observations

In each year the first observation period was scheduled 2-3 weeks before the grazing season started to obtain the behaviour of the animals without having access to pasture as a starting point. During the grazing season 3-4 observation periods were scheduled with intervals of 4 to 8 weeks. Direct observations were conducted during 48 hours using the scan sampling method (Altmann, 1974) with 15 min intervals. Usually one person observed during 8 hours, followed by the next person. In some cases when pasture distance was large or the herd was spread out over pasture as well as in the barn, observations were carried out by two people. All observers were instructed at the beginning of each observation period to minimize subjectivity.

Location of animals and standing or lying position were observed. Locations in the barn were: in the AMS, in the waiting area, in the feeding compartment, at the feeding gate, at the concentrate feeder, by the water troughs, within the first five cubicles on each side near the waiting area, within other cubicles, in the corridor of the lying compartment and the selection gate (see figure 1). The waiting area was defined as the corner area near the AMS entrance. In a pilot observation it was found that in some occasions cows seemed to be part of (and close to) the group waiting for the AMS but not standing in the defined waiting area. It was decided to count those cows as if they were in the waiting area. It also became clear that sometimes cows used the first five cubicles near the AMS to stand in while waiting. Therefore standing (and lying) in these cubicles was scored separately from standing and lying in the other cubicles in the barn. In the analysis it was added to the scores in the waiting area and labelled 'extended waiting area'.

Locations outside the barn were the path to the pasture site, an area of about 10 m around the water through at the pasture entrance and the pasture site.

All behavioural elements were observed at herd level as well as at individual level. At herd level the number of animals per behavioural element was scored, at individual level the element itself was recorded.

Technical data

Herd size, average number of AMS visits, pasture time and frequency of going to pasture was recorded by the AMS (and selection gate) management software. Numbers of fetched cows were recorded by the farm manager daily.

Analysis and presentation of the results

Behavioural data were analyzed with Genstat (Genstat, 1993). Number of scores per behavioural element or per cow and total number of scores were used in the analysis. Predicted means and standard deviations of these traits were first calculated under the assumption of being normally distributed. The presented percentages represent the mean duration of the behavioural elements per day, as well as the relative number of cows per behavioural element. Differences between percentages of the first observation period (referred to as winter season) and the observation periods during the grazing season were analyzed over years (n=3 for the winter season and n=10 for the grazing season) and per year (n=1 vs. n=3, 4 and 3 for 2001, 2002 and 2003, respectively for the winter season vs. the grazing season) using the GLM procedure. No block-effect for year was found.

Results

Time budget

The number of cows in the herd was 61, 56 and 68 respectively in 2001, 2002 and 2003 in the winter season and on average 59, 62.1 and 63.1 respectively in the grazing season.

The average number of fetched cows during the restricted grazing periods (all except the last period in 2003) was 16.8, ranging from 0-38, and average pasture time of the herd was 7.7, 7.3 and 7.3 hr in 2001, 2002 and 2003, respectively. The average pasture time of the selected cows was 7.5, 7.2 and 7.6 h in 2001, 2002 and 2003, respectively. The average daily frequency of going to pasture was 1.1 in all 3 years.

Table 1 shows the mean time budget of the herd in the winter and the grazing season. The time budget of the individual selected cows was comparable with that of the total herd.

Table 1. Mean time budget (%) of the herd in the winter and grazing season.

	Winter season	Grazing season
AMS	2.3	2.0
Waiting area	3.5	2.9
Standing in feeding area	7.2	4.1
Near roughage/concentrate/water	21.7	10.6
Standing in lying area outside cubicles	5.5	3.5
Standing in cubicles	14.1	12.1
Lying in cubicles	45.7	30.8
Standing/walking at cow path	0	1.1
Standing in the pasture	0	20.0
Lying in the pasture	0	12.9
TOTAL	100	100

Use of AMS and waiting area

Table 2 shows the mean percentage of the herd in the AMS and waiting area during the winter and the grazing season in 2001, 2002 and 2003. The mean percentage of cows on herd level in the extended waiting area was on average 2% higher than in the waiting area.

Table 2. Mean percentage of the herd in the AMS and waiting area during the winter and grazing season in 2001, 2002 and 2003.

Location	Winter season		Grazing season	
	Mean %	S.d.	Mean %	S.d.
AMS 2001	2.4	1.2	2.1	1.4
AMS 2002	2.6	1.4	2.0	1.3
AMS 2003	2.0	1.2	2.0	1.8
Waiting area 2001	2.9	3.0	3.6	4.7
Waiting area 2002	3.6	3.1	2.3	2.6
Waiting area 2003	4.0	2.4	3.0	3.5

Over years the mean percentage of cows in the AMS was significantly higher in the winter season than in the grazing season (2.3 ± 1.2 vs. 2.0 ± 1.4%, respectively; P<0.05). No difference was found for mean presence in the waiting area (respectively 3.5 ± 2.9 vs. 2.9 ± 3.6%; P>0.05). Interactions between year and winter/grazing season were not statistically identified.

The mean percentage of 2.3 vs. 2.0 in the AMS equals a mean of 1.4 vs. 1.2 cows and a total of 17.0 vs. 14.7 hr occupancy of the AMS per unit per day. The mean percentage of 3.5 vs. 2.9 in the waiting area equals a mean of 2.2 vs. 1.8 cows and 50.4 vs. 42.0 min per cow per day.

Data from the AMS showed on average 2.9 and 2.7 milkings per cow per day in the winter season and in the grazing season, respectively. The AMS visiting frequency of the selected cows was in agreement with those on herd level.

During the winter season a peak in mean percentage of cows in the waiting area or visiting the AMS was seen between 7-9 am. In the grazing season no clear peak could be found, but there was an off-peak in AMS visits in the afternoon (between 1-3 pm).

Lying behaviour

Table 3 shows the mean lying behaviour during the winter and the grazing season in 2001, 2002 and 2003. Although not statistically proven, it is interesting to notice the decrease in mean percentage of lying in cubicles in 2003 in favour of lying at pasture.

The mean percentage of lying in the cubicles was significantly different (P<0.05) between the winter and grazing season. The sum of the lying duration in cubicles and at pasture did not differ (P>0.05) between seasons on herd level as well as on selection level.

Discussion

The first observation, referred to as in the winter season, was usually carried out in April/May. In the first experimental year (2001) it was delayed because of the occurrence of Foot and Mouth Disease in The Netherlands till mid June.

On herd level 2.3% of the cows was in the AMS in the winter season compared to 2.0% in the grazing season. This equals 1.4 vs. 1.2 cows or an occupancy of 17.0 vs. 14.7 h per

Table 3. Mean lying behaviour during winter and grazing season in 2001, 2002 and 2003.

	Winter season		Grazing season	
	Mean %	S.d.	Mean %	S.d.
Lying in cubicles 2001	43.0	3.5	33.4	4.9
Lying in cubicles 2002	47.7	3.0	32.7	5.0
Lying in cubicles 2003	46.3	3.9	25.4	4.7
Lying at pasture 2001	-	-	11.3	2.5
Lying at pasture 2002	-	-	9.6	2.2
Lying at pasture 2003	-	-	18.8	3.8

day of the AMS in the winter vs. the grazing season. This indicates that in the system used (2-units-in-row) there was sufficient free time. The difference in mean percentages between winter and grazing season was in agreement with the AMS data on milking frequencies.

During the winter season cows spent on average 3.5% of their time in the waiting area. In the grazing season this was less, although not significantly. It equalled 50.4 min in the winter season compared to 42.0 min in the grazing season. In the grazing season the number of AMS visits was less than in the winter season, which probably caused this difference. Another reason may have been the forced cow traffic in 2001 and partly in 2002. To get to the feeding gate the cows had to pass the waiting area and the AMS first. In 2003 there was free cow traffic within the barn. Cows were free to move from the lying compartment to the feeding compartment and vice versa, or to go to pasture when milking intervals allowed them to. The mean percentages of lying in cubicles and at pasture in 2001, 2002 and 2003 indicate that with free cow traffic cows seem to prefer lying at pasture, as is found by Ketelaar-de Lauwere *et al.* (1999). The total mean lying percentage did not differ between winter and grazing season. It was within the range mentioned in other studies (Albright and Arave, 1997, and references therein). Durations of 10.1 to 12.5 h for lying at pasture compared to in a barn were seen. The mean percentage in our study (45.7 vs. 43.7% in the winter vs. the grazing season equals 11.0 vs. 10.5 h lying per day, respectively.

Conclusions

Herd behaviour as well as behaviour of the selected individual cows showed comparable results. There was a difference in mean percentage of cows in the AMS between the winter and grazing season: it was significantly lower but still on an acceptable level during the grazing season. No significant differences were found for mean presence in the waiting area in front of the AMS or for the total lying behaviour. Cow behaviour was not affected dramatically over the grazing seasons. However, these results apply to the management practices described. Further research on the effect of grazing systems, pasture distances, fetching regimes, cow traffic systems or occupancies of the AMS and their interaction on herd level as well as individual level is needed to optimise the management of dairy cattle offered grazing combined with automatic milking.

Automatic milking – A better understanding

Acknowledgements

The EU and the Dutch Ministry of Agriculture, Nature and Food Quality are acknowledged for funding this project. Special thanks go to all observers and farm personnel of Nij Bosma Zathe for help with the collecting of data, Petra Lenskens for analyzing the data and Albert Meijering and Andrea Ellis for their useful comments on earlier versions of this manuscript.

References

Albright, J.L. and C.W. Arave, 1997. The behaviour of cattle. CAB International, Oxon UK, p. 38.

Altmann, J., 1974. Observational study of behaviour: sampling methods. Behaviour 49: 229-267.

Genstat 5 Committee, 1993. Genstat 5 reference manual; version 3. Clarendon Press, Oxford, 796 pp.

Ketelaar-de Lauwere, C.C., A.H. Ipema, E.N.J. van Ouwerkerk, M.M.W.B. Hendriks, J.H.M. Metz, J.P.T.M. Noordhuizen and W.G.P. Schouten, 1999. Voluntary automatic milking in combination with grazing of dairy cows; milking frequency and effects on behaviour. Applied Animal Behaviour Science 64: 91-109.

Ruis-Heutinck, L.F.M., H.J.C. van Dooren, A.J.H. van Lent, C.J. Jagtenberg & H. Hogeveen, 2001. Automatic milking in combination with grazing on dairy farms in The Netherlands. In: J.P. Garner, J.A. Mench and S.P. Heekin (Eds.), Proceedings of the 35th International Congress of the ISAE, The Center for Animal Welfare at UC Davis, Davis (CA), USA, p. 188.

INTRODUCTION OF AMS IN ITALIAN DAIRY HERDS: WELFARE ASSESSMENT BASED ON METABOLIC AND ENDOCRINE RESPONSES IN PRIMIPAROUS DAIRY COWS

F. Abeni [1], L. Calamari[2], M. Speroni[1] & G. Bertoni[2]
[1]Animal Production Research Institute, Dairy Cattle Section, Cremona, Italy
[2]Institute of Zootechnics, Faculty of Agriculture, Piacenza, Italy

Since animal welfare is a significant issue that has an impact on the public acceptance of new technology, the automatic milking system (AMS) must demonstrate excellent technical performance without causing stress responses above the levels that are generally accepted in conventional milking system. Our aim was to assess the animal welfare in primiparous cows, conventionally or automatically milked, through the evaluation of the metabolic and endocrine responses during lactation.

Twenty-nine Friesian heifers, selected for similar expected calving date, were divided into two groups, homogeneous for pedigree index, after calving. The first group (C) of 15 cows was housed in a free stall barn with a traditional herring-bone milking parlour, and milked twice daily. The second one (AMS) of 14 cows was housed in a free stall barn provided with a single box De Laval Voluntary Milking System (VMS™), with a one-way traffic system from the resting area to the feeding area, passing through the milking area. Both groups were fed once daily with the same total mixed rations (TMR). The AMS group was added with concentrate (nearly isoenergetic and isoproteic to the TMR) during milking, at the rate of 1 to 4 kg/head/d. Controls were performed on: 1) feed and diet characteristics together with daily dry matter intake (DMI) of each group; 2) metabolic (Piacenza Metabolic Profile and energy profile) and endocrine (cortisol) responses from blood samples took at 7.30 AM, before TMR distribution, at -14 d relative to expected calving date, and at 1, 7, 14, 28, 42, 70, 98, and 154 days in milk (DIM); 3) nutritional conditions, with the same frequency of blood sampling, through the evaluation of the Body Condition Score (BCS); 4) health status. The statistical analysis was performed only on data of 10 cows from each group, excluding cows that experienced problems just after calving not related to the milking system.

Only during lactation, plasma cortisol was higher in AMS ($P < 0.05$ at 7, 42 and 98 DIM) and more cows with spike values were observed in this group. These results seems to indicate some more difficulties in AMS to adapt to the system and could suggest a chronic stress situation with a stronger responsiveness to any stress-like stimulus. The stress situation could have also affected the non esterified fatty acids (NEFA) levels, which were slightly higher in AMS. This situation has also been confirmed by the observed difficulties to adapt to the feeding pattern; the AMS cows had in fact less meals in the first few days that could - together with the higher stress situation - justify the higher lipolysis. This adaptation period, at least for the feeding habit, was however very short; in fact, the BCS showed a similar pattern in both groups. The blood parameters related to the inflammatory conditions increased in both groups after calving, but ceruloplasmin values were slightly higher in AMS at 7 DIM ($P < 0.05$). The γ-glutamil transferase (GGT) values were higher in AMS at 28 DIM ($P = 0.06$). Other clinical diseases were not observed in relation to the different milking system.

We therefore conclude that the welfare at the beginning of first lactation seems equally acceptable in conventional and automatic milked cows; nevertheless, the slight stress situation suspected in AMS seems to require further researches.

INDICATORS OF STRESS IN SIMMENTAL AND BROWN SWISS COWS DURING MILKING IN A ROBOTIC SYSTEM COMPARED TO A HERRINGBONE PARLOUR

K. Hagen[1], D. Lexer[1], R. Palme[2], J. Troxler[1] &S. Waiblinger[1]
[1]Institute of Animal Husbandry and Welfare, University of Veterinary Medicine Vienna, Vienna, Austria
[2]Institute of Biochemistry, University of Veterinary Medicine Vienna, Vienna, Austria

Automatic milking systems need to be evaluated for their effect on cow welfare both with regard to milking itself and with regard to the altered management. The aspect on which we focus here concerns the milking process itself. The study was conducted in two groups of dairy cows at the Landwirtschaftliche Bundesversuchswirtschaften GmbH in Austria. One group was milked in a Lely Astronaut robot, the other was milked twice a day in a Happel 2x6 herringbone parlour with automatic stripping and cluster remover. Both groups were fed a total mixed ratio and given some additional concentrate in the milking robot and in concentrate feeders, but not in the parlour. The groups were matched for lactation, milk yield and pregnancy and consisted of 30 cows each: 15 Austrian Simmental and 15 Brown Swiss. A total of 42 healthy lactating cows were included in the observations: 12 Simmental and 11 Brown Swiss in the robotic milking group, 10 Simmental and 9 Brown Swiss in the milking parlour group. Observations were made of two to six milkings per cow, allowing repeated measures analysis of variance as well as nonparametric analysis of individual cows' average responses. The behaviour of the cows and details of the milking procedure were recorded directly by two observers. Heart rate was recorded as interbeat intervals using a telemetric system (Polar™). Samples of composite milk were frozen and later analysed for cortisol using an enzyme-immuno-assay.

Milking lasted longer in the HMP than in the AMU (general mixed model: $F_{1,39}$=12.06, P=0.0013), after significant effects of milk yield, day of lactation and time of day had been taken into account. Location of the teats by the robot took longer in Simmental than in Brown Swiss cows (Mann-Whithey-U-test: $U_{11,12}$=32, P=0.037). Kicking and stepping with the hind legs was less frequent in the AMU than in the HMP ($U_{23,19}$=76.5 for kicks; $U_{23,19}$=85 for steps; P<0.001 in both cases). Brown Swiss cows stepped less than Simmental cows in the AMU ($U_{11,12}$=32, P=0.036). In the HMP, kicks occurred more frequently during the time period from the beginning of udder cleaning until cluster attachment was completed, than at other times (Friedman test: $F_{19,4}$=32.41, P<0.001; Bonferroni-corrected posthoc-comparisons: P<0.01). There was no difference between groups in heart rate during milking. The AMS-group had higher milk cortisol values than the HMP-group ($U_{23,19}$=130.5, P=0.026).

In conclusion, cows showed more behavioural signs of discomfort during milking in the HMP-group than in the AMS-group, but they had higher milk cortisol levels in the AMS-group. The latter was probably not related to milking itself, but to other aspects of the system. Robotic milking needs to be evaluated further on the basis of the overall evidence from different studies, also taking into account conditions and factors that vary on farm-level.

EFFECTS ON FEED INTAKE, MILKING FREQUENCY AND MILK YIELD OF DAIRY COWS OF SEMI-FORCED OR FREE COW TRAFFIC IN AMS

D.J. Haverkamp[1,2], G. Pettersson[2] & H. Wiktorsson[1]

[1]Department of Animal Nutrition and Management, Swedish University of Agricultural Sciences, Uppsala, Sweden

[2]Department of Animal Nutrition and Management, Swedish University of Agricultural Sciences, Kungsängen Research Centre, Uppsala, Sweden

Introduction

The present study focused on the effect of the cow traffic system on the milking frequency (MF) and cows' milk and feed related performances. The hypothesis was that semi-forced cow traffic (SFct) would give less variation in MF's between cows, but both SFct and free cow traffic (Frct) allow cows to have sufficient feeding area visits.

Material and methods

The study involved 45 Swedish Red and White dairy cows at an automatic milking barn, but only data of 35 cows were used. During preceding SFct (30 days) the cows were forced to enter the feeding area either by passing the milking unit (MU) or one of the two controlling gates. During Frct (25 days) the feeding area was freely accessible. The cows were split up in low, middle and high MF groups on basis of their mean individual MF. During both cow traffic systems roughage was fed ad libitum.

Results and discussion

Differences between MF groups in visits to the MU and the feeding area were smaller during SFct versus Frct, due to regulation effects of the selection gates. After correction for days postpartum no significant differences in milk production were found between SFct and Frct. In general the number of MU visits was lower during Frct (P<0.001). Compared to older cows the heifers showed an increase in MF from SFct to Frct (P<0.001). The authors assume that less queuing in front of the MU during Frct, as observed in behavioral studies on similar cow traffic systems, reduces the chances of interaction between heifers and higher ranking cows. Therefor the heifers might enter the MU relatively more frequently.

Eating patterns of the cows changed from SFct to Frct. Total time spent eating roughage did not differ between both cow traffic systems, but the number of feeding area visits was higher during Frct (P<0.001). The DM intake was higher during Frct for both low and high MF groups (P<0.001). This could be in disagreement with the hypothesis that cows had sufficient feeding area visits. The milk production expressed per kg DM intake was higher during SFct (P<0.001). During both cow traffic systems the low MF groups had less efficient milk production per kg DM intake (P<0.01).

BEHAVIOR OF HIGH- AND LOW-RANKED DAIRY COWS AFTER REDIRECTION IN SELECTION GATES IN AN AUTOMATIC MILKING SYSTEM

G.G.N. Hermans, M. Melin, G. Pettersson & H. Wiktorsson
Department of Animal Nutrition and Management, Swedish University of Agricultural Sciences, Uppsala, Sweden

Cow traffic in automatic milking systems can be controlled with gates at the entrances to feeding areas, where cows are either let through (gate opens) or redirected (gate remains closed) to the milking unit (MU). The motivation and possibility for a cow to enter the MU after a redirection can be measured as redirection time, which is the elapsed time from a redirection in a gate until the next registration in the MU. Melin and Wiktorsson (unpublished) observed a considerable individual variation in redirection time; individual averages ranged from 22 to 172 minutes. In order to increase the understanding for this individual variation in redirection time, a behavior study was performed at the University Cattle Research Center Kungsängen in Uppsala, Sweden. Two groups of 12 cows each (high-ranked and low-ranked) in a controlled cow traffic system were used as focal cows in a herd of 46 cows. After a redirection in gates, the behaviour and the location of the focal cows were continuously observed until entrance in the MU or during a maximum observation time of one hour. This was repeated five times for each focal cow, generating 120 observations in total. The dominance order was established from registrations in roughage stations. Dominant cows spent on average 13 minutes in the waiting area during the observation period, to be compared to 19 minutes for subdominant cows. In 22% of all observations the cow entered the cubicles during the one-hour observation period, and when they did, they on average spent 41 minutes of totally 60 in the cubicles. No individual cow entered the cubicles in all five observation sessions, and the maximum individual average of time spent in cubicles was 25 minutes (Table 1). Individual differences were observed; 9 of the totally 24 cows never entered the cubicles, and 9 other cows were responsible for 80% of all cubicle observations.

Table 1. LS-means (minutes) of time spent in different barn areas within 1 hour after a redirection in gates for high-ranked and low-ranked cows, respectively (N=116). The range of individual means (minimum-maximum) is given within parenthesis.

Barn Area	High-ranked cows	Low-ranked cows
Waiting area in front of MU	13[a] (1-25)	19[b] (11-34)
Resting area with cubicles	14 (0-25)	10 (0-25)
Passage area in front of gates	6 (2-18)	5 (2-14)

[a] Significant difference between LS-means within rows

A long time since last feeding decreased the time spent in the resting area. The number of cows in the waiting area at the time-point of redirection was in positive relationship to the time until entrance in MU. It was concluded that there exist individual differences as well as differences due to dominance order in the behaviour after a redirection in gates. These behavioural differences could to a part explain the individual variation in redirection time observed by Melin and Wiktorsson (unpublished).

IMAGE ANALYSIS FOR MONITORING UNREST OR DISCOMFORT OF COWS DURING MILKING

A.H. Ipema, L. de Haan, J. Hemming & W.G.P. Schouten
Agrotechnology and Food Innovations B.V., Wageningen UR Wageningen, The Netherlands

Several measuring techniques are known to acquire information about the cow's behavioural and physiological status during the application of new technologies. These techniques diverge from behavioral observations till blood sampling. Behavioral observations are time consuming and often affected by the observer. Blood sampling has negative aspects for the animal welfare and should therefore be restricted.

In this research we looked for possibilities to use video imaging for analyzing the level of unrest or discomfort (movements) of cows, which were exposed to new technologies. In this case the new technology was a robotic milking system. The movements of the cows during a milking stall visit were recorded with a videorecorder. The analog signal from the videoreorder was converted in a digital video-signal and analysed with LabView-software on a PC. Different recording positions for the videocamera were tested. The best position seemed to be from above, because then the presence of not relevant moving objects in the images could be restricted. During 4 days from all milkings of 38 cows video recordings were made. With the before mentioned method these recordings were analysed. Each milking stall visit was divided into four elements (Figure 1).

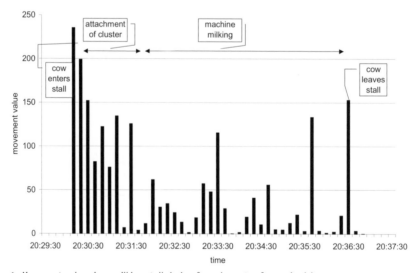

Figure 1. Movement values in a milking stall during four elements of a cow's visit.

First the entrance of a cow is detected. The staying time in the milking stall is then divided in the cluster attachment phase (here 1.5 mins) and the milk removal phase (here 4 mins). Finally the cow leaves the milking stall. For the cluster attachment and the milk removal phase movement values per unit of time (10s) were obtained. From these movement

Automatic milking – A better understanding

values means per phase were calculated. The mean values per milking stall visit (attachment and milk removal phase) varied between cows from 22 to 61. A statistical analysis of the obtained movement values showed that there was a significant difference between the attachment and the milking phase. The movement values in these phases were respectively 46 and 39 ($p<0.05$). This proves that cows were moving more during attachment than during milk removal. It was concuded that image analysis could be a useful tool for studying effects of (milking) technology on animal behaviour without any adverse effect on animal welfare.

CLINICAL EXAMINATIONS IN WELFARE ASSESSMENT IN HERDS OPERATING WITH AN AUTOMATIC MILKING SYSTEM (AMS)

I.C. Klaas, C. Fossing, M.K. Bonde & J. Hindhede
Danish Institute of Agricultural Sciences, Department of Animal Health and Welfare, Foulum, Denmark

Health is an important aspect in welfare assessment. An evaluation of health based on farm treatment records only is biased by the farmer's threshold when to initiate treatment. Systematic clinical examinations may reveal directly how system and management affect the individual cow's health and welfare as well as the herd health. The aim of the study was to develop and test a protocol for systematic clinical examinations that could be used as an applicable tool to assess health independently of farm treatment records.

The protocol for clinical examination included assessment of the locomotion system, udder health, pressure sores, skin condition, cleanliness and the overall health status as well as body condition scoring. Systematic clinical examinations were carried out in 8 Danish dairy herds operating with an AMS.

Herd sizes were 60-70 milking cows (1 automatic milking unit (AMU), n = 5 herds) or 110-130 milking cows (2 AMUs, n = 3 herds). During 2002 a veterinarian visited the farms six times. Prior to each visit, 40 to 50 cows per farm were randomly chosen for clinical examination. These cows were examined within 1.5 to 2 hours starting at feeding time.

The prevalence of lameness was 13.3%. 62.5% of cows had pressure sores on the hock, 1.6% of cows had severely inflamed hock lesions. In 37.0% of cows the hock was swollen, in 2.7% of cows the swelling was severe. Symptoms of clinical mastitis, defined as swelling or nodes in at least one quarter, were found in 3.7% of cows. Only 0.3% of cows had acute teat lesions. Asymmetry between left and right quarters was observed in 12.0% of cows, 6.7% of cows had a dry quarter. Lameness and hock lesions were the most important welfare problems detected by clinical examinations.

Systematic clinical examinations can supply valuable information in animal welfare assessment regarding the chronic disorders lameness, pressure lesions and chronic mastitis (nodes in the quarter). But due to a low prevalence of acute diseases, their impact on animal welfare on herd level cannot be evaluated by systematic clinical examinations alone.

MEASURING WELFARE IN DAIRY COWS - BEHAVIOURAL ASPECTS

Jan Olofsson
Department of animal nutrition and management, Swedish university of agricultural sciences, Uppsala, Sweden

Introduction and objectives

Finding methods to measure animal welfare is important but also relatively difficult, since the methods should be both accurate and suitable for practical use. Many attempts have been made to assess different welfare protocols to estimate the well-being of animals in a herd. However, few protocols are being used in practice. A research team at the Danish institute of agricultural sciences at Foulum has developed a protocol for measuring welfare in dairy cattle and adapted it to fit the conditions of Automatic milking systems (AMS). A part of the evaluation of the protocol was done with studies in the AMS barn at SLU cattle research centre, Kungsängen in Uppsala.

Research procedure

The AMS herd at Kungsängen was studied according to the welfare protocol at two different occasions (summer and winter). Behavioural aspects of the protocol was applied according to the instructions but somewhat expanded regarding the number of observations. The studied behaviours included for instance resting behaviour, getting up, location in the barn (activity), distance of avoidance to man, and behaviour during milking. The aim of the study was to find out how applicable the methods in the protocol were and how the number of observations affected the results.

Results and discussion

The variation in the collected data determines the number of observations needed to certify an acceptable mean value. The time it takes for a cow to get up is an example of a behaviour showing a high coefficient of variance. In theory, many observations are needed. However, fewer observations may be needed for practical use. The study at Kungsängen also detected behaviours with small variation, such as the distance when cows showed fear when the testing personnel approached them. Almost all cows in the study allowed physical contact during the test, resulting in an average distance of avoidance close to zero.

Getting-up behaviour, resting behaviour and the avoidance test are examples of low time-consuming observations, on condition that a sufficient number of cows are available. Cow activity is quickly recorded by counting the number of individuals in separate parts of the barn. However, since these recordings have to be repeated in order to get an idea of the diurnal rhythm of the heard, it ends up taking rather long time. Studying behaviour during milking is also very time-consuming, due to long milking times and the need for many observations.

Generally, animal behaviour is quite easy to measure but often difficult to fully understand. For instance, health problems such as disorders in legs or hoofs may be indicated by a long time for a cow to get up. However, the cause may also be a poor functioning cubicle (narrow, short or slippery). On the other hand, a long getting-up time may also be an effect of a cow feeling comfortable when resting or by the fact that the cow is fearless and do not respond to commands. More knowledge is certainly needed to fully understand the complexity in animal behaviour.

IMPROVED ANIMAL WELFARE IN AMS?

Jan Olofsson & Kerstin Svennersten-Sjaunja
Department of animal nutrition and management, Swedish university of agricultural sciences, Uppsala, Sweden

Introduction and objectives

Automatic milking systems (AMS) give enhanced possibilities to meet the need of the individual cow. However, has animal welfare really been improved by the introduction of AMS? A study was conducted in order to investigate whether or not the welfare of an individual was affected by its social rank. Earlier studies have indicated such results. Animals low in rank often loose in competitive situations and are sometimes referred to less attractive hours of the day for their activities. Studies have also shown that animals both have a stress- and an anti-stress system. The later increase the tolerance when strained and is stimulated by smoothing stimuli. The pituitary hormone oxytocin seems to be crucial in the activation of this system.

Research procedure

The study consisted of two experiments, both conducted in the AMS barn at SLU cattle research centre, Kungsängen in Uppsala. The social dominance order in the herd was recorded by studying the succession order at the roughage feeding stations. The 6 highest and the 6 cows of lowest rank were then studied regarding their behaviour, physiology and performance. Blood samples were drawn during milking and resting and were analysed for oxytocin and cortisol. The automatic feeding and milking system generated data on behaviour and performance. The cows were also studied by video recordings, resulting in notes of their behaviour and location in the barn. The second experiment concentrated on the behaviour and physiology in low and high ranked cows during milking and in the area close to the milking station. Additionally, a comparison with conventional "herringbone milking" was done. The last experiment is not yet analysed whereby those results are not included.

Results and discussion

It turned out that low ranked animals had higher levels of oxytocin and lower levels of cortisol, both during resting and milking. No differences were noted in behaviour during milking (tripping, kicking, manuring or urinating). Nor did feed consumption differ between high and low ranked cows. However, low ranked cows spent less time in the feeding area and had less visits to the feeding stations. They spent longer time standing in cubicles and were more often found closer to the milking station, especially during resting.

The results indicate that the low ranked cows showed no physiological signs on stress. On the contrary, their hormonal status indicated anti-stress. The low ranked cows did not have any problems supporting themselves nutritionally, but they had a more effective feeding behaviour. They probably avoided confronting higher ranked cows by standing in

dermatitis. The agreement between days when evaluated by the same person was good for BCS and OHS. The HY scores varied somewhat, most likely due to true day-to-day variation. The number of cows with PS was similar at the two registrations, but the number of observations varied. In general, the conformity between days was considered acceptable given the type of examination. The agreement between the two veterinarians was good for OHS, hoof condition and lameness. Some discrepancies were found in BCS and HY, while the numbers and degree of PS and UH remarks varied more. However, the total impression of the herd was similar. Some between-person variation is difficult to avoid but can be minimized with proper training. The presented data is based on a pilot study and the results must be interpreted with care. Further evaluation of the protocol is needed.

INTRODUCTION OF AMS IN ITALIAN HERDS: EFFECTS ON ANIMAL BEHAVIOUR

Marisanna Speroni, Susanna Lolli & Giacomo Pirlo
Animal Production Research Institute (ISZ), Dairy Cattle Section, Cremona, Italy

Diurnal behaviour pattern and usage of stable of dairy cows milked by AMS were recognised to be indicators of animal welfare (Hindhede *et al.*, 2002) and possible consequences on social synchronization of feeding and resting behaviour arise some concern. However the use of different areas in the barn depends on many other environmental factors than milking system. Two groups of Italian Friesian cows were observed seasonally by video recording in order to compare the behaviour of cows reared in identical condition and milked by AMS or twice a day by milking parlour (MP).The two groups were housed in the two opposite sides of the same free stall barn with cubicles. On one side there was an 8+8 herring-bone MP, on the other side there was a single box AMS (Delaval VMS™). Cows had access to outdoor areas. The one-way traffic was selectively guided by a selection gate, which allowed cows to access directly to the feeding area or through the AMS if interval from last milking was >5h. Once a day, at about 7.00, both groups were fed the TMR; AMS group received a concentrate supply during milking. Because our previous experience with AMS which showed a dramatic decrease of frequency of milking and passages to feeding area in the hot season, all cows were encouraged to move to the milking area twice a day, at about 5.00 and at 16.00. Cows in different areas of stable at different hours of the day were counted using videorecorded observations. Proportions of cows eating, lying or standing were also measured. The overall mean % of cows lying in cubicle did not differ between the systems but an interaction with season was found (p<0,001). Diurnal pattern of cows lying in cubicles was reported in figure. Average % of cows observed in the feeding area was affected by the season; the overall % per group were 17,84% in AMS and 21,72% in the MP (p<0,05); diurnal pattern of MP group (see figure) indicates that between 7.00 and 8.00, after the TMR distribution, 68.52% of cows were in the feeding area; between 17.00 and 19.00, after the afternoon milking, 35,79% of cows were found in the feeding area on average. Even in AMS group feeding activity was rather synchronised: at 7.00 and at 16.00 more than 35% of cows was found in the feeding area. Proportion of cows eating (with head through the feeding gate) was 12,61% and 14,36% for AMS and MP group respectively. Average of cows standing, including when milked, was smaller (p<0.05) in AMS than in MP

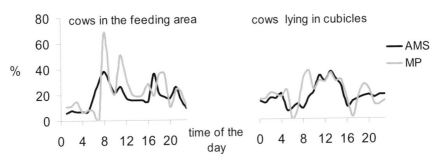

system, therefore proportion of cows total lying, including lying outdoor, was greater. On summary the behaviour patterns of the two groups resulted rather affected by the season and were consistent with human activity in the barn (feeding, cleaning stalls) and with the milking routine in PM or with the diurnal milking pattern in the AMS. The high synchronization in the AMS can be also explained by the strategy adopted in fetching cows. It could be interesting repeating the experiment without fetching cows and/or with free traffic.

Hindende J. *et al.*, 2002. Report D23, EU project, QLK5-2000-31006.

Automatic milking – A better understanding

FARM AND HERD MANAGEMENT

DEMANDS AND OPPORTUNITIES FOR OPERATIONAL MANAGEMENT SUPPORT

Wijbrand Ouweltjes & Kees de Koning
Applied Research, Animal Sciences Group, Wageningen UR, Lelystad, The Netherlands

Abstract

It is generally recognised that with automatic milking operational farm management is considerably changed compared to conventional machine milking, and is a key issue to fulfil requirements in practise. Automatic milking causes a reduction of physical labour, but an increase in decision making tasks of the dairy farmer. Decision making is supported by information obtained with sensors. Each manufacturer has different solutions for management support and performance of automatic tasks. The concept of management by exception and the distinction between normal and abnormal parameter values are discussed. Difficulties that arise when detection of abnormalities is based on correlated traits are outlined. Characteristics of sensors and their consequences are discussed for both automated decision making and decision support.

The paper also describes differences between conventional and automatic milking with respect to operational farm management. A clear difference with conventional milking is that milking intervals have to be controlled for individual cows. Feeding strategy is a key element in this, especially when grazing is applied. With regard to health, in general much more information is available, especially detection of mastitis partly depends on abnormalities detected with sensors. However, visual inspection of the animals remains an important method to control health. Because milkings are unattended, regularly abnormalities have to be checked. Due to the limited reliability of the alarms, and the fact that these are not yet integrated, this requires specific skills from the farmer. Furthermore the automatic milking system (AM system) has to be maintained and its cleaning and functioning must be controlled more or less constantly.

In order to be able to comply with legislation some of the currently existing regulation needs to be adapted, avoiding double standards. Farmers must have affinity with automation. They have to work with secure procedures, for instance first enter treatment data in the computer and then treat sick cows, and respond as should on alarms for system and animals. Udders and teats must be kept clean. Farmers using AM systems in general are satisfied with the current possibilities. Despite this, not all of the demands for automatic milking are yet fulfilled by the current systems but this is also due to inadequate legislation. Especially automatic separation of abnormal milk and secured teat cleaning should be realised. Furthermore, monitoring of the animals and equipment can be further improved. With these improvements automatic milking certainly has advantages with respect to milk quality and food safety compared to conventional milking.

Introduction

Basic requirements with regard to automatic milking are determined both by animal physiology, milk quality, food safety and farmers demands. Farm management is a key issue to fulfil requirements in practise. Compared to conventional machine milking, farm management is considerably changed with automatic milking. Some actions or decisions are completely automated and purely based on information gathered by the system. Most decision making however requires farmer's involvement. Decision making is based on information. With conventional milking a lot of this information is obtained visually around and during milking, but this is impossible with automatic milking. Therefore automatic milking systems are equipped with sensors and software. Each manufacturer has different solutions for management support and performance of automatic tasks, and users can adapt various settings to their own preferences. An overall picture of changes in management when milking with an automatic milking system however is not available.

The objectives of this paper are to describe demands and opportunities for operational management with automatic milking systems relative to conventional milking, compare these with the possibilities of currently available systems, signal shortcomings and indicate possibilities to overcome these.

Operational farm management

Management is generally regarded as a cyclical process of planning, implementation and control (Pietersma *et al.*, 1998). It contains both physical activities and decision making, for which the farmer is primarily responsible. On more automated large farms proportionately less time is spent on physical work, and more time is spent on decision making (Tomaszewski, 1993). Traditional automation is focused on automation of physical activities, but to some extent decision making can be automated as well, especially when it occurs often and is relatively simple.

What needs to be planned, implemented and controlled? In general management is divided in strategic, tactical and operational management. In this paper only operational management, referring to short term management activities, is considered. Although in The Netherlands most dairy farms produce roughage and raise their own replacements, these tasks are not worked out in this paper. It is assumed that there are no essential differences between farms with conventional or automatic milking systems with respect to operational management of these tasks. With this limitation the following operational management functions and underlying sub functions (as distinguished by Brand *et al.* (1996)) are listed in table 1.

Each sub function can be divided further in a number of processes. The herd replacement function suggests that replacement heifers are bought from elsewhere, but it is meant as a within farm purchase. Although in principle each function also contains an assessment sub function, these are only mentioned if they are to be used frequently. Welfare is regarded as part of health care and not as a separate operational function.

Table 1. Operational management functions.

Function	Sub functions
Nutrition	Grassland utilization, ration composition, control of feed supply, grazing/feeding, body condition scoring, assessment
Health care	Observation, examination, prevention, treatment
Reproduction	Observation, insemination, examination/treatment, calving assistance
Milk production	Milking, storage, milk testing, assessment
Herd replacement	Sale (including selection), purchase
Fixed assets and labour	Acquisition, maintenance, hiring
Cash management	Borrowing/investing, payments/receipts

Sensors and information

As mentioned above, automatic milking causes a reduction of physical labour and an increase of decision making tasks for the dairy farmer. Detection and diagnosis of problems or opportunities is part of decision making, as is development and analysis of alternatives and selection of follow up action (Brand *et al.*, 1996). Thus decision making is based on information. Traditionally farmers obtain information for decision making from various sources: eyes, ears, nose, feeling but also from experts and figures from management systems and milk recording. With automatic milking some information normally used for decision making is not available.

Information can also be obtained from sensors, for instance measuring milk conductivity. This information can be meant to replace otherwise missing information, but it can also be additional information meant to improve decision making. Another reason to use sensors in automatic milking systems is to be able to automate the milking process. This is because automation of the milking process also includes some on-line decision making. As an example, teats have to be detected first before teat cups can be attached. During milking teat cup attachment must be controlled, for instance by monitoring air leakage. This information is essential to automate the milking process, but is not directly useful for decision making by the farmer. As signalled by Tomaszewski (1993) and Pietersma *et al.* (1998) a growing amount of available data only improves decision making if sensor systems are reliable and the data is utilized appropriately. If data from the AM-system is combined with data from for instance milk recording, feeding and the veterinarian, this will support correct interpretation and utilization.

Management by exception

Automatic milking systems partly rely on sensor systems. With their sensors AM systems can collect enormous amounts of data, which can be processed with appropriate software. Normally most of the information when it would be studied carefully would lead to the conclusion that there is no need to take specific action, but situations that require action should be indicated. Besides that it is considered too time consuming to go through all of the information every day, this could result in lowered attention for these figures. An obvious disadvantage of this is that real problems could be overlooked. The challenge for manufacturers is to detect the abnormalities or at least help the farmer to detect them, so that action can be taken. Because abnormalities are rare this is called management by exception.

One of the acute problems that has received much attention so far is clinical mastitis, especially in relation to abnormal milk. By definition, milk of cows that suffer from mastitis has an abnormal visual appearance. It is also one of the most frequently occurring diseases in dairy cattle, and is responsible for the majority of abnormal milk. Despite this, milking a cow with abnormal milk is a rather exceptional event on most dairy farms. As an example, assume that abnormal milk is always caused by mastitis, that 50% of all cows have one case of mastitis each year, and that each mastitis case causes 10 milkings with abnormal milk. For a 100 cow herd with 310 days in milk per cow per year and 2.5 milkings per cow per day, only 0.64% of all milkings are abnormal. This clearly indicates that, even with a high mastitis frequency, the percentage of abnormalities is very low.

Detection of abnormalities

Abnormal milk is defined as milk with clots or blood (Rasmussen, 2003). Blood causes a change in colour of the milk from white to reddish, depending on the concentration. Colour of normal milk however is not fixed, and very low concentrations of red blood cells in milk do not visibly change colour. Thus the distinction between normal and abnormal milk based on colour is not sharp but to some extent arbitrary. Also measurement errors can occur, these do make it even more difficult to distinguish normal and abnormal situations. For conductivity this has been a relevant problem, because potentials that were shown in lab experiments were not repeated under practical circumstances in conventional parlours (Hamann and Zecconi, 1998; De Mol et al., 2001). Therefore sensors used to detect abnormalities for individual cows must have sufficient precision.

Correlated traits

As mentioned above, the distinction between normal and abnormal values to some extent is arbitrary for most parameters. It depends on the threshold which percentage of the observations is regarded as abnormal. There is also another reason why detection of abnormalities is difficult. For abnormal milk the reference is the presence of clots or blood (Rasmussen, 2003). Currently detection of abnormalities in milk in AM systems is largely based on conductivity. It could well be that conductivity is truly abnormal, but no clots or blood are observable. This is also known for milk with high cell counts. Although in such situations there certainly are abnormalities, alarms for abnormal milk would be considered false positive. The reference trait, visual inspection, and conductivity are not identical but are positively correlated. This means that with "normal" conductivity values the likelihood that milk is visibly abnormal is low, but with increasing conductivity this likelihood also increases. The stronger the correlation the better visible abnormalities can be detected. With correlated traits four different situations are possible, this is illustrated in figure 1.

In this figure it is assumed that both visual inspection and conductivity have a range of possible values. A visual inspection value of 80 or higher is considered abnormal, a conductivity of 80 or higher causes an alarm. Because the traits are correlated positively

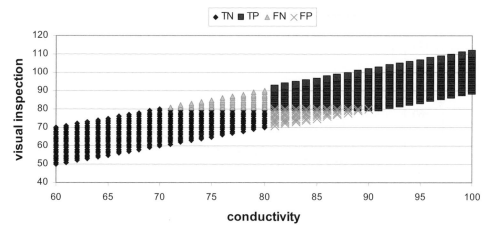

Figure 1. Detection of abnormalities with correlated traits.

visual inspection values increase (become more abnormal) with increasing conductivity. As is clearly indicated situations do exist where conductivity is below 80, but visual inspection indicates abnormality. These situations are called False Negative (FN). The other situations without alarms are called True Negative (TN), this will normally be the majority of the observations. On the other hand in some cases conductivity is ≥80, but visual inspection does not indicate abnormality. These situations are called False Positive (FP). Situations where both conductivity and visual inspection indicate abnormality are called True Positive (TP).

When the threshold is shifted to the right, more FN situations will occur and a smaller percentage of the true abnormalities will be detected. The advantage is that a larger proportion of the alarms is TP. The closer the trait(s) measured resemble the reference trait, the better the performance of the sensor will be. The ideal sensor only has TN and TP values, but then the reference trait has to be measured directly and without error. As abnormal milk is defined as milk with clots or blood (Rasmussen, 2003), sensors that measure homogeneity or colour will have a better detection performance than sensors that measure invisible chemical or physical parameters. If these parameters are reliable indicators for physiological changes they can however be very useful to improve health care.

Sensitivity and specificity

Sensor systems are often characterised by their values for sensitivity (% of abnormalities detected: TP/(TP+FN)*100) and specificity (% of non-problems classified as such: TN/(TN+FP)*100). As indicated above for each sensor these two characteristics are related, and increasing sensitivity will reduce specificity for the same sensor. For automatic separation of abnormal milk FP are costly and thus should occur rarely, whereas when alarms are generated for the farmer FP are less problematic. Because as indicated the percentage of abnormalities is usually very low, even a rather high specificity is not acceptable for automatic separation. This can be seen in figure 2.

This figure clearly shows that especially with low incidences of abnormalities a very high specificity is required to get a reasonable percentage of correct alarms.

Automatic milking – A better understanding

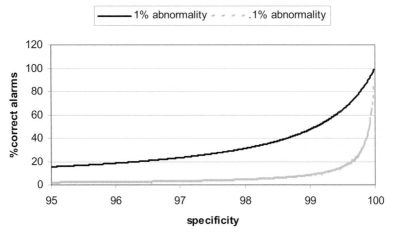

Figure 2. Relationship of specificity and percentage of correct alarms for 2 frequencies of abnormalities.

Demands

Demands are addressed to legislators, farmers and manufacturers. The demands for manufacturers are described following the operational management functions from table 1.

Legislators

In conventional milking visual inspection of animals and milk are important to ensure food quality and food safety in dairy production. Therefore according to legislation visual inspection is mandatory, although it is hardly controllable in practise. According to Rasmussen (2003) visual detection of abnormal milk at time of milking by a skilled milker has a sensitivity of ±80% and a specificity of almost 100%. With automatic milking visual inspection at time of milking is not possible. The wording of legislation should allow other solutions to control food quality and food safety, and avoid double standards. Required sensitivity of detection and testing protocols must be described in order to be able to test these solutions. Specificity is a matter of practical concern, only with a very high specificity the amount of falsely separated milk is limited.

Farmers

With AM systems an important part of the farmers daily working schedule is no longer fixed. Therefore it is considered important that management activities are planned. As described, management with AM systems to a large extent relies on the use of information gathered by these systems. To work with AM systems farmers must have faith in their AM system, and feel comfortable with management by exception.

Only a small amount of the information obtained from AM systems is really necessary because it cannot be obtained otherwise, but to a large extent it is produced to improve herd management. As described by Verstegen and Huirne (2001) the value of this information arises from improved decision making. Because it is the farmer who is responsible for decision making, the value of the information depends on the skills of the farmer to use it. This also could mean that not all farmers need the same degree of sophistication of their AM system. The extra options could be of no added value for some users, but worthwhile for others. Farmers with interest in computer technology are likely to be able to handle information from AM systems (Meskens et al., 2001). Verstegen and Huirne (2001) concluded that farmers with high management levels get more added value from management information than farmers with low management levels. Management levels were deteminded by education and training, modernisation, farm policy, tactical and operational planning and social aspects. Rougoor et al. (1998) argue that management capacity has personal aspects such as education, experience, goals and age and decision making aspects related to planning, execution and control. It is likely that the same management factors that determine milk quality with conventional milking also are relevant for automatic milking, because milk quality before and after introduction of AM systems are positively related (Van der Vorst and Ouweltjes, 2003).

AM systems

One of the major differences between conventional and automatic milking is that cows visit the milking unit voluntary. Achieving planned milking frequencies is therefore an important aspect of herd management. A challenge for the farmer is to encourage the cows to visit the milking unit regularly, without having to restrict them in access to feed or water. In general all AM systems give alarms for cows that have too long milking intervals. Manufacturers advise farmers not to fetch all cows from the list, but use these together with other available information, such as lactation stage and yield, to decide which cows will be fetched. In contrast to other alarms for individual cows interval alarms are not based on normal variation of milking intervals. Although health problems are one of the possible causes for not showing up, the alarms do not necessarily indicate problems.

With respect to the operational management functions other than health care and milk production, automatic milking does not have specific demands with regard to management information. With conventional milking important management information is obtained during milking on udder health and to a lesser extent on general health. According to legislation this information seems to be needed at time of milking, because it states that cows with inadequate health should not be milked in the bulk tank. This indicates that in certain situations milk should be separated automatically. It is however questionable whether conventional farms ever separate visually normal milk from animals that are (possibly) ill. For more severe problems it is likely that the animals do not visit the system voluntarily and appear on interval alarm lists.

Most of the demands are directly related to milking. As mentioned before, teat detection is not used for decision making by the farmer, but it is very important for the performance of AM systems. The teat cleaning device should ensure clean teats and inform about failures. According to current legislation teats should be clean, cleaning itself is not required. Before milk is directed to the bulk tank it should be checked for abnormalities, and milk that is considered abnormal should be separated. For conventional milking foremilk is to be inspected, because during milking inspection is impossible and often milk is directly sent to the bulk tank. With automatic milking and pumping the milk to the bulk tank after milking is finished it is not necessary to inspect foremilk. If teat cups are kicked off the system must intervene in such a manner that no dirt will enter the milk, and the farmer must be able to get information on the occurrence of such events. After detachment of the teat cups the system must be able to spray individual cows, preferably as efficient as possible.

The demands described above all replace activities that are normally done by the farmer during milking. There are also demands that are additional, and specific for automatic milking. While the bulk tank is emptied or cleaned, milk entry into the bulk tank must be blocked. With automatic milking due to its continuous milking this requires special provisions. Furthermore the farmer must have information available to monitor the performance of the AM system, both with regard to capacity and technical milking parameters. Preferably software does indicate potential problems and not just give figures, because problems are likely to be exceptions. The farmer also has to monitor system cleaning, both with regard to occurrence and with regard to its results. The manufacturer must give a list of checkpoints. Because some of the sensors are essential for good decision making, these must be monitored automatically and if there are doubts about their functionality this should be indicated. Service and maintenance instructions also should address sensors. Data that is essential to operate the system must be backed up in a safe way. If the system breaks down quick service must be available.

Opportunities

For farmers with health problems automatic milking might enable them to continue dairy farming. The attractivity of farm jobs could be increased, thus improving the possibilities to hire labour. As indicated above, with automatic milking also increased amounts of management information can be generated. This could improve decision making. However, as indicated by Van Asseldonk (1999), the more skills are required to use the information correctly, the less of the potential benefits are realised in practice. Because detection of abnormalities in conventional milking is subjective (Rasmussen, 2003), automatic milking systems have the potential to outperform conventional milking with regard to food quality and food safety. The chance of misidentification of cows in the milking unit is neglectable, whereas this can occur in conventional parlours. Thus, if the management database is up to date, errors in milk separation are unlikely. The onset of diseases possibly can be detected in an early stage, before animals show visible abnormalities. If adequate protocols for follow up are available, the severity of health problems can be reduced. The information from sensors could also be used to monitor recovery of animals after detection and treatment of diseases.

Especially with regard to teat end condition, the possibility to detach teat cups per quarter is a potential advantage in comparison to conventional milking (Neijenhuis and Hillerton, 2002, 2003). Compared to visual inspection, which only allows foremilk to be inspected, with sensors all milk can be inspected. If the decision to separate milk is made after the milking is finished, for instance milk yield can also be taken into account to decide. The use of additional sensors can further enhance the reliability, but also could enable more specific alarms. Continuous monitoring of sensors can be used to quickly detect and repair defect parts. Further improvement of the decision support not only is beneficial for the technical performance of the farm, but it also could help to reduce labour requirements.

Current situation and future perspectives

Automatic milking is a reality nowadays, but it is important to realise that it is still a rather new technology which is developing rapidly. Both technical aspects and user friendliness are more or less continuously improved. Farmers that are currently using AM systems are satisfied with the possibilities of their systems, and do not bother too much about false alarms. Probably this is because they do not automatically separate abnormal milk. Management of milking intervals and detection of mastitis are not solely based on the attention lists generated by the AM system, but they combine available information. The farmers needed some time to get used to this way of working, but since they are used to it they can handle all information and do not feel a need for more integrated alarms. Improved exchange of information between AM software and farm management programs however would strongly be appreciated.

Demands for automatic milking are not all fulfilled yet. Legislation is not yet adapted, especially with regard to ensure food quality and food safety. However, the EU is currently working on a new version that is also suited for automatic milking (Chalus, 2002), and a suitable definition of abnormal milk is discussed by Rasmussen (2003). As long as the new legislation is uncertain, it is difficult for the manufacturers of AM systems to focus on solutions that will fulfil the mandatory requirements. Not surprisingly, none of the available systems currently separates abnormal milk automatically. Thus it might occur occasionally that visibly clearly abnormal milk is directed to the bulk tank, although AM farms usually produce good quality milk (Van der Vorst et al., 2002). Detection of abnormalities in milk nowadays is largely based on conductivity, but with conductivity alone it is considered impossible to enable automatic separation with sufficient specificity. This is supported by results of Rasmussen & Bjerring (2004). It is expected that signals that are closer related to clots or blood, for instance colour or viscosity, will result in better performance than conductivity. Thus, automatic separation can become possible when such sensors are available.

Conductivity can also support the farmer with the detection of mastitis. De Mol and Ouweltjes (2001) showed that adding milk yield information and improving the calculation model substantially improved the detection of mastitis compared to detection based on conductivity alone. Major problem that remained was the large number of false positives. By applying fuzzy logic the number of FP values could be reduced with 95% (De Mol and Woldt, 2001). This improvement certainly is beneficial for herd management support. The reliability of alarms is increased enormously, and also the number of alarms is much less and thus easier to handle. It is assumed that further improvement is possible by increasing the number of signals used for detection. If one sensor indicates abnormality, but several other sensors give no such indication, the alarm is likely false. In order to be able to interpret all the information, it should be analysed with appropriate software. Measurement of invisible physiological parameters is only useful for decision support if there is a clear advice what to do in case of alarms. Early warnings for mastitis based on conductivity for instance so far do not really improve decision making, because treatment is usually not started unless clinical signs of mastitis are observed. A reduction of the amount of false positive alarms such as realised by De Mol and Woldt (2001) makes early treatment more realistic.

Other areas where automatic milking does not fulfil requirements according to current legislation are the inspection of animal health at time of milking and control of teat cleaning. The legislation must be updated in such a manner that these requirements could also be fulfilled with automatic milking, and manufacturers have to develop solutions to achieve the results that are intended by the legislation. It is likely that the purpose of the legislation is to increase food safety and food quality. The farmer remains responsible for the whole farm operation. This includes monitoring of the AM system, both with regard to capacity, technical performance and cleaning. The support of these tasks by automation can be improved.

Besides some unfulfilled demands, automatic milking also does have opportunities compared to conventional milking. Apart from economic perspectives resulting from further technological improvement, and enabling people with health problems to continue dairy farming, it potentially also enhances food safety and food quality and could enable improved health care.

Conclusion

Automatic milking is a reality, but both legislators and manufacturers have work ahead to enable further improvement. Automatic milking not only automates physical tasks, but to some extent also automates decision making. Currently automatic milking requires experience and skills from the farmer, especially because the information is not presented in an integrated way. Integration could reduce time required for management tasks and facilitate introduction of automatic milking on farms. Most of the demands are fulfilled, but several opportunities compared to conventional milking still have to be realised.

References

Asseldonk, M.A.M.P. van, 1999. Economic evaluation of information technology applications on dairy farms. PhD-thesis Wageningen Universiteit.

Brand, A., J.P.T.M. Noordhuizen and Y.H. Schukken, 1996. Herd health and production management in dairy practice, Wageningen pers, 543pp.

Chalus, T., 2002. Current legislation and thoughts on the coming Hygiene Directive. In: DIAS intern report no. 169, Foulum, DK.

Hamann, J. and A. Zecconi, 1998. Evaluation of the electrical conductivity of milk as a mastits indicator. Bulletin of the International Dairy Federation no 334, 23 pp.

Meskens, L., Vandermersch, M. and E. Mathijs, 2001. Implication of the introduction of automatic milking on dairy farms. Literature review on the determinants and implications of technology adoption. Deliverable D1, EU-project QLK5-2000-31006. www.automaticmilking.nl

Mol, R.M. de and W. Ouweltjes, 2001. Detection model for mastitis in cows milked in an automatic milking system. Preventive Veterinairy Medicine 49:71-82.

Mol, RM. De, W. Ouweltjes, G.H. Kroeze and M.M.W.B. Hendriks, 2001. Detection of estrus and mastitis: field performance of a model. Applied engineering in agriculture 17:399-407.

Mol, R.M. de and W.E. Woldt. 2001. Application of fuzzy logic in automated cow status monitoring. Journal of Dairy Science 84: 400-410.

Neijenhuis, F. and J.E. Hillerton. 2002. Health of dairy cows milked by an automatic milking system. Review of potential effects of automatic milking conditions on the teat. Deliverable D21, EU-project QLK5-2000-31006. www.automaticmilking.nl

Neijenhuis, F. and J.E. Hillerton. 2003. Health of dairy cows milked by an automatic milking system. Effects of milking interval on teat condition and milking performance with whole udder take off. Deliverable D22, EU-project QLK5-2000-31006. www.automaticmilking.nl

Pietersma, D., R. Lacroix and K.M. Wade, 1998. A framework for the development of computerized management and control systems for use in dairy farming. Journal of Dairy Science 81: 2962-2972.

Rasmussen, M.D., 2003. Definition of normal and abnormal milk at time of milking. Consequences of definitions of acceptable milk quality for the practical use of automatic milking systems. Deliverable D6, EU-project QLK5-2000-31006. www.automaticmilking.nl

Rasmussen, M.D., 2004. Definition of normal and abnormal milk at time of milking. Possibilities of automatic milking systems to detect and separate milk based on quality. Deliverable D7, EU-project QLK5-2000-31006. www.automaticmilking.nl

Rougoor, C.M., G. Trip, R.B.M. Huirne and J.A. Renkema, 1998. How to define and study farmers' management capacity: theory and use in agricultural economics. Agricultural Economics 18:261-272.

Tomaszewski, M.A. 1993. Record-keeping systems and control of data flow and information retrieval to manage large high producing herds. Journal of Dairy Science 76:3188-3194.

Verstegen, J.A.A.M. and R.B.M. Huirne, 2001. The impact of farm management on value of management information systems. Computers and electronics in agriculture 30:51-69.

Vorst, Y. van der, K. Knappstein and M.D. Rasmussen, 2002. Milk quality on farms with an automatic milking system. Effects of automatic milking on the quality of produced milk. Deliverable D8, EU-project QLK5-2000-31006. www.automaticmilking.nl

Vorst, Y. van der and W. Ouweltjes, 2003. Milk quality and automatic milking; a risk inventory. Report 28, Research Institute for Animal Husbandry, Lelystad, The Netherlands.

UDDER CISTERN EVALUATION TO HELP THE DECISION MAKING PROCESS OF ASSIGNING THE APPROPRIATE MILKING FREQUENCY IN AUTOMATICALLY MILKED COWS

M. Ayadi[1], I. Llach[1], G. Caja[1], I. Busto[2], A. Bach[2] & X. Carré[3]

[1]Grup de Recerca en Remugants, Departament de Ciència Animal i dels Aliments, Universitat Autònoma de Barcelona, Bellaterra, Spain
[2]ICREA and IRTA-Unitat de Remugants, Barcelona, Spain
[3]Diputació de Girona, Semega, Monells, Spain

Abstract

To take the decision on the number of milkings per day for each cow is a key factor to optimize the use of an automatic milking system (AMS). Currently, this decision is mainly taken from yield and stage of lactation data, but no udder capacity is taken into account. In this aim, external (teats, fore and rear udder) and internal (fore quarters cistern area) udder traits of a group of 30 Holstein cows milked by a 'DeLaval VMS' were evaluated and correlated with daily milk yield and average milk flow traits. External udder traits were evaluated as linear data on left side and back pictures and cistern area (left- and right-front quarters) measured by ultrasonography with a 7 to 10 h milking interval. Cows were scanned in duplicate before and after (5 min) an intramuscular oxytocin injection (OT) to evaluate cistern elasticity by difference. Significant and positive correlations were observed between milk yield and udder depth (r = 0.62) and cistern area before OT (r = 0.65). Cistern area also significantly correlated with milk flow (r = 0.49). Cistern area per quarter dramatically increased after OT injection (0 min, 7.9 cm^2; and, 5 min, 15.8 cm^2; $P < 0.001$) and cows individually differed in cistern elasticity (mean: 10.8 ± 0.8 cm^2; range: 3.18 to 19.6 cm^2). Cistern elasticity significantly correlated with milk yield (r = 0.72). Despite these observations, assigned milking frequency during the experimental period in the cows with smallest and greatest cistern elasticity were close (2.2 vs 2.5 milkings/d). The results from this study point out the interest of evaluating the area of udder cisterns, as a helping criterion of udder milk storage capacity, when establishing cow milking frequencies in an AMS.

Introduction

A key factor to optimize the use of an automatic milking system (AMS) is to take the right decision about the number of milkings per day for each cow. The decision making process is currently taken from yield and stage of lactation data, but no storage capacity of the udder is taken into account. One of the expected benefits from the use of an automatic milking system (AMS) is the increase in milk yield as a consequence of a more frequent milking. Increasing milking frequency from 2 to 3 times per day increases milk yield by 3-39% in conventional farms (Pearson et al.,1979; Klei et al.,1997; Smith et al., 2002). When milking frequency rises up 3 times per day, milk yield increases by 6-14% (Van der lest and Hillerton, 1989; Bar-Peled et al., 1995; Henshaw et al., 2000). A fixed effect of number of milkings on milk yield have been proposed at long term (3.5 kg/milking) by Erdman and Varner (1995) and at short term (1.6 l/milking) by Caja *et al.* (2000) and Ayadi *et al.* (2004), in Hostein dairy cows.

Daily milking frequency averages 2.3 to 2.8 milkings/cow in most AMS farms and number of milkings per cow range from <2 to >3.4 (De Koning and Ouweltjes, 2000; Wendl *et al.*, 2000). Nevertheless, the increase in milking frequency obtained with AMS does not confirm the expected milk yield results in most cases. Svennersten-Sjaunja *et al.* (2000) reported an increase in milk yield of 7% in Sweden, and Veysset *et al.* (2001) reported increased milk yield by only 3% in the first year, and by 9-13% after two years of AMS installation in French dairy farms.

Total number of milkings per day is also a main limiting factor in the capacity of AMS farms, and in consequence most farmers try to maximize the amount of milk milked daily by optimizing the number of cows milked per day and by taking the right decision on the number of milkings per each cow. Available knowledge recommends to use high milking frequencies in early lactation (3-4 milkings/day), to let the cow to express their milk yield potential, and to reduce milking frequency thereafter. Milking frequency in mid lactation ranges between 3 and 2 milkings/day, according to AMS capacity, and less than 2 milkings/day are usually applied during late lactation.

Negative effects of milking frequency reduction are related to udder cistern size, with the large cisterned cows being less affected by long milking intervals (Knight and Dewhurst, 1994; Ayadi *et al.*, 2003b) and, on the contrary, lower increases in milk yield should be expected when milking frequency is greater in these cows.

Udder morphology and cow behavior are also taken into account to improve the adaptation of cows to AMS (De Koning and Ouweltjes, 2000; Caja *et al.*, 2002) and to maximize the number of cows milked daily. Moreover, milk flow has a positive effect on both number of milkings and capacity of AMS in dairy cows (De Koning and Ouweltjes, 2000).

The aim of this work was to study the use of udder morphology traits as complementary criterion to assist in decision of establishing the adequate number of milkings per day for each cow in an AMS.

Materials and methods

Animals, feeding and routine milking

The study was carried out using 30 multiparous Holstein Friesian cows from the experimental farm of Semega in Monells (Girona, Spain). Milk yield and days in milk during the experimental period averaged 31.8 ± 1.9 l/d and 178 ± 10 d, respectively. Cows were in a single group of 65 milking cows, with loose housing and fed ad libitum a total mixed ration containing 1.53 MJ EN_l and 17.2% CP. Milking was conducted automatically with an AMS single unit ('DeLaval VMS') with a free traffic of cows. Milkings per cow ranged between 1.75 to 3.30 milkings per day.

Experimental procedures

External and internal udder morphology was evaluated at random for each cow during an experimental period of 2 weeks. Selected udder morphological traits (udder depth and teat length) were evaluated as linear data on individual digital pictures (Sony MpegMovie

HQX, Cyber-shot 5.0 Megapixels) by using a breakdown system in wich independent traits were pointed according to a linear scale of 9 points (Pérez-Cabal and Alenda, 2002).

Internal udder morphology was evaluated by measuring cisternal udder area by ultrasonography of the left and the right front udder quarters. Udder scans were performed according to the methodology proposed by Ayadi et al. (2003a) using a real time B-mode ultrasonograph (Ultra Scan 900, Ami Medical Alliance Inc., Montreal, Canada) with a 5 MHz sectoral probe (80° scanning angle). Images were transmitted to a portable computer and processed in triplicate by using an image processing software (MIP4 Advanced System, Microm España, Barcelona, Spain). Area in pixels were converted to cm^2 (1 cm^2 = 1,024 pixels) as Ayadi et al. (2003b).

Small groups of cows (2-4 cows) were locked in the feed bunk 7 to 10 h after milking and scanned in duplicate before and after (5 min) an intramuscular oxytocin (OT) injection (20 IU/cow). Cisternal elasticity was calculated by difference between the cisternal area measured before (0 min) and after (5 min) the OT injection.

Cows were brought to the AMS immediately after the last scan and milked. Individual milk yield and milk flow were recorded at each milking during the experimental period and averaged for calculations.

Statistical analysis

Data were processed by the PROC MIXED for repeated measurements of SAS (v. 8.1). The model included the fixed effects of the treatment time (0 and 5 min) and the specific milking interval (7 to 10 h); the random effects of animal (1 to 30) and quarter side (left and right); the correspondent interactions; and the residual error. Significance was declared at $P < 0.05$. When the probability of the interaction term was not significant ($P > 0.20$), it was removed from the model. Peason's correlation coefficients between parameters were also calculated.

Results and discussion

Ultrasonography images obtained in our work were similar to those previously reported in dairy cows (Bruckmaier and Blum, 1992; Bruckmaier et al., 1994; Ayadi et al., 2003a). No differences were observed between left and right front udder quarters and therefore, their values were averaged and discussed jointly.

Values of cistern area per quarter varied largely between cows (2.9 to 36.3 cm^2) and differed according to time after OT injection. Cows also differed in initial value of cistern area and were categorized in three groups according to cisternal size (small, medium and large) as shown in table 1. Cisternal area increased dramatically after OT injection, reaching its maximum value at 5 min vs 0 min (15.8 vs 7.9 cm^2; $P < 0.001$). The increase in cisternal area after OT injection was due to milk transfer from alveoli to cistern when the myoepithelial cells contracted by effect of OT (milk letdown) as previously reported (Bruckmaier and Blum, 1992; Caja et al., 2003). As a result, cows differed in cistern elasticity (overall mean: 10.1 ± 0.8 cm^2; range: 3.2 to 19.6 cm^2). Moreover, increase in cisternal area after OT injection varied according to animal, and 4 cows (13%) of the small-cisterned group showed a great enlargement of the cisterns reaching the same final area as the cows of the large-cisterned group. Elasticity value of the small-cisterned cows (78% of increase) also indicated that these cows did not reach their maximum cisternal size (114 and 118% in medium- and large-cisterned cows, respectively) and, as a consequence, they should be able to resist a reduction in daily milking frequency.

Daily milk yield : cistern area ratio decreased as cistern size increased in all cow groups (Table 1) and cows in the large- and medium-cisterned groups showed a lower milk holding capacity than small-cisterned cows ($P < 0.01$).

Positive correlations ($P < 0.01$) were observed between milk yield and udder depth (r = 0.62). These results agree with those reported by Normen and Van Vleck (1972) and Labussière and Richard (1965), who reported values of 0.27 and 0.50, respectivelly). Moreover, positive correlations were also observed between daily milk yield and cisternal area before (r = 0.65) and after (r = 0.68) the OT injection, and with cistern elasticity (r = 0.72). These values are similar to the correlation found between udder volume and milk yield in dairy cows (Labussière and Richard, 1965; Knight and Dewhurst, 1994). The obtained results indicate that animals with large and more elastic cisterns should be more efficient producers of milk and better adapted to reduced milking frequencies. Moreover, the evaluation of cistern area at 5 min can be considered as a methodology to estimate the maximum size of cisternal area in dairy cows, as previously indicated by Caja et al. (2003).

The positive correlation between milk flow and cistern area measured by ultrasonography before (r = 0.49) and after OT injection (r = 0.40) agrees with those reported by Labussière and Richard (1965) who observed a negative correlation (r = - 0.27) between milk flow and conjunctive tissue, indicating that animals with great udder cisterns should be milked faster. A negative correlation (r = - 0.51) was also observed between milk flow and teat length in our study indicating that cows with short teats milked more quickly in AMS. Similar results

Table 1. Milking characteristics of dairy cows in an AMS according to udder cistern size.

Item	Initial size of udder cistern (0 min)		
	Small	Medium	Large
Cows, n	9	11	10
Milk yield, l/d	25.1 ± 1.5[a]	30.0 ± 1.4[b]	34.0 ± 1.4[c]
Lactation stage, d	193 ± 13	185 ± 20	158 ± 16
Daily milking frequency	2.2 ± 0.1[ab]	1.9 ± 0.1[b]	2.5 ± 0.2[a]
Milking interval, h:min	8:33 ± 0:31	9:30 ± 0:30	8:10 ± 0:19
Cisternal area, cm²/quarter:			
0 min	7.2 ± 1.0[b]	9.5 ± 0.8[ab]	13.3 ± 1.6[a]
5 min	12.7 ± 1.3[c]	20.3 ± 1.9[b]	29.0 ± 1.7[a]
Elasticity	5.6 ± 0.2[c]	10.8 ± 0.2[b]	15.7 ± 0.1[a]
	(78%)	(114%)	(118%)
Milk yield:Cisternal area[1] ratio, ml/cm²			
0 min	873 ± 10[a]	790 ± 5[b]	640 ± 3[c]
5 min	495 ± 5[a]	370 ± 2[b]	293 ± 2[c]
Elasticity	1113 ± 12[a]	695 ± 4[b]	543 ± 3[c]

[1]Cisternal area estimated as the quadruple of average front quarters; [a, b, c]: Different letters in the same column indicate differences at $P < 0.05$.

were previously reported in conventional milking (Stallkup *et al.*, 1963; Labussière and Richard, 1965; Le Du *et al.*, 1994). No correlation was observed between cistern area, either before or after OT injection, and teat length. Milking frequency during the experimental period in the cows with smallest and largest cistern elasticity were similar (2.2 vs 2.5 milkings/day; $P > 0.05$) as shown in table 1, indicating that number of milkings were not optimized for each cow in our case.

In conclusion, we propose to use the evaluation of the udder cisterns area, as a helping criteria of udder milk storage capacity, when establishing milking frequencies for each cow in an AMS.

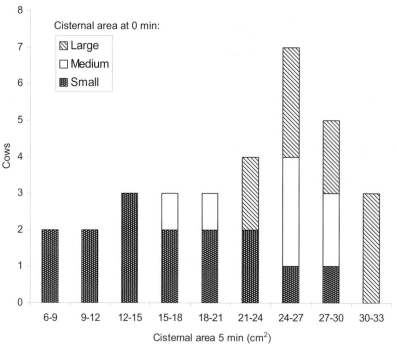

Figure 1. *Distribution of maximum cisternal area of dairy cows according to initial size of front quarters udder cistern.*

References

Ayadi, M., G. Caja, X. Such, and C.H. Knight, 2003a. Use of ultrasonography to estimate cistern size and milk storage at different milking intervals in the udder of dairy cows. Journal of Dairy Research 70: 1-7.

Ayadi, M., G. Caja, X. Such, and C.H. Knight, 2003b. Effects of omitting one milking weekly on lactational performances and morphological udder changes in dairy cows. Journal of Dairy Science 86: 2352-2358.

Ayadi, M., G. Caja, X., Such M., Rovai, and E. Albanell, 2004. The effect of different milking intervals on milk composition of cisternal and alveolar milk in dairy cows. Journal of Dairy Research (in press).

Bar-Peled, U., E. Maltz, I. Bruckental, Y. Kali, H. Gacitua, A.R. Lehrer, C.H. Knight, B. Robinzon, H. Voet, and H. Tagari, 1995. Relationship between frequent milking or suckling in early lactation and milk production of high producing dairy cows. Journal of Dairy Science 78:2726-2736.

Bruckmaier, R.M. and J.W. Blum, 1992. B-mode ultrasonography of mammary glands of cows, goats and sheep during α- and β-adrenergic agonist and oxytocin administration. Journal of Dairy Research 59:151-159.

Bruckmaier, R.M., E. Rothenanger, and J.W. Blum, 1994. Measurement of mammary gland cistern size and determination of the cisternal milk fraction in dairy cows. Milchwissenschaft 49:543-546.

Caja, G., M. Ayadi, C. Conill, M. Ben M' Rad, E. Albanell, and X. Such, 2000. Effects of milking frequency on milk yield and milk partitioning in the udder of dairy cows. In: H. Hogeveen and A. Meijering (editors), Robotic Milking. Wageningen Pers, The Netherlands, pp. 177-178.

Caja, G., M., Ayadi, X. Carré, and M. Xifra, 2002. Los robots de ordeño en España: Situación actual y perspectivas. In: Ordeño Robotizado. Ed. Agrícola Española S.A., Madrid, Spain, pp. 11-16.

Caja, G., M. Ayadi, and C.H. Knight, 2003. Evidence of cisternal recoil after milk letdown in the udder of dairy cows. Journal of Dairy Science 86(Suppl. 1):117 (Abstr).

De Koning, K. and W. Ouweltjes, 2000. Maximising the milking capacity of an automatic milking system. Automatic milking: reality, challenges and opportunities. In: H. Hogeveen and A. Meijering (editors), Robotic Milking. Wageningen Pers, The Netherlands, pp. 38-46.

Erdman, R.A. and M. Varner, 1995. Fixed yield responses to increased milking frequency. Journal of Dairy Science 78:1199-1203.

Henshaw, A.H., M. Varner, and R.A. Erdman, 2000. The effects of six times a day milking un early lactation on milk yield, milk composition, body condition and reporoduction. Journal of Dairy Science 83(Suppl. 1):242 (Abstr.).

Klei, L.R, J.M. Lynch, D.M. Barbano, P.A. Oltenacu, A.J. Lednor, and D.K. Bandler, 1997. Influence of milking three times a day on milk quality. Journal of Dairy Science 80:427-436.

Knight, C.H. and R.J. Dewhurst, 1994. Once daily milking of dairy cows: relationship between yield loss and cisternal milk storage. Journal of Dairy Research 61:441-449.

Labussière, J. and P. Richard. 1965. La traite mécanique: aspects anatomiques, physiologiques et technologiques. Mise au point bibliographique. Annales de Zootechnie (Paris) 14:63-126.

Le Du, J., F.A. Chevalerie, M. Taverna, and Y. Dano. 1994. Aptitude des vaches à la traite mécanique : relation avec certaines caractéristiques physiques du trayon. Annales de Zootechnie (Paris) 43:77-90.

Norman, H.D. and L.D. Van Vlek, 1972. Type appraisal: III. Relationships of first lactation production and type with life time performance. Journal of Dairy Science 55:1726-1734.

Pearson, R.E., L.A. Fulton, P.D. Thompson, and J.W. Smith, 1979. Three times a day milking during the first half of lactation. Journal of Dairy Science 62:1941-1950.

Pérez-Cabal, M.A. and R. Alenda, 2002. Genetic relationships between lifetime profit and type traits in Spanish Holstein cows. Journal of Dairy Science 85:3480-3491.

Smith, J.W., L.O. Ely, W.M. Graves, and W.M. Gilson, 2002. Effect of milking interval in DHI performance measures. Journal of Dairy Science 85:3526-3533.

Stallkup, J.M., J.M. Rakes, and G.L. Ford, 1963. Relationship between milk flow and anatomical characteristics of udder. Journal of Dairy Science 46:624-625.

Svennersten-Sjaunja, K., G. Pettersson, and I. Bergulund, 2000. Evaluation of the milking process in an automatic milking system. In: H. Hogeveen and A. Meijering (editors), Robotic Milking. Wageningen Pers, The Netherlands, pp.196.

Van der Iest, R. and J.E. Hillerton, 1989. Short-term effects of frequent milking of dairy cows. Journal of Dairy Research 56:587-592.

Veysset, P., P. Wallet, and E. Prugnard, 2001. Le robot de traite: pour qui? Pour quoi? Caractérisation des exploitations équipées, simulation économiques et éléments de réflexion avant investissement. INRA Productions Animales 14:51-61.

Wendl, G., J. Harms, and H. Schon, 2000. Analysis of milking behavior on automatic milking. In: H. Hogeveen and A. Meijering (editors), Robotic Milking. Wageningen Pers, The Netherlands, pp. 143-151.

EFFECTS OF MILKING MACHINE PARAMETERS ON THE MEAN MILK FLOW RATE IN A ROBOTIC MILKING SYSTEM

A.H. Ipema & P.H. Hogewerf
Agrotechnology and Food Innovations B.V., Wageningen UR Wageningen, The Netherlands

Abstract

In this experiment the focus was on the machine on time needed for the milk removal process. This experiment aimed to examine the effect of four milking machine treatments on the mean milk flow rate. In the four test situations the milking machine settings were: A) milking- and pulsation vacuum 42 kPa; pulsation ratio 65/35; pulsation rate 58 cycles per minute; detachment level 200 g per minute; B) Settings as in A except the detachment level was increased from 200 to 350 g per minute; C. Settings as in B except the vacuum level was increased from 42 to 45 kPa; D. Settings as in C except the pulsation ratio varied between 55/45 and 70/30 depending on the actual milk flow rate.

Treatments A, B, C and D were applied during three 16-d periods. In a 16-d period each treatment was used for 4 consecutive days. The order of the test situations differed in each 16-d period. The experiment was carried out in a two-box robotic milking system (Prolion). Cows were accepted for milking 6 hours after the previous milking visit. When intervals became longer than 12 hours cows were fetched. During each milking visit 0.5 till 2.0 kg of concentrates were fed. Data from 54 cows that were on the system during the three test periods were analyzed.

The mean milk flow rates differed significantly ($p<0.05$) between all treatments and were 1.65, 1.81, 1.90 and 1.84 kg per minute for respectively the A, B, C and D treatment. With higher milk yields (range 3-18 kg) and longer attachment durations (range 45-120 s) per milking milk flow rates increased significantly. Reactions on the treatments of individual cows sometimes showed clear deviations from the general trends.

Introduction

High investment and maintenance costs require an efficient usage of robotic milking systems. Aspects that should be considered in this respect are for example the features of the robotic milking system, the milking technology, the herd and milking management, the housing situation and the milk yield and milkability properties of the herd.

The aim of this research was to analyze effects of different milking technology treatments. However, the automated attachment of the teat cups implies variable durations of this procedure. This leads to differences in the time between attachment of the first teat cup, which evokes milk ejection, and the moment all teat cups are attached and the milking process starts. The differences in attachment durations will affect the milk removal efficiency. Ipema *et al.* (1997) found a significantly increased mean milk flow rate for up to three minutes waiting time.

Further, milking management in robotic milking systems implies that cows will be milked with variable intervals. In this research intervals for acceptance of cows for milking was set at 6 hours, but because of the voluntarily visiting regime the intervals between milking varied. Maximum milking intervals were determined by the fetching regime. The consequence of the variable intervals is that milk yields per milking also show a wide range. De Koning & Ouweltjes (2000) showed that the milk flow rates increased with higher milk yields per milking.

The impact of attachment duration and milk yield per milking will therefore have to be taken into account when effects of milking technologies are studied.

Finally, it is well known that the milkability can differ largely between cows. These properties are hereditary, but will also show some changes with parity and lactation stage.

Materials and methods

The research was carried out on the experimental farm 'De Ossekampen' in Wageningen, The Netherlands. The cows were milked in a 2-stall robotic milking system (Prolion). There were at least 74 cows on the system during the experimental period. In the barn forced cow traffic was applied, which means that the cows had to pass a selection unit when they move from lying to feeding area. In the selection unit was decided whether a cow should be send to the waiting area in front of the milking stall or to the feeding area. The minimum interval for accepting a cow for milking was set at 6 hours for all cows in the herd. Fetching was done at fixed times (7:00 and 16:00 h) for cows with milking intervals longer than 12 hours. The milking installation was cleaned three times a day. During each milking stall visit at least 0.5 and maximal 2.0 kg of concentrates were fed. The amount per cow depended on the milk yield level. The composition of the forage ration fed at the feeding gate was not changed during the experiment.

After a cow had entered a milking stall, the milking robot system started to locate the position of the individual teats after which the attachment of the teat cups was carried out one by one. After attachment during a short period (10-15 s) water was injected in the teat cup in order to clean the teat and to stimulate milk ejection. Cleaning water, dirt and some first milk were separated into a dump line. When all teat cups were attached the milking system started a short stimulation function. At the end of a milking an overmilking prevention function was applied.

The experiment focused on the effects of four milking-machine treatments:
A. Reference situation: settings for the milking machine parameters were: milking- and pulsation vacuum levels of 42 kPa, a pulsation ratio of 65/35, a pulsation rate of 58 cycles per minute and a detachment level of 200 g per minute;
B. Increased detachment level: settings as in A, except the detachment level was increased from 200 to 350 g per minute;
C. Increased vacuum level: settings as in B, except the vacuum level was increased from 42 to 45 kPa;
D. Milk flow dependent pulsation: settings as in C, except the pulsation ratio varied between 55/45 and 70/30 depending on the actual milk flow rate.

Each milking-machine treatment was applied during a period of 4 consecutive days and was repeated three times. The order of the treatments within a repetition was drawn.

Data about the milking process were automatically recorded by the robotic milking system. For each milking-machine treatment information from all milkings of all cows during 12 days (4 days per treatment and 3 repetitions) was available. In the analyses data from 54 cows, which were at the system during the complete experimental period, were used.

Results and discussion

For a proper evaluation of different milking technologies besides effects on the milk removal efficiency also possible influences on milk yield and udder and teat health are of importance. However, in this experiment the treatment periods were too short for detectable effects on udder and teat health. Therefore, here we will only discuss milk yield and milk removal.

Milk yield
The daily milk yield for each cow is calculated from the sum of the yields of all milkings in a day divided by the sum of all preceding intervals (in hours) of that milkings multiplied by 24. The mean daily milk yield of the 54 analyzed cows was 28.9 kg and varied between 15.0 and 46.5 kg for the individual cows.

The mean daily milk yields for the milking machine treatments A, B, C and D were respectively 28.9, 29.0, 28.9 and 28.7 kg. There were no significant differences between these yields.

Milk removal
In this experiment 4609 milkings were analyzed for aspects that affect the milk removal efficiency. The milk yields in the analyzed milkings varied between 3 and 18 kg. The distribution of these yields is given in figure 1.

The average milk yield per milking was about 10 kg. About two third of the milkings had a yield between 6 and 12 kg. The average milk yield per milking for a single cow depends on her daily yield and milking frequency.

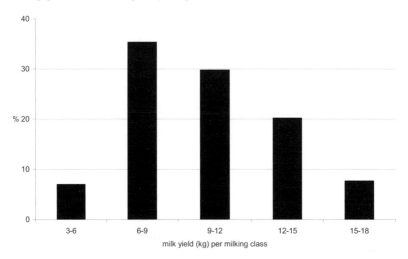

Figure 1. Frequency distribution of milk yields (kg) per milking.

Automatic milking – A better understanding

Only milkings with attachment durations less than 2 minutes were analyzed. It took at least 45 s after the start of the procedure for locating the teats before the robotic system attached the first teat cup. In more than 50% of the analyzed milkings all teat cups were attached within one minute. In about 12% of the milkings this procedure took between one and a half and two minutes.

The effects of the milking technology treatments on the mean milk flow rate were analyzed together with the effects of milk yield per milking and attachment duration per milking (Table 1). For the milk yield per milking the classes as shown in figure 1 were used. For the attachment duration the classes were 45-6-, 60-90 and 90-120 s.

In formulae the used statistical model was: $y = \beta + u$

In which:

y: mean milk flow rate

β: a vector with the fixed effects of the milking technology treatments (A, B, C, D), milk yield per milking classes (3-6, 6-9, 9-12, 12-15 and 15-18 kg) and attachment duration classes (45-60, 60-90 and 90-120 s)

u: a vector with random effects of cow (1,..., 54)

Table 1. Effects of milking technology treatments, milk yield per milking and attachment duration on the mean milk flow rate.

Milking technology treatment	A	B	C	D	
Mean milk flow rate (kg/min)	1.65 [a]	1.81 [b]	1.90 [d]	1.84 [c]	
Milk yield (kg) per milking	3-6	6-9	9-12	12-15	15-18
Mean milk flow rate (kg/min)	1.33 [a]	1.61 [b]	1.83 [c]	2.01 [d]	2.23 [e]
Attachment duration (s) per milking	45-60	60-90	90-120		
Mean milk flow rate (kg/min)	1.73 [a]	1.79 [b]	1.88 [c]		

[abcde] values with different subscripts in the same row are significantly different (p<0.05)

With an increase of the detachment level from 200 to 350 g per minute (milking technology treatment A vs. B) the mean milk flow rate increased significantly (p<0.05) from 1.65 to 1.81 kg per minute (10%). This increase corresponds with results from earlier research by Ipema & Hogewerf (2002).

When in addition the vacuum level was increased from 42 to 45 kPa the mean milk flow rate also significantly increased to 1.90 kg per minute. Compared with the reference treatment A the increase was 15%. The mean milk flow rate at the milk flow dependent pulsation ratio (treatment D) was significantly (p<0.05) lower than at treatment C, but larger than at treatment B.

With increasing milk yield classes the mean milk flow rates were significantly higher. The largest increase was found between the classes '3-6' and '6-9'. The relation between milk yield per milking and milk flow rate is already known from literature (Ipema et al., 1997, de Koning & Ouweltjes, 2000).

The attachment duration classes also had significant effects on the mean milk flow rates. The effect from class '60-90' to '90-120' was larger than from class '45-60' to '60-90'. This was probably caused by a better milk removal from the beginning of the milking process, because of a larger amount of available cisternal milk. The effect of the attachment duration or time delay between first contact of teat cleaning or milking equipment with teats or udder was also found in other researches (Ipema *et al.*, 1997, Macuhova & Bruckmaier, 2000).

In further analyses, it was found that there were interactions between the milking technology treatments and the milk yield per milking classes and the attachment duration classes. The results are summarized in figures 2 and 3. In these figures the increases in the mean milk flow rates are given for the comparison of treatment A with B (effect of increased detachment level), treatment B with C (effect of increased vacuum level with fixed pulsation ratio) and treatment B with D (effect of increased vacuum level with milk flow dependent pulsation ratio).

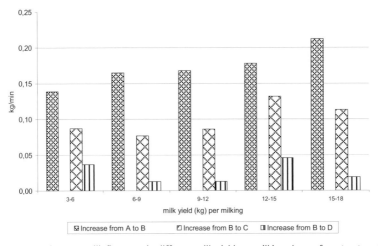

Figure 2. Increases in mean milk flow rate in different milk yield per milking classes from treatment A to B, B to C and B to D.

The absolute effect of an increased detachment level was larger with increased yields per milking. In larger milk yields the descending phase of the udder milk flow pattern will be longer. This explains the larger effect of the increased detachment level at higher milk yields.

The increases of the mean milk flow levels caused by an increased vacuum level with a fixed pulsation ratio showed a trend to be larger with higher milk yields per milking. An increased vacuum level with a milk flow dependent pulsation ratio showed a small significant increase in the mean milk flow rate (Table 1). A clear effect of the milk yield level on this increase was not found (Figure 2).

When the detachment level was increased from 200 to 350 g per minute the increases in the mean milk flow rates were larger with longer attachment durations. The chance that the milk removal of some of the teats already started before all four teats were attached will be increased when the attachment duration was longer. This will result in larger differences in the milk yield of the quarters and will have lead to a longer descending phase of the milk flow pattern from the whole udder. The effect of an increased detachment level will then be larger.

Automatic milking – A better understanding

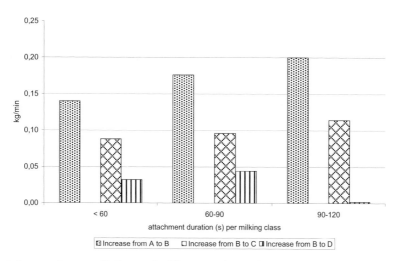

Figure 3. Increases in mean milk flow rate in different attachment duration per milking classes from treatment A to B, B to C and B to D.

The increases of the mean milk flow levels caused by an increased vacuum level with a fixed pulsation ratio or with a milk flow dependent pulsation ratio were not significantly affected by the attachment durations.

In figure 4 the effects of the milking machine treatments for three individual cows are compared with the effects for the whole herd.

The mean milk flow rate of the three cows in the reference treatment A was at the average level for the whole herd (all). Figure 4 shows that the reactions of the three cows on the milking machine treatments widely differ from the herd average and among each other.

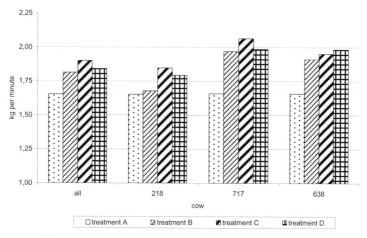

Figure 4. Mean milk flow rates at the four treatments for the herd and for three cows.

Conclusion

The daily milk yield per cow was not significantly affected by the milking machine treatments. The mean milk flow rates differed significantly between all four milking machine treatments. An increase of the detachment level from 200 to 350 g per minute resulted in a 10% higher mean milk flow rate. Increasing the vacuum level from 42 to 45 kPa gave another 5% higher mean milk flow rate. When at the increased vacuum level a flow dependent pulsation ratio was applied the increase of the mean milk flow rate was only 1%.

The results showed that there are possibilities to improve milk removal efficiency by changing milking machine settings. The differences in the effects between cows and milk yield levels showed that a more controlled application depending on actual and historic (quarter) milk flow information is advisable.

References

Ipema, A.H., C.C. Ketelaar-de Lauwere, C.J.A.M. de Koning, A.C. Smits and J. Stefanowska, 1997. Robotic milking of dairy cows. Beiträge zur 3. Int. Tagung Bau, Technik und Umwelt in der landwirtschaftlichen Nutztierhaltung, Kiel, Deutschland, p. 290-297.

Ipema, A.H and P.H. Hogewerf, 2002. Detachment criteria and milking duration. Proceedings of the First North American Conference on Robotic Milking, Toronto, Canada, P. II-33-44.

Koning, de K. and W. Ouweltjes, 2000. Maximising the milking capacity of an automatic milking system. Proceedings of the International Robotic Milking Symposium, Lelystad, The Netherlands, p. 38-46.

Macuhova, J. and R.M. Bruckmaier, 2000. Oxytocin release, milk ejection and milk removal in the Leonardo multi-box automatic milking system. Proceedings of the International Robotic Milking Symposium, Lelystad, The Netherlands, p. 184-185.

CONSIDERATIONS AT ESTABLISHMENT OF AUTOMATIC MILKING SYSTEMS IN EXISTING HERD FACILITIES

Troels Kristensen & Egon Noe
Danish Institute of Agricultural Sciences, Department of Agroecology, Foulum, Denmark

Abstract

The aim of this study was to describe the process going from the first considerations about buying an automatic milking system (AMS) to the short-term results (1/2 to 3/4 year). Focus was on feeding, working conditions, and herd management. In addition to this, it was a specific aim to discover conditions of vital importance for the functioning of the AMS on the farm. We conducted qualitative interviews with six farmers. This approach was chosen to get an in-depth insight in how these farmers have experienced the change to AMS and the considerations they have been though.

The farmers involved were minded for the learning process necessary for the implementing of AMS on the farm. The implementation of AMS calls for a complete reconstruction of the labour processes on the farm. A higher degree of flexibility benefits the social family life on the other hand it is important to have a network of skilled people around the farm that can look after the system when the family is away. AMS can be fitted into many types of existing dairy production systems as seen in relation to type of stable, fodder supply, and yield level in the herd.

The work pointed out three problem complexes of special relevance for research and development of dairy production systems with AMS: 1) AMS challenges our knowledge about the relationship between feeding, milking, health and animal behaviour, and the abilities for controlling and optimising these factors. 2) There is a need for focusing the data from the AMS to the condition in an AMS herd, including user friendly presentations, before these data are transformed into a useful decision support tools for the farmers. 3) The present models for economic optimising are not sufficiently specific, when an investment in an AMS is being considered. AMS influences and is influenced by a series of production factors on the farm, such as labour, fodder supply, and utilization of quotas.

Introduction

Farmer's motivation for investment in automatic milking systems (AMS) is of interest both for advisors and research in order to meet the demand from this rapid increasing group of dairy farmer. The AMS differs from most other agricultural technologies in that it is not only a new way of doing an operation, milking, at the farm, but it has impact on many other aspects of the production and management at the farm, including the more social dimensions (Koning & Vorst, 2002). The discussion of AMS as a technology is therefore not sufficient without taken these elements into consideration. Most of the AMS are introduced in existing stables (Meskens *et al.*, 2001), and for a major part of the farmers are the investment not based on economic, but social reasons (Meskens *et al.*, 2002). These facts mean that the understanding of introduction of AMS at a dairy farm and the impact of AMS on the farm has to be seen in a holistic context.

The aim of this study was to describe the process going from the first considerations about buying an automatic milking system (AMS) to the short-term results (1/2 to 3/4 year). Focus was on feeding, working conditions, and herd management. In addition to this, it was a specific aim to discover conditions of vital importance for the functioning of the AMS on the farm, and by that to identify problem complexes of special relevance for research and development of dairy production systems with AMS.

Material

The study was based on qualitative interviews conducted with six farmers. The farms were chosen from a pool of 24 farms that participated in other research activities. The criteria for selection of the farms were that AMS was implemented into an existing system for about 1/2 a year before the visit. This was done in order to get information that was as closely related to AMS as possible, and not, as in some cases, a mix of AMS introduction and increase of herd size or new buildings.

Results and discussion

Noe & Kristensen (2002) presents each of the six cases, and some of the observations are summarised in table 1. Each farm is presented with the information given by the farmers during the visits. This approach was chosen in part because it was not possible to make follow-up registrations on the farms, in part - and not least - to get an insight in how these farmers have experienced the change to AMS and the considerations they have been though.

The production level in the herd before AMS introduction was high, from 7600 kg energy corrected milk (ECM) to 11500 kg. Four of the herds used grazing before AMS, but only one herd, expected the same level (40 to 60 percent of DM intake) of grazing after AMS.

Motivation and decision support
Social aspects and lower workload are the only motivations that were general given at all farms (Table 1). By information given at the time of interview was the workload reduced from 10 to more than 100 percent, within milking, feeding and management of the cow herd. Also other study finds very variable effects of AMS introduction on the work load (Meskens *et al.*, 2001; Raun & Rasmussen, 2002). This is due to factors like function of the AMS, frequency of involuntary milking and capacity of the old milking parlour. More than 80 percent of the AMS farmers in Denmark have invested in AMS due to social reasons (Meskens *et al.*, 2002), which support our findings.

Only two farms had an expectation to higher income due to the investment in AMS, or in higher income compared to other way of changing the production. Nielsen & Vestergaard (2003), found that only 13 farms out of 41 dairy farms with AMS had a positive return to farmer. Compared to other investments most of the farmers had made the investment in AMS due to other aspect than expected profit, and often without support from the traditional farm consultants. AMS consultants and farmers that had invested in AMS had been their main backup.

Automatic milking – A better understanding

Table 1. Productionsystem and motivation for introduction of AMS at six Danish dairy farms.

	Herd number					
Productionsystem	1	2	3	4	5	6
Herd size, no. of cows	55	70	100	65	105	110
Production level, kg ECM per cow	11500	9400	9000	7600	9500	8000
Grazing, before / after AMS	+ / -	+ / -	- / -	+ / +	- / -	+ / (+)
Motivation for, expectation to AMS (0 none, +++ high)						
• higher income	+++	0	+	0	0	0
• higher social welfare	+	+	++	+	+	+
• lower work load	+	++	+	+++	+++	+
• no hired labour needed	+	0	0	0	++	0
• new technology	+	0	++	++	0	++
• higher yield per cow	+++	++	+	0	+	+

Barn layout

All the herds were housed in loose housing system with slattered floor, and the AMS was placed in an existing barn lay out. Two of the farms used one-way cow traffic, while the others had free cow traffic. But the six farms also showed that there are many ways for an overall layout of the barn, like grouping of the cows, feeding space per cow and concentrates parlours, that can be combined successfully with AMS.

The AMS need to be placed with respect for the good access for both the cow and the manager (Koning & Vorst, 2002). Several of the farms had put too less attention to the last part. There has to be optimal access to inspection and cleaning, and with good facilities for computer, printer and other management's tools, close to the AMS.

Feeding

Aspects of feeding only played a minor role in the decision before investing in AMS, which is surprising, as all the farms after introduction of AMS have made adjustment in the feeding. Digestibility, protein content, bulk factor and time and frequency of feeding of the roughage have been changed in order to optimise the system. Two herds had tried with fresh grass in the stable, but the experience was negative, as the variation in drymatter - and energycontent, had major impact on herd production and use of the AMS.

Several of the herds have had problems with ketosis and other metabolic diseases, due to unbalance between the intake of roughage and the amount and type of concentrates given in the AMS. Knowledge of the strategy used for adjustment of the amount given in the AMS, due to production and stage of lactation, was among several of the farmer's weak. They referred to the standards, without actually knowing the strategy.

Working condition and management

The change in work situation, both in time and type, has to most off the farmers been even more successful than expected. Several of the farmers emphasized the importance of skilled labour for back up, both short time and during holidays. Most of the farmers really had not changed their way of herd management significant. Data from the AMS was used primarily for first indication for mastitis, while data from the AMS was used little in reproduction and general health observation.

Perspectives for research

We have identified three major problem complexes that need to be given high priority in the research. The first problem complex relates to the strategic management of the dairy farm. This is related both to optimisation of an existing production and to development of the production. Within an existing production there is a need for methods to optimise the use of the AMS with focus on selection of cows for replacement and for optimising feeding and access to milking at individual cow level. Development of a farm often is a continuous process, where herdsize and quota is gradually increased. As the milking facilities at an AMS farm is based on units with step of 40 to 60 cows, there is a need for models to calculate the most efficient way of development. These models has to take several of the aspects in figure 1 into consideration. As feedsupply has a major impact on farm economy and environment impact, it is important to consider how AMS influence the choice of feeding, not only from a herd perspective, but also from a farm perspective.

Feeding need to be given far more attention as it influence and are influenced by some of the major elements in a successful use of AMS, as illustrated in figure 2. The problems can be addressed at both herd- and cow level. Several investigation (Lauwere, *et al.*, 2000; Sporndly & Wredle, 2002) has focused on grazing in combination with AMS, as a sever example of the conflict between the ideas behind voluntary milking and access to high digestible roughage (Weidemann *et al.*, 2002). Regulation of feeding to the individual cow interacts with cow health and production. Some of the case farmers had expired problems with ketosis, due to unbalance between feedintake and milking in periods with high milking frequency. The relations between the elements in figure 2, becomes even more complex than in systems with traditional milking, while the elements to some extent becomes uncontrollable and are more influenced by each other. It is important to combine knowledge from physiology and nutrient, with animal behaviour and welfare in order to meet the demands from these new types of productions system.

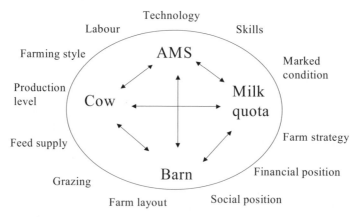

Figure 1. Illustration of elements to be taken into a strategic planning at AMS farms.

Automatic milking – A better understanding

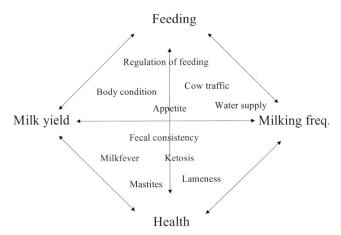

Figure 2. *Illustration of major elements in relation to production and health for a successful AMS.*

Management, or more specific the use of information from the AMS, in daily herd management, is the third problem complex we have identified. The use of the data has to be oriented against two dimensions, which refer to different styles of farm management (Noe, 1999). Observation can in theory start with information from the computer or with observation done in the stable. Based on these information can the actions taken, be either determinated by a cow specific or a herd specific tress holds.

There is also a general need for more advanced ways of treating data, and combine data from different sources in a high tech dairy unit, like data from feedintake, live weight, milking, milkcontent and activity, in order to get more exact information.

In conclusion, research within AMS has to change from research of a technology to research into systems including this technology. This calls for both system oriented research and for experimental work within areas of herd health and production, based on a system analyse.

References

Koning, K & Vorst, Y. 2002. Automatic milking - changes and chances. In. Proc. British Mastitis Conference, 68-80.

Lauwere, C.C.K-de., Ipema, A.H., Lokhorst, C., Metz, J.H.M., Noordhuizen, J.P.T.M., Schouten, W.G.P., Smits, A.C. 2000. Effect of sward height and distance between pasture and barn on cows visits to an automatic milking system and other behaviour. Livest. Prod. Sci. 65, 1-2, 131-142.

Meskens, L., Vandermersch & Mathijs, E. 2001. Implication of the introduction of aoutomatic milking on dairy farms. Literature review on the determinants and implications of technology adaption. Http://www.automaticmilking.nl

Meskens, L. & Mathijs, E. 2002. Socio-econimic aspects of automatic milking. Motivation and characteristics of farmers investing in automatic milking systems. Http://www.automaticmilking.nl

Nielsen, A.H. & Vestergaard, M. 2003. Erfaringer og økonomiske resultater med automatisk malkning. In Produktionsøkonomi - Kvæg 2003. Dansk landsbrugsrådgivning, 28 -31.

Noe, E. 1999. Værdier, rationalitet og landbrugsproduktion. Ph.d.-afhandling. KVL. 198 pp.

Noe, E. & Kristensen, T. 2003. Driftsledelsesmæssige udfordringer ved etablering af AMS i eksisterende mælkeproduktionssystemer. DJF rapport Husdyrbrug, nr. 47, 24 pp.

Raun, C & Rasmussen, J.B. 2002. Arbejdsforbrug ved automatisk malkning - Afsluttet Farm Test. Farm Test - Kvæg nr. 8, LR.

Spörndly, E. & Wredle, E. 2002. Management and animal husbandry. The effect of distance to pasture and level of supplementary feeding on visiting frequency, milk production and live weight of cows in an automatic milking system. In. The First North American Conference on Robotic Milking. March, 2002. Toronto, Canada. 76-77.

Wiedemann, M., Wendl, G. & Schön, H. 2002. Management and animal husbandry. Effects of energy and protein content in basic feed on milking and cow behaviour in automatic milking systems. In. The First North American Conference on Robotic Milking. March, 2002. Toronto, Canada. 93-96.

INTEGRATION OF COW ACTIVITY IN INDIVIDUAL MANAGEMENT SYSTEMS: ANALYSIS OF INDIVIDUAL FEEDING PATTERNS IN AUTOMATIC MILKING SYSTEMS

Martin Melin & Hans Wiktorsson
Department of Animal Nutrition and Management, Swedish University of Agricultural Sciences, Uppsala, Sweden

Abstract

With increasing possibilities to obtain on-line information on an individual cow, systems for individual management can be developed. Cow activities such as feeding patterns from automatically obtained records can be valuable input information in these systems. With the aim of analyzing individual feeding patterns, records of 30 fresh cows from feeding stations (n=111176) were extracted. To obtain a reliable method of grouping feeding visits into meals, intervals between visits were modeled in a mixture of lognormal densities. Meal criteria, i.e. the longest interval between two feeding visits to not separate two meals, ranged from 49 to 76 min, which were longer than what has been reported for dairy cows in loose housings with conventional milking systems. The major part (83 to 96%) of the random variation in feeding patterns of the cows was due to differences between individual cows. It was concluded that feeding patterns are characteristic for individual cows and consistent over time, making them valuable in individual management control systems. Further, meal criteria differed between individual cows as well as cow traffic systems. Arbitrary chosen meal criteria may be misleading in analyses of feeding patterns.

Introduction

Information about cow activity can be obtained by register when cows report at milking units, control gates and feeding stations. We hypothesized that feeding patterns can be analyzed and serve as valuable input information in a management control system. In such a system, automatically obtained on-line information about individual cows continuously is processed and gives early warnings of disturbances and may suggest management changes. Naturally for dairy cows, feeding (Metz, 1975; Tolkamp et al., 1998) are not totally random processes, i.e. the feeding visits are not randomly spread over a time period. Instead, feeding visits occur in distinct bouts or meals. Little is known about feeding patterns in automatic milking systems, where access to feeding areas often is somehow restricted. Tolkamp et al. (1998) described the occurrence of meals in feeding behavior of cows as the presence of clusters in data of feeding intervals. The time intervals within each cluster were assumed to be lognormally distributed, and a mixture of two normal densities was used as a model for the frequency distribution of LN-transformed (Natural Logarithm) feeding intervals. These densities represented different types of intervals between feeding visits. When two densities were included in the model, the intervals were separated into one density with short within-meal intervals and another density with long between-meal intervals. Meal criterion, i.e. the longest interval between two feeding visits to separate two meals, was then identified as the point where an interval was assigned to both densities with equal probabilities.

The purpose of this study was to analyze feeding patterns of cows in an automatic milking system with respect to individual differences. Information of feeding patterns was extracted from individual on-line registrations in roughage and concentrate stations, and modeled in a mixture of lognormal densities.

Material and methods

Animals, housing and experimental design

A total of 30 cows of Swedish Red and White dairy bred (SRB), of parities 1-7 were in the study that extended over the period September 2001 to June 2002. Cows were offered ad lib. of mixed grass silage:concentrate (ratio 75:25) and concentrates supplied in milking unit and in concentrate feeding stations. Average daily dry matter intake was 19.2 kg for cows in first lactation and 24.7 kg for multiparous cows. The average total amount of cows in the barn during the period was 46 (min: 37 max: 52). The study lasted until the 19:th week of lactation and, throughout the study the cows were subjected to controlled cow traffic by assessing a time limit for milking permission. The cows were divided into two groups allowing a high (HF) or a low (LF) milking frequency. The HF group was given a minimum time limit for milking permission of 4 hours and the LF group was given a minimum time limit for milking permission of 8 hours. The experiment was performed at the University Cattle Research Centre Kungsängen in Uppsala, Sweden. The barn was a loose-housing system including a resting area, two separate feeding areas and a milking unit (DeLaval VMS™). The barn was equipped with a data acquisition system for concentrates and roughage.

Individual feeding patterns

The individual feeding patterns were studied in a feeding visit analysis. Feeding intervals were calculated as the elapsed time between two subsequent visits to either roughage or concentrate stations. Feeding intervals were LN-transformed and modelled in a mixture of lognormal densities. Tolkamp et al., 1998 showed that this statistical method estimated meal criteria that was biologically relevant for cows held in an ordinary loose-housing. Melin and Wiktorsson (unpublished) confirmed this biological relevance for cows held in an automatic milking system. Two normal densities were fitted to LN-transformed feeding visit intervals. For each density three parameters were estimated; density mean μ_i, density variance σ^2_i and the proportion parameter p_i. Meal criterion was calculated as the point where the functions of density one and two crossed.

Results and discussion

Meal criteria, i.e. the longest interval between two feeding visits to not separate two meals, were on average 74.3 and 50.6 minutes for heifers and multiparous cows of the HF group, respectively. The corresponding values for cows of the LF group were 38.5 and 47.3 minutes. Meal criteria estimated in this study were longer than what has been reported for dairy cows in loose housings with conventional milking systems. Even though meal criteria in this study were longer than meal criteria that have been reported by others, meal criteria

Automatic milking – A better understanding

were within the range of 20 to 60 minutes where Metz (1975) recorded few between feeding intervals for tied-up dairy cows.

Estimated functions of the mixture density models are shown in figure 1 for six of the totally 30 cows. Density means, meal criteria and confidence intervals of meal criteria are shown for the same six cows in table 1. The first three cows (Figure 1a-1c) are cows from

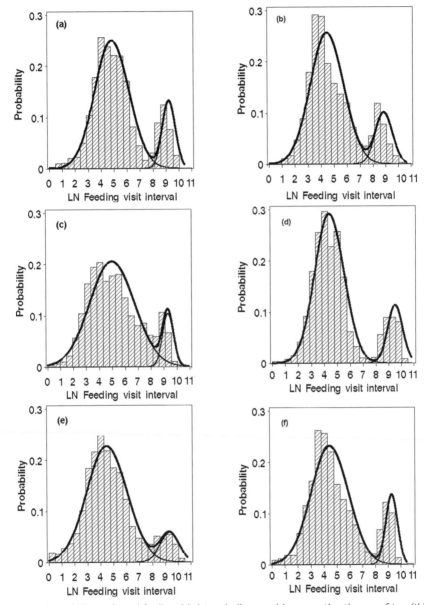

Figure 1. Functions of LN-transformed feeding visit intervals (in seconds) representing the sum of two (thick line) and separate densities (thin lines). Frequency distributions of intervals are also presented in bar charts with bar widths of 0.5 LN-units. Each graph represents data for individual cows.

Table 1. Proportion parameter of density one (P1), meal criteria, means of density one (μ_1) and density two (μ_2) for feeding intervals modelled in a mixture of two densities of six individual cows. Values for means and criteria are given in minutes within parenthesis (exp(LN)/60). Given is also confidence intervals (CONF) of meal criteria and calculated number of meals per 24 hours.

Figure	MF[1]	P1	μ_1	μ_2	Meal criteria	CONF	Meals/24hrs	N
1a	LF	0.83	2.2	173	58	53-62	5.6	3586
1b	LF	0.83	1.5	121	32	29-35	7.0	4493
1c	LF	0.89	2.5	186	91	81-102	4.7	4003
1d	HF	0.83	1.4	229	46	42-50	4.5	2987
1e	HF	0.89	1.5	191	58	51-66	5.4	4250
1f	HF	0.84	1.5	184	62	57-67	5.6	3794

[1]Milking frequency group (LF or HF).

the HF group and the other three cows (Figure 1d-1f) are cows from the LF group. The feeding patterns between individual cows were considerably different between cows within the same milking frequency group. For all six cows the proportion of within meal intervals (p1) ranged from 0.83 to 0.89. The average within-meal feeding visit interval (μ1) ranged from 1.4 to 2.5 minutes for the six individual cows, and the average between meal interval (μ2) ranged from 121 to 229 minutes for the six individual cows. Based on calculated confidence intervals, meal criteria could be estimated with a low uncertainty. Meal criterion of cow (c) was estimated with a confidence interval ranging from 81 to 102 minutes, which was the widest of all six cows. This uncertainty can be seen as a big overlap between densities and a generally bad fit of model functions to the frequency bar charts (Figure 1c). Figure 1d shows two very distinct densities with a small overlap between fitted model functions, and meal criterion for this cow is estimated with a confidence interval ranging from 42 to 50 minutes. Cow (e) had many feeding visits within each meal, which resulted in a high proportion of intervals in density one (p1=0.89).

The parameter estimates of the mixture density models were included in a statistical analysis, where the total random variation (σ^2, variation not due to age category or milking frequency group) in these parameters was divided in random variation within individual cows (σ^2_{within}) and variation between individual cows ($\sigma^2_{between}$). $\sigma^2_{between}$ explained most of the random variation in the parameter estimates of the mixture density models, and it was always significantly different from zero. For all parameters in all periods, $\sigma^2_{between}$ was greater than 0.83, leaving a small part of the random variation to be explained by variation within cows (Table 2).

Calculated number of feeding meals per 24 hours showed a wide variation between individual cows in the same frequency group (Table 1). Further, the estimated model parameters of the mixture density models showed a wide variation between individual cows. This suggests that individuality plays an important part in how cows interact with an automatic milking system. Others have stated that individual cows behave differently in the same conditions. Ketelaar-de-Lauwere (1999) noticed individual differences in how cows used the selection unit in an automatic milking system, and Hopster and Blokhuis (1999) found that cows responded to a social stressor in a way that was characteristic for the

Automatic milking – A better understanding

Table 2. Proportion of the total random variation in mixed density model parameter estimates and meal criteria for feeding intervals that was explained by variation between individual cows ($\sigma^2_{between}$). Data is given separately for four different periods of lactation.

Parameter[1]	Lactation week			
	2-6	7-12	13-16	17-19
P1	0.91	0.89	0.83	0.92
μ1	0.96	0.96	0.96	0.93
μ2	0.96	0.95	0.95	0.96
MC	0.92	0.90	0.84	0.94

[1]Parameter labels explained in material and methods section.

individual animal. The fact that individual cows exhibit different reactions to the same situation, suggests that management change taken on group level is an inefficient management strategy. In this study, the major part of the variation in feeding patterns was due to differences between individual cows while differences within individuals explained a smaller part of the variation. This suggests that cows develop feeding patterns in the automatic milking system that is characteristic for each individual and is consistent over time. The consistency over time of cow behaviour increases the possibility to use feeding patterns or other cow activities as input information in an individual management system.

Conclusions

Modelling feeding intervals in a mixture of normal densities gave estimations of meal criteria with low confidence intervals. Due to differences in meal criteria both on group and on individual level, arbitrary chosen meal criteria is likely misleading in studies of feeding behaviour. Because of the finding that the major part of the variation in feeding patterns can be explained by differences between and not within individual cows, it can be concluded that cows develop feeding patterns that are characteristic for the individual cow and consistent over time. Based on this, feeding patterns likely have the potential as input information for purposes as monitoring and decision-making in individual control management systems.

References

Hopster H. and Blokhuis H. J., 1994. Consistent individual stress response of dairy cows during social isolation. Appl. Anim. Beh. Sci. 40(1), 83-84.

Melin and Wiktorsson. (unpublished).

Metz, J. H. M. 1975. Time patterns of feeding and rumination in domestic cattle. Meded. Landbouwhogesch. 75-12, Agric. Univ., Wageningen, The Netherlands, 66pp.

Tolkamp, B J. Allcroft, D J. Austin, E J. Nielsen, B L. Kyriazakis, I., 1998.

Satiety splits feeding behaviour into bouts. J. Theor. Biol. 194(2), 235-250.

Ketelaar-de-Lauwere-CC. 1999. Cow behaviour and managerial aspects of fully automatic milking in loose housing systems. Wageningen Agricultural University. Landbouwuniversiteit Wageningen; Netherlands, 189 pp.

Automatic milking – A better understanding

ILLUMINATION OR GUIDING LIGHT DURING NIGHT HOURS IN THE RESTING AREA OF AM-BARNS

Gunnar Pettersson & Hans Wiktorsson
Department of Animal Nutrition and Management, Swedish University of Agricultural Sciences, Uppsala, Sweden

Abstract

It is a common practice to have full illumination day and night in AM-barns. It is however questioned whether cows prefer/need guiding light/darkness during night hours. The aim was to investigate if dairy cows actively choose light or dark area for resting during night hours.

The study was conducted in an AM research barn with controlled traffic and 2 by-pass gates. The resting area of the barn was divided into 2 identical sections by screening. The study was divided into 3 periods. Period 1 with full lightning in both parts of the resting area. Period 2 and 3 with one half of the resting area with full lighting (app. 200 lux), while the other had guiding light (5 - 7 lux) from 23:00h to 05:00h. After 3 weeks the lighting in the two sections was reversed.

The cows distributed equally in both sections of the resting area during the whole experiment and no effect of level of illumination on production parameters could be seen on herd level. However there are indications that some cows preferred the light or the dark side and followed the changes, but no relation between ranking order and choice of light was found.

Introduction

Shortly after automatic milking was introduced in Sweden some commercial herds reported a decrease in number of milkings if the illumination was reduced to guiding light in the barn during night hours compared to full illumination. Therefore it has become a common practice to have full illumination day and night in AM-barns.

Effects of the length of the photoperiod have been studied in several experiments. Dahl et al. (2000) discuss, in a review article, the relations between production, endocrine response, hormonal status and the length of the photoperiod and they found positive effects on daily milk yield up to 16h illumination. However, photoperiods longer than 16h were not discussed. Reksen et al. (1999) found in a survey study positive effects on fertility parameters when the night time illumination in the barn was reduced to guiding light compared with herds without illumination at night.

It is however questioned whether cows prefer guiding light or darkness during night hours. In Sweden and other Scandinavian countries the length of the photoperiod varies greatly over the year with very short nights or even without sunset during the summer. The cows adjust their feeding periods accordingly with grazing late in the evening and again at sunrise. The aim of this study was to investigate if dairy cows actively choose between light and dark areas for resting during the night hours, and if so whether they prefer a light or a dark area.

Material and methods

The study was conducted in an AM research barn with 4 rows of cubicles, 1 milking station DeLaval VMS™, 20 feeding stations for roughage and 2 concentrate stations (Figure 1). The cow traffic system during the study was controlled traffic with 2 by-pass gates between the resting and feeding area. The cows received milking permission 6h after the previous milking, and before that time they could directly enter into the feeding area through the by-pass gates. The resting area of the barn was divided into 2 identical sections by screening between the two central rows of cubicles and towards the feeding area and milking unit, but the passages remained unaffected. Each section of the resting area had 25 available cubicles. 46 cows, Swedish Red and White, were used in the study.

The study was divided into 3 periods. In period 1 full lighting (app. 200 lux) was applied in both sections of the resting area. In period 2 and 3 one half of the resting area had full lighting, while the other had guiding light (5 - 7 lux) from 23:00h to 05:00h. After 3 weeks the lighting in the two sections was reversed during another 3 weeks.

Manual identification of every cow in the resting area was performed every 12:th minute during 48 consecutive hours in period 1 and 2 x 48h in period 2 and 3. The observations were performed during the last 2 weeks of each 3 weeks period. Dominance values were calculated automatically on data from the roughage feeding stations (Olofsson, 2000).

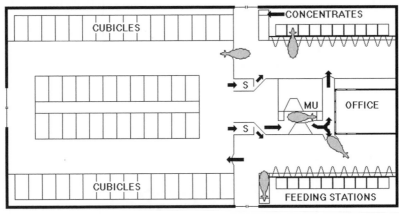

Figure 1. Schematic drawing of the experimental barn layout. Abbreviations: S=Selection by-pass gate, MU=Milking Unit.

Results

Distribution of the cows in the resting area

The cows distributed equally in both sections of the resting area during the whole experiment and no effect of level of illumination could be seen on herd level (Table 1). In all periods there were free cubicles in both sections during both day and night time (Figure 2, 3 and 4), and there were at least 4 free cubicles on each side at every observation during the observation periods. However during period 2 and 3 there is a slight increase in the maximum number of occupied cubicles in the section with guiding light (p=0.09).

Table 1. Average distribution of the cows during night (23:00 - 05:00) and daytime with full illumination in the hole resting area compared to full illumination in one half of the resting area and guiding light in the other half.

Time of the day	Full illumination in both sides of the resting area		Full illumination on the left side and guiding light on the right side of the resting area		Full illumination on the right side and guiding light on the left side of the resting area	
	Left side	Right side	Left side	Right side (5 - 7 lux)	Left side (5 - 7 lux)	Right side
05:00-23:00	50%	50%	50%	50%	51%	49%
23:00-05:00	51%	49%	53%	47%	50%	50%

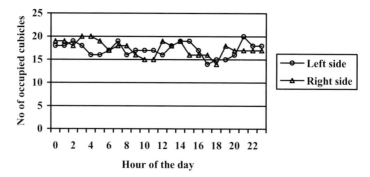

Figure 2. Maximum no. of occupied cubicles during each hour of the day in period 1, full illumination in both sections of the resting area.

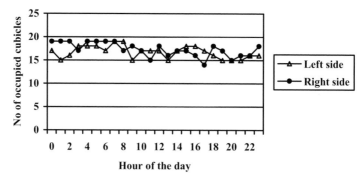

Figure 3. Maximum no. of occupied cubicles during each hour of the day in period 2, full illumination on the left side and guiding light on the right side of the resting area.

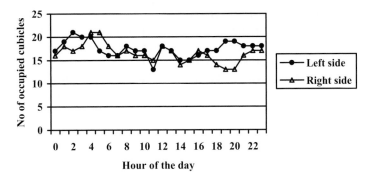

Figure 4. Maximum no. of occupied cubicles during each hour of the day in period 3, guiding light on the left side and full illumination on the right side of the resting area.

Although no effect could be seen on herd level, a number of individual cows seemed to prefer the light or the dark side and followed the changes (Figure 5), but there were no relation between ranking order and choice of illumination. When the cows left the resting area for milking, feeding or drinking water they returned to the same side of the resting area in 60.5% of the cases as an overall mean. During the period 23.00h to 05.00h the return percentage was lower with guiding light than with full illumination (Table 2).

Table 2. Frequency of cows returning to the same side in the resting area after having left it during different degree of illumination over the day.

Period of the day	Full illumination 00:00 - 24:00	Full illumination 05.00 - 23.00 Guiding light 23.00 - 05.00
23.00 - 05.00	67	59
05.00 - 11.00	58	61
11.00 - 17.00	53	61
17.00 - 23.00	58	61

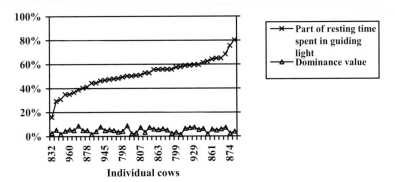

Figure 5. Resting time from 23:00 to 05:00, distribution between fully illuminated or guided light in individual cows.

Milking frequency and feeding visits

One way to measure if the illumination during night-time had any influence on cow activity is the number of milkings per day and in different periods of the day. The number of milkings was not negatively affected when the illumination in one half of the resting area was reduced (Table 3). On the contrary there was a slight increase in the number of milkings, 2.61 milkings/cow/day compared to 2.53 ($p<0.05$). This effect might be caused by other factors as the distribution of all milkings over the day was not affected ($p<0.98$).

Table 3. Number of milkings/cow and day and the distribution of the milkings over the day.

	Full illumination 00:00 - 24:00	Full illumination 05:00 - 23:00 Guiding light 23:00 - 05:00
Milkings/cow/day	2.53	2.61
Milking distribution over the day		
23.00 - 05.00	25.4%	25.3%
05.00 - 11.00	24.0%	24.7%
11.00 - 17.00	25.3%	25.3%
17.00 - 23.00	25.3%	24.9%

Feeding visits per cow and day is another way to describe how active the cows are. During the study the mean number of feeding visits per cow and day was 7.6 with full illumination and 7.7 when guiding light was applied in one half of the resting area. The distribution of the feeding visits over the whole day is presented in figure 6 as the total number of cows in the feeding areas over the day. The degree of illumination has not influenced the number of cows in the feeding areas during different parts of the day. There is a decrease in number of cows in the feeding areas during the late night hours in both systems, but the decrease is somewhat smaller when guiding light was applied in one part of the resting area. In the experiment it is noted that lying periods starting before 05:00h and ending at 05:00h or later are shorter for cows lying in a cubicle with guiding light than for cows in a cubicle with full illumination. The increase in illumination at 05:00h might have an activating effect on the cows.

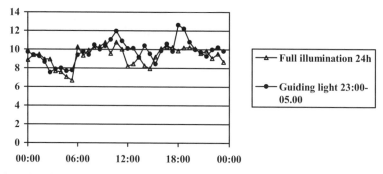

Figure 6. Total number of cows in the feeding areas during the day.

Discussion

The results of the studies did not reveal obvious preference for a particular degree of illumination. However, the increased maximum number of occupied cubicles in the sections with guiding light and the fact that some cows "followed" one level of illumination when it shifted, although there were free cubicles on both levels, indicate that individual cows may have preference for a lighter or darker resting area. This behaviour was not related to the calculated dominance value.

Conclusion

The conclusions of the study are:
- reduced illumination to guiding light in one half of the resting area during night-time had no negative influence on the overall milking and feeding frequency, nor did it influence the milking and feeding pattern.
- no difference between full lighting or guided light could be seen in the distribution of the cows in the resting area during night-time on herd level
- indications on individual preference of full lighting or guided light could be observed, but was not due to ranking order.

References

Dahl, G.E., B.A. Buchanan and H.A. Tucker, 2000. Photoperiodic effects on dairy cattle. A review. Journal of Dairy Science 83:885-893.

Olofsson, J., 2000. Feed Availability and Its Effects on Intake, Production and Behaviour in Dairy Cows. Doctoral dissertation. Agraria 221, Department of Animal Nutrition and Management, Swedish University of Agricultural Sciences, Uppsala, Sweden.

Reksen, O., A. Tverdal, K. Landsverk, E. Kommisrud, K.E. Bøe and E. Ropstad, 1999. Effects of photointensity and photoperiod on milk yield and reproductive performance on Norwegian red cattle. Journal of Dairy Science 82:810-816.

SYSTEM CAPACITY OF SINGLE BOX AMS AND EFFECT ON THE MILK PERFORMANCE

Rudolf Artmann
Institute of Production Engineering and Building Research, Federal Research Centre (FAL), Braunschweig, Germany

For planning and economic decisions are the knowledge of the milking capacity and also the effects of varying milking intervals as well as more frequent milking on the milk yield important parameters. Apart from working time requirement and social effects the more cows at each unit can be milked and the higher the milk yield increases, the more win the milking automatic systems on competitiveness.

For the determination of the system capacity and the effect of varying milking intervals on the milk production the data from two larger dairy farms were analyzed.

The closely co-operating farms have now seven automatic milking systems (AMS) with single boxes (Astronaut) in operation. All data, available since spring 2000, were collected and analyzed.

The cows hold in rebuilt L-203 stalls was firmly assigned to the individual AMS. Unsuitable cows as well as problem cows were milked always or temporarily in the old herringbone parlor.

An estimation of the capacity boundaries of the AMS can be derived from the variation of the milking frequency when the AMS are loaded with different number of cows with varying milk yield. This dependency shall be estimated with statistical analysis (GLM). Provisional analyses show a clear decrease of the number of milking when more than 45 high performance cows are milked per AMS.

The influence of automatic milking on the milk performance is also statistically analyzed. Laktationsnummer, milking interval (MI) and day in lactation are the most important

Results from the MS (daily means).

AMS	1	2	3	4	5
Number of cows on the AMS	46.6	45.1	47.3	45.9	50.2
Total number of milking *10_	89.4	85.1	93.6	95.6	90.0
Milking per day	132.5	126.1	138.7	141.7	133.4
Milking interval [h]	8.5	8.7	8.3	7.9	9.2
Milking per cow and day	2.82	2.76	2.89	3.04	2.61
Milk yield [kg] per milking	11.8	8.2	11.5	9.8	7.5
Time with milk flow [min/d]	636.9	454.4	612.3	586.1	473.3
Time without milk flow [min/d]	70.9	66.9	70.7	81.2	71.4
Visits without milking	38.6	91.4	37.0	74.6	112.7
Milking with problems	5.6	7.7	5.6	6.2	6.0

variables. On the basis of estimations the influence of the MI on the milk yield per lactation are calculated and represented. First results show that with the frequency and uniformity of the MI the milk production rise. But with only two milkings per day and unfavorably MI less performance can occur.

Table 1 shows some preliminary characteristic results of five already longer used AMS. For the presentation the analyzed time series should be as long as possible. Perhaps an influence of the time in use on parameter of performance can be found.

UDDER STIMULATION WITHOUT MILKING DECREASES MILK PRODUCTION AT THE FOLLOWING MILKING BUT HAS NO FURTHER CONSEQUENCES ON MILK PRODUCTION

A. Bach[1], I. Busto[2] & X. Carre[2]
[1]ICREA and IRTA-Unitat de Remugants, Barcelona, Spain
[2]Diputació de Girona, Semega, Girona, Spain

Increasing milking frequency has a positive effect on milk production (Wagner-Storch and Palmer, 2003). With an automatic milking system (AMS) it is not uncommon that in occasions (0.8 and 1.6% in the front and rear quarters respectively) some units fail to be attached to the udder causing the udder to be milked in all quarters except that or those which the AMS failed to attach the units. The objective of this study was to quantify the consequences of a prolonged milking interval in a given quarter when the rest of quarters of the udder have been milked.

A total of 15,796 milkings from an average of 73 cows during 90 days were traced and recorded for later statistical analyses. The database contained information about milk yield from individual quarters and interval between milkings (for both the entire udder and individual quarters in case of a failure of unit attachment failure). An analysis of variance was conducted using cow as a random effect and days in milk and the failure of attaching a milking unit plus their corresponding interaction as fixed effects. Both fixed effects and their interaction had a significant effect ($P < 0.001$) on quarter milk production per milking.

With similar milking intervals, the quarters that had undergone stimulation but had not been milked yielded less ($P < 0.001$) milk than those quarters that had not been stimulated (Table 1). Despite the lower milk production obtained when the AMS failed to milk a given quarter, subsequent productions from that quarter were not negatively affected ($P = 0.60$). It is likely that stimulation of a quarter without a subsequent emptying stops milk synthesis but does not affect death rate of secretory cells.

Table 1. Effect of udder stimulation with or without milking on milk production at the following milking.

	Milking interval, h	DIM	Yield, l/quarter/milking
Front left quarter			
No milking stimulus	19.9[a]	184	4.00 [a]
With Milking stimulus[1]	19.4 [a]	180	3.76 [b]
Front right quarter			
No milking stimulus	19.7 [a]	188	4.02 [a]
With Milking stimulus[1]	19.9 [a]	186	3.66 [b]
Rear left quarter			
No milking stimulus	19.6 [a]	174	5.98 [a]
With Milking stimulus[1]	19.6 [a]	181	5.31 [b]
Rear right quarter			
No milking stimulus	19.9	185	5.68 [a]
With Milking stimulus[1]	20.0	178	4.99[b]

[1]Consequence of a unit attachment failure in the corresponding quarter.

[a,b]Different superscripts between the two rows of each udder indicate significant differences ($P < 0.001$).

THE IMPACT OF MILKING INTERVAL REGULARITY ON MILK PRODUCTION

A. Bach[1], I. Busto[2] & X. Carre[2]
[1]ICREA and IRTA-Unitat de Remugants, Barcelona, Spain
[2]Diputació de Girona, Semega, Girona, Spain

Milking intervals with conventional milking parlors are usually kept constant. However, with automatic milking systems (AMS), it is expected that the milking intervals within cows will be different across days and larger than those obtained with conventional milking parlors. The greater variation in milking interval with the AMS is mostly due to the fact that cows will not visit the AMS the same number of times nor at the same time of the day along their lactation. Therefore, the udders of these cows will have to store different amounts of milk depending on the consistency of their milking intervals. The objective of the present study was to assess the impact of milking interval regularity on milk production with an AMS.

Milk production, number of daily milkings, and milking intervals of a total of 83 cows were traced for a period of 10 months (a total 35,291 milkings). The cows had free access to a DeLaval Voluntary Milking System. The coefficient of variation of the milk interval was calculated for each cow within day and within week. The data were analyzed following an analysis of variance with cow as a random effect and lactation number, days in milk and the coefficient of variation of daily milking intervals as fixed effects. All fixed effects were coded into discrete variables. Both, days in milk and coefficient of variation of milk intervals were divided into four categories, coinciding with the four quartiles of their respective distributions (each category held 25% of the observations). The data from cows with equal or greater than 4 milkings per day were excluded from the dataset due to their relative low frequency of appearance (less than 1.3% of the data). The number of milkings per day increased ($P < 0.001$) total daily milk production and apparent milk synthesis rate (kg/h). Cows milked one, two, three and four times a day produced 16.5, 28.0, 34.1, and 37.6 kg/d, respectively. These production increases were larger than the ones previously reported in the literature with traditional milking parlors (Erdman and Varner, 1995) but similar to those reported with an AMS (Wagner-Storch and Palmer, 2003). There was no relationship between interval between milkings and the coefficient of variation of the milking intervals within day or week. However, uneven frequency (large coefficient of variation of milk intervals) decreased ($P < 0.001$) milk yield (Table 1). It is concluded that uneven milking intervals (with coefficient of variation within day above 34%) have a negative impact on milk production in dairy cows.

Table 1. Total daily production, milk production per milking, and apparent milk synthesis rate as affected by regularity of milking intervals.

	Coefficient of variation of milk interval, %			
	<11.7	11.7-21.9	21.9-34.4	>34
Milk production, kg/d	31.3[b]	32.1[a]	31.6[b]	28.7[c]
Milk production, kg/milking	16.35[a]	16.14[a]	16.12[a]	14.98[b]
Milk synthesis rate, kg/h	1.13[a]	1.13[a]	1.1[b]	1.0[c]

a,b,c Different superscriptis within row indicate significant differences at $P < 0.05$.

CHANGE OF UDDER CONFORMATION TRAITS BY LACTATIONS IN A HOLSTEIN FRIESIAN HERD

E. Báder[1], I. Györkös[2], P. Báder[1], M. Porvay[3], E. Kertészné Györffy[1] A. Kovács[1] &
L. Pongrácz[1]
[1]University of West Hungary, Faculty of Agricultural and Food Sciences, Institute of Animal Breeding and Husbandry, Mosonmagyaróvár, Hungary
[2]Research Institute for Animal Breeding and Nutrition, Herceghalom, Hungary
[3]Enyingi Agricultural Corporation, Kiscséripuszta, Hungary

Many authors state that the number of undergone lactations (or age) affects the size and shape of the udder. Multiparous cows had a significantly larger volume of the udders than primiparous cows. Similar rate of changes in udder volume with age was observed in other studies. The aim of this study is to establish the change of udder conformation traits by lactations because it is essential for robotic milking.

Material and methods

We made our investigations on the dairy farm of Agrár Rt, Enying at Kiscséripuszta. The conformation traits based on minimal two consecutive scorings with a 50-point scale were evaluated in a Holstein Friesian herd consist of 739 cows. The two evaluations were performed at the first and second lactations.

Results and discussion

During the lactations the scores of fore udder attachment were decreased slightly (with 1.6 points) and for rear udder height increased with 0.9 point (by 20-24 cms). The scores for udder width were better with 1.1 points but were under by 15 cms, and udder suspension were better with 0.3 point. As for udder depth, the udder floors were above the hocks at the first scoring by 7.5 cms and by 5 cms at the second. At the first evaluation the rear teat placements were scored as normal but during the second lactations the scores for trait of side teat placements increased. The udder system generally was evaluated for "good" level but later its score (77.6-77.0 points) decreased with 0.6 point.

Conclusions

By the distribution of scores we can established that the number of cows with weak udder attachment were slightly increased but also increased the number of cows with strong attached udder. The number of evaluated cows with intermediate scored were decreased by 10% and there were a few cows with very strong udder.

Teat placement also changed during lactations, centrally placed were decreased (by 5%) and inside teats were increased (by 3%).

References

Báder, E., Györkös, I., Báder, P., Porvay, M., Kertészné, Györffy E., (2001) Change of udder conformation traits during the first two lactations in a holstein friesian herd, 3rd International Conference of PhD students,Miskolc, 7-9.

Kuczaj, M. (2003). Analysis of changes in udder size of high-yelding cows in subsequent lactations with regard to mastitis Electronic Journal of Polish Agricultural Universities, Animal Husbandry, Volume 6, Issue 1.

MANAGING THE CHANGE TO AUTOMATIC MILKING

J.R. Baines

Fullwood Ltd, Ellesmere, United Kingdom

Moving cows to a new milking facility, whether conventional or robotic, can be stressful and disruptive to the performance of any herd of cows. Time taken to accustom both cows and personnel to the new facility can result in impaired milking time management and aggravation of underlying sub-clinical mastitis problems.

Where the change is from milking in a conventional parlour to the use of an automated system, the challenge of change is even greater. In addition to the potentially stressful period of training, for both cows and personnel, milking no longer occurs at predetermined times and regular intervals. Milking frequency depends on management strategies relating to housing and feeding together with their effects on the willingness of cows to visit the milking station.

Studies in a number of countries indicate that milk quality and mastitis levels tend to deteriorate, at least for a time after the introduction of a robotic milking system. The extent of the deterioration is often related to the overall effectiveness of management and mastitis control in the herd prior to conversion.

In addition to the changes or routine facing the cows, milkers and herd managers lose their regular opportunity for close contact with cows which occurs through twice or thrice daily milking. For many people, this means that they must learn to use the information provided by the robotic system. Instead of relying on visual and tactile perception of the cows welfare and condition, reliance must be placed on the production of adequate information by the system and its interpretation. In turn, a successful changeover is entirely dependent on milkers/herd managers making correct management decisions and taking the appropriate actions.

The adoption of automatic milking in larger herds, and those with seasonal calving patterns, can present even greater challenges during the introductory period.

This paper discusses the practical implications of the changeover period for both cows and people. Advice, based on field experience, is given on ways of recognising and addressing the potential problems.

TRANSFER OF DATA FROM MANAGEMENT SYSTEMS TO CENTRAL DATABASES

M. Bjerring[1], T. Lind[2], F. Skjøth[2] & M.D. Rasmussen[1]
[1]Danish Institute of Agricultural Sciences, Foulum, Denmark
[2]Danish Cattle, Aarhus N, Denmark

Automatic milking systems create tons of data, which besides being crucial for the on-farm management system may be very useful for extension and research purposes. The transfer of data and the merging with external databases, are however, not simple matters. This poster focuses on the technical details and important steps that should be taken in order to ensure good data quality and useful analyses.

Cow number with 11 digits is a primary key in the official Danish database, whereas in the AMS-system the cow is identified by a 4-digit number. Purchased animals may have the same last four digits as a cow in the herd and consequently this number cannot be used unequivocally. Farmers are very creative in developing new numbers for the AMS, which is acceptable for the daily use. This attitude has no consequences for data filing as long as data are not transferred directly between the AMS and the official databases. Data transfer between management programs and official cattle databases will serve many purposes with the most important being no double registration. The establishment of such an interface would motivate farmers to keep precise records of all animals and treatments in their herds.

Data from the automatic milking systems are designed only for management use. The different management systems store data for a variable number of days: 10 days, until new lactation, until culling, or 3 months. Sometimes data are lost at the installation of new software. Consequently, valuable information is lost. A solution is to ensure that data are downloaded regularly to a safe environment. The maintenance of these programs is not a concern of the farmer and regular visits have to be paid to the farms in order to avoid missing data. Electronic transfer of data is possible in some cases but occupies the network for a long time. The ultimate benefit of a frequent data transfer is back-up of the system. However, official databases are not designed to store the large amount and very specific data needed for the all technical functions of the robot but some key numbers could be useful.

The different variables in the AMS databases do not have the same meaning in the different companies and it is often difficult to get a precise description of the variables. The ISO group on AMS standards has started some of this work concerning definition of machine-on time, milk yield, and successful milking etc. However, many variables unnecessary the ISO-standard should be defined. This could be done in a vocabulary from IDF.

There is a need to motivate the companies to follow specific standards for data recording and subsequently to show farmers the benefits of data transfer, in terms of more stable systems, a better extension service and production control.

EFFECTS OF MILKING FREQUENCY AND LACTATION STAGE ON MILK YIELD AND MILK COMPOSITION OF TEST DAY RECORDS IN ROBOTIC MILKING

I. Llach[1,2], A. Bosch[2], M. Ayadi[1], G. Caja[1], M. Xifra[3] & X. Carré[4]

[1]Grup de Recerca en Remugants, Departament de Ciència Animal i dels Aliments, Universitat Autònoma de Barcelona, Bellaterra, Spain
[2]Escola Superior Agricultura, Universitat Politècnica de Catalunya, Barcelona, Spain
[3]Mas Gener, Riudellots de la Selva, Girona, Spain
[4]Diputació de Girona, Semega, Monells, Girona, Spain

A total of 74 Holstein cows (average milk yield, 33.1 kg/d; days in milk: 132), milked by a single unit of 'Lely Astronaut' AMS (automatic milking system) with free traffic of cows, were used during 3 consecutive months to study milk performance data obtained in the official test day milk recordings conducted monthly. Milk sampling was done by using a 'Lely Shuttle' sampling device during a 48-h extended period. Samples (50 ml milk with azidiol) were analysed for milk fat and protein by infrared analysis (MilkoScan, Fosselectric) in the following 48 h. Daily milk yield was calculated by adding the records of all milkings and the portion of milk proportionally produced during the 24-h test day period, as indicated by De Mol and Ouweltjes (2000; In: Robotic Milking, Wageningen Pers, pp. 97-107). Fat and protein contents in milk were calculated from the proportional amount of fat and protein yielded during 24 h. Milking frequency during the 48 h recording period ranged from 3 to 9 milkings per cow and data for all milk variables were standardized to a 24 h interval.

Different evaluation models were tested by using the PROC MIXED for repeated measures of SAS (v. 8.1) to ascertain the influence of fixed and random effects on the milk performance dependent variables (milk yield, milk fat and milk protein). Retained fixed effects included: lactation stage (early-, first 84 d; mid-, 85 to 168 d; and, late-lactation, after 168 d), recording month (1st, 2nd, and 3rd), and cow (1 to 74). Random effects included either milking interval (h), or their inverse number of milkings in the last 24 h, and the residual error. Overall means for milk yield, and fat and protein milk contents, along with effects of main variation factors are summarized in table 1, from which it was concluded that:

- Stage of lactation did not showed any significant effect on the estimation of milk yield and milk composition data in AMS.
- Milking interval in AMS significantly affected the estimation of all the milk performance variables studied, in agreement with available knowledge on milk synthesis. Greater standardized daily milk yields were obtained for shorter milking intervals. On the contrary, greater milk fat and protein contents were obtained for longer milking intervals.
- Daily number of milkings had a significant effect on estimated milk yield in AMS; the lesser frequent milkings producing lower daily milk yield. The value of the slope (+1.3 kg/milking) agreed with the hypothesis of a fixed short term effect of number of milkings on milk yield, as previously reported (Caja et al., 2000; In: Robotic Milking, Wageningen Pers, pp. 177-178). However, not significant effect was evidenced for milk fat and milk protein contents.

Table 1. Milk performance (mean ± SE) and effects in AMS test day records.

Item	n	Yield (kg/d)	Fat (%)	Protein (%)
Overall mean	164	33.1 ± 0.3	3.51 ± 0.03	2.92 ± 0.02
Lactation stage:				
1st	49	33.1 ± 1.4^{ns}	3.63 ± 0.14^{ns}	2.87 ± 0.06^{ns}
2nd	59	33.3 ± 1.2^{ns}	3.52 ± 0.12^{ns}	2.91 ± 0.05^{ns}
3rd	56	31.4 ± 1.7^{ns}	3.56 ± 0.17^{ns}	2.99 ± 0.07^{ns}
Milking frequency slope:				
/hour	164	-0.9 ± 0.2*	0.04 ± 0.02*	0.02 ± 0.01*
/milking	164	1.30 ± 0.65*	-0.05 ± 0.06^{ns}	-0.04 ± 0.03^{ns}

^{ns}: Not significant ($P > 0.05$), * Significant at $P < 0.05$.

In conclusion, milk recording in AMS farms should take into account the variability among days and the effects of different number of milkings on milk yield and milk composition and, consequently, milk dilution effects should be corrected for cow lactation standardization.

INTRODUCTION OF AMS IN ITALIAN HERDS: INTEGRATION AND PERFORMANCES IN A CONVENTIONAL BARN OF NORTH ITALY

M. Capelletti[1], C. Bisaglia[2] & G. Pirlo[1]
[1]Istituto Sperimentale per la Zootecnia (ISZ), Sezione di Cremona, Cremona, Italy
[2]Istituto Sperimentale per la Meccanizzazione Agricola (ISMA), Sezione di Treviglio, Treviglio (BG), Italy

Abstract

In North Italy, the design of conventional, milking parlor-oriented barns have to take the climatic conditions into account with particular attention to high summer temperatures and relative humidity. Mainly for this reason the lay out of most of the Italian conventional barns are long and narrow, North-South oriented and provided for high volumes of natural ventilation. With this backgroud the integration of an AMS (Automatic Milking System) encounters definite constraints and relies, up to now, on few experiences. With the aim to evaluate the possibilities of a correct integration of a single-box automatic milking unit in a conventional barn, an investigation was carried out at the ISZ research farm, Cremona, Italy, over a period of two years. Two aspects of the research project are presented: i) the daily utilization of the milking area (i.e. waiting area and automatic milking unit) in order to evaluate the correct position of the AMS (a DeLaval Voluntary Milking System, VMS™) in the existing barn related to the cow traffic, waiting time and milking time; and ii) the frequency of successful single teat-cup attachments in order to assess the functional aspects of the robot with a given, not selected herd. A random group of 110 dairy cows (breed *Italian Holstein-Friesian*) with different lactation status was analysed in the period February 2001 - July 2003 and 38.8 ± 3.3 of them, on average, were constantly present in the barn for milking. All the data were recorded with the AMS database. 88253 milkings were registered during the investigation period. On average, each cow was milked 2.5 ± 0.24 time per day for 8.32 ± 3.27 min and visited on average 4.0 ± 0.81 times the milking box. Two peaks from 4:00 to 8:00 a.m. and from 3:00 to 5:00 p.m. were observed with a consequent fluctuation of the waiting times (14 to 86 min). The average daily utilization of the milking unit varied from 30.0 to 82.7%. Moreover, high successful attachment frequency of each teat-cup is a good parameter as well to evaluate the correct operation of the system. Data recorded show a low frequency of attachment failures (5.88%) with different behaviour among quarters: rear quarters reveal the higher frequency of failures with a slight difference between left and right quarter; also the attachment time varyes with longer values recorded for the rear ones. Furthermore, a slight difference between left and right attachment time has been noticed in relation, probably, to the relative greater distance covered by the attachment arm from the teat-cup storage to the left quarters. The study offers a designing experience with conventional, milking parlor-oriented barns integrated with an AMS in order to establish i) the threshold of convenience in maintaining cows with high frequency of attachment failures and ii) the real utilization of the milking area during the 24 hours in order to avoid excessive waiting times or, on the other hand, an underutilization of the system. A project for the allocation of a second automatic milking system in the ISZ research farm will be also briefly presented.

WORKING TIME STUDIES IN FARMS WITH CONVENTIONAL AND AUTOMATIC MILKING

Mats Gustafsson

JTI - Swedish Institute of Agricultural and Environmental Engineering, Uppsala, Sweden

When a farmer changes system from conventional milking to an automatic milking system (AMS) it is stated that there is an considerable change in the amount of work and the alteration within the work. Therefore the present study was to evaluate the indicated differencies in working time and working tasks between conventional and automatic milking.

Time studies were made on four farm with conventional milking (tandem parlour 2x3) and four farms with one AMS unit. The conventional herds varied from 59 to 71 cows and the AMS herds from 55 to 67 cows. The staff was observed during one entire working day. Only stable related work was studied. The work was divided into the follwing categories; cleaning, AMS, management, animal health, feeding, young cattle, milking and other work. Each category was divided furthermore, for example the category AMS has 11 subgroups.

When comparing the conventional and the AMS stables there was no difference in the work concerning animal health, management, young cattle and other work. Factors such as barn lay-out, floors, housing and feeding type of young cattle affected work requirements more than type of milking system. Work connected to conventional milking consumed 4,3 (4,0-4,6 (min-max)) minutes per milking cow and day (mpmc). This should be compared with the AMS work which consumed 1,5 (1,1-1,7) mpmc. AMS farms consumed more labourtime for feeding, 0,75 (0,6-0,9) mpmc, than conventional milking 0,45 (0,3-0,7) mpmc. This difference is mainly caused by a more frequent feeding. Cubicles were cleaned and littered more frequently in the AMS farms (0,6 mpmc) than in the conventional farms (0,4 mpmc).

The conclution of this study is that working time can be approximately 2,0 minutes shorter per milking cow and day in an AMS, when compared with a conventional stable of the same size. A farm with 55 milking cows would in totalt save approximately two hours a day which can be used for other farm work, labour savings, other occupations or leisure.

Automatic milking – A better understanding

MANAGING AN AUTOMATIC MILKING FARM: MINIMIZING THE AMOUNT OF CONCENTRATES IN THE ROBOT

Ilan Halachmi

Agricultural research organization (ARO), The Volcani Center, Bet-Dagan, Israel

Under hot climate conditions such as in Israel, the new milking technology might interfere with the well-proven local management practices: (1) Feeding total mixed ration (TMR) is unfeasible - part of the concentrates is subtracted from the TMR and is eaten in the robot. (2) Cooling all the cows together while gathering just before milking in the parlor's waiting yard is impracticable since the cows arrive to the robot one-by-one spread all over day and night and wait only few minutes before the robot. The common way of holding the entire group in the yoke, after milking, for few hours showering contradicts the robotics' management concept. (3) Full control of milking frequency for a specific group is not 100% possibly - the system is more depending on cow behavior.

The cow cooling aspect was reported in a previous research (Halachmi 2004), the present paper investigates the possibility to minimize the amount of concentrate in the robot in order to stick into a balanced TMR almost unharmed.

Two hypotheses were investigated: (1) that reducing the concentrate allocation in the feed to a certain amount (1.2 kg) would not impair the cows' readiness to visit the robot; and (2) that the fat yield would not decrease if we chose for high milking frequency only those cows at the beginning of lactation.

There were eight robots and 453 cows on the farm from which 100 cows were selected randomly for the experiment but pairs, characterized by their equal numbers of visits to the robot, had one member allocated to the control and one to the experimental group. The cows were divided into two control groups and two experimental groups. Two feeding regimes were used: the "candy concept", with only the minimum feed concentrate provided at each visit to the robot, 1.2 kg per visit, with the aim of attracting the cow; or alternatively, a switch from TMR to completely individual feeding management with allocation of a maximum of 7 kg of feed concentrate per day.

The cows in the first treatment consumed approximately 3.5 kg of concentrate per day and those in the second treatment approximately 5 kg per day. No significant differences were observed in milk yield, in milk fat or protein content, or in cow behavior as expressed in voluntary visits to the robot.

Halachmi, I. 2004. Designing the automatic milking farm in a hot climate, Journal of Dairy Science, In Press

INFLUENCE OF INCOMPLETE MILKINGS ON MILK YIELD UNDER AUTOMATIC MILKING CONDITIONS (VMS®)

J. Hamann, H. Halm, R. Redetzky & N.Th. Grabowski
Department of Hygiene and Technology of Milk, School of Veterinary Medicine Hannover, Hannover, Germany

Goal of the study

Depending on the frequency of occurrence and the extent of milk left in the udder, incomplete milkings (ICM), are generally assumed to cause a marked reduction in milk yield. This study was performed to evaluate the influence of ICM on milk yield under the automatic milking conditions.

Material and methods

A group of 40 high-yielding German Holstein Frisian cows at different lactation stages and lactation numbers was milked with an automatic milking system (Voluntary milking system VMS®; DeLaval, 42 kPa vacuum, 60 cycles/min pulsation rate, 65% pulsation ratio). VMS® identified a milking as incomplete when the actual milk yield did not reach a certain level of expected yield (60% at quarter and/or 80% at cow level). Over 400 days, survey was carried out every 20 days (test day; TD), and each session included 24 hours of sampling where clusters were attached manually. Test day milkings were compared with 35,762 recorded milkings from which 4,560 were treated as incomplete.

Results

The analysis showed that in 70% of all cases, only one quarter of the udder was incompletely milked. Strikingly, 50% of incompletely milked quarters were rear ones. This data suggest that the majority of ICM was caused due to limited accessibility of the rear quarters (technical difficulties). On the other hand and considering cow-individual factors, it was seen that on average, 5 cows out of 40 (= 12.5%) were responsible for 44% of all ICM. During TD, the cluster attachment was done manually, so that no ICM did occur then. The evaluation of the influence of ICM on average milk yield was done by comparing calculated weekly yields WY (TD values by seven days) with the actually recorded ones. Table 1 summarizes the comparisons for the first and second 10 TD with the recorded milk yield of the related weeks.

The results did not indicate any significant impairment of ICM on milk yield.

Table 1. Comparison of mean weekly yields (WY; [kg]) between test days (TD) and corresponding weeks.

TD	WY calculated	WY recorded	weekly ICM	p values
1 - 10	5891 ± 570	5998 ± 635	9.7%	0.349 (n.s.)
11 - 20	6357 ± 998	5763 ± 906	12.1%	0.090 (n.s.)

Implications

The results suggest that the occurrence of ICM is conditioned by both technical (reduced accessibility to rear quarters) and cow-individual circumstances. In our case, the effects caused by a mean of approx. 10% ICM were compensated by the increase in the milking frequency (2.86 milkings/cow/day) and shorter inter-milking intervals so that no significant differences in milk yield could be detected.

INFLUENCE OF COW TRAFFIC ON MILKING AND ANIMAL BEHAVIOUR IN A ROBOTIC MILKING SYSTEM

Jan Harms & Georg Wendl
Bavarian State Research Center for Agriculture, Institute of Agricultural Engineering, Farm Buildings and Environmental Technology, Germany

An important aim when applying automatic milking systems is to reach an optimal milking frequency combined with a minimum of costs and work load. At the same time animal behaviour should be impaired as less as possible. To reach these aims cow traffic is very important. The aim of this project was, to analyse whether it is possible to combine advantages of free and guided cow traffic which were normally used. Therefore "selectively guided cow traffic" was developed in 1998 (access to the feeding area via the milking box and two selection gates between lying and feeding area) and analysed in the following years. To estimate whether the found results are repeatable, the same investigation was carried out in a second research farm with different hardware and breeds. Both farms are described below (as farm A/farm B):

48/45 cows with a milk yield of 7000/8200 kg/a were milked in single box systems (Fullwood/DeLaval). On each farm roughage intake was acquired exactly using 24 weighing troughs. Cows received concentrates in the milking box and additionally in two concentrate feeders on farm B.

To evaluate the effects of the different forms of cows traffic, the following parameters were investigated: Feed intake (per day and per eating period), feeding periods, milkings and visits (number and distribution), milking/visiting intervals, fetched cows (number and distribution), waiting cows in front of the milking box (number and distribution), influence of yield and day in lactation, usage of the different gates/passages.

In both farms (farm A/farm B) comparable results were obtained. When using free cow traffic every cow was milked 2.3/2.2 times per day and visited the milking box 0.6/0.6 times additionally. More milkings (2.6/2.6) were registered with guided cow traffic, but the number of additional visits increased too (1.4/1.5). With selectively guided cow traffic the milking frequency was as high as with guided cow traffic (2.6/2.5), whereas the number of additional visits was nearly as low as with free cow traffic (0.7/0.9). Big differences were found in the number of cows, which had to be fetched for milking: With free cow traffic this was necessary 15.2/24.4 times per day, with guided cow traffic 3.8/3.4 times and with selectively guided cow traffic 3.9/9.8 times. One of the main reasons for the choice of free cow traffic is the assumption that cows will visit the feeding area more often, so that feed intake increases. This could only partly be confirmed in this investigation. Even though the highest number of visits was found with free cow traffic (8.9/7.5) and the lowest with guided traffic (6.6/5.0), the roughage intake (DM) did only show small differences. Cows probably were able to compensate less visits by a higher feed intake per visit. To identify possible reasons for these differences all passes between the resting and the feeding area were counted. With free cow traffic cows changed the area 7.9/9.8 times, with guided cow traffic 4.2/4.0 times and with selectively guided cow traffic 4.7/5.7 times.

In total this investigation showed, that selectively guided cow traffic combines advantages of free and guided cow traffic. These results were confirmed independently on two research farms. For the future, additional improvements in the configuration of the gates, like special rights to pass them at certain times or for certain cows, may help to benefit even more from this form of cow traffic.

EFFECT OF A SIMPLIFICATION IN THE ENTRANCE AND EXIT PROCEDURE FOR A MULTIPLE STALL ROBOTIC MILKING SYSTEM ON THE COW'S TIME BUDGET FOR MILKING

A.H. Ipema & P.H. Hogewerf

Agrotechnology and Food Innovations B.V., Wageningen UR Wageningen, The Netherlands

A milking visit contains two main parts: the handling time and the machine on time covering the milking process. Total handling time in single stall systems is in average between 2-2.5 minutes. In multiple-stall systems arranged one behind the other, the handling time is increased resulting in a less than proportional increase in the milking capacity. For example in a two-stall-system the increase in performance amounted to approx. 60%. The less than proportional increase in milking capacity could be explained by longer waiting times before the robot arm is available (robot arm has to serve more stalls) and less efficient cow traffic because cows in the entrance/exit alley sometimes block the milking stalls.

The effect of a simplification in the entrance and exit procedure in a two-box robotic milking system (Prolion) was tested. In the barn forced cow traffic was applied. Cows could get access to the waiting area by a selection stall placed in the passage between lying and feeding area. In the selection stall it was decided based on the milking interval whether a cow should be sent to the waiting area in front of the milking robot or to the feeding area. When the interval was longer than 6 hours a cow was guided to the waiting area. At shorter intervals a cow sent to the feeding area. To return to the lying area in three passages between feeding and lying one-way-gates were mounted. From the waiting area the cows could get access to the milking stalls. In the experiment two different ways of access from waiting area into the milking stalls were tested.

Original situation: Access from the waiting area into the milking stalls through an alley and exit from the milking stalls into the feeding area also through an alley. The length of the entrance alleys was about 2 m for the first stall and 5 m for the second. The exit alley had a length of 8 m for the first stall and 11 m for the second stall. New situation: Direct access from the waiting area into the individual stalls of the robotic miking system and exit from these stalls back into the waiting area. Milked cows had to leave the waiting area trough the selection stall, where they then had a milking interval less than 6 hours and were sent to the feeding area.

The activities of 21 cows in the original and the new 'access-exit' situation were studied (Table 1).

In the original situation the mean waiting time before entering a milking stall was 30 min per visit. In 12% of the visits cows stayed more than one hour in the waiting area. In the new situation the mean waiting time before entering a milking stall decreased to 11 min per visit. In only 1.4% of the visits the waiting time was longer than one hour. Because the cows came back into the waiting area after leaving a milking stall the total staying

Access/exit situation	Number of visits	Staying duration in waiting area (min/visit)			Milking stall visit duration (min/visit)	Milk yield (kg/visit)
		before	after	total		
Original	167	30	-	30	9.6	10.8
New	146	11	14	25	9.5	11.2

duration was 25 minutes, which was 5 min less than in the original situation. In about 16% of the visits in the new situation cows entered a milking stall for a second time. These visits without milking generally lasted less than 30 s. Only in 2.7% of the visits the total (before and after) staying duration was longer than one hour. The milk yield and duration per milking stall visit were at the same levels in both situations.

It was concluded that milking in the new situation was much more efficient for the cows because the duration for the complete process was decreased. Moreover, the number of milkings with long waiting durations, which is bad for the animal welfare, was significantly lower. The change from the original to the new situation was to the entire satisfaction of the farm and was used ever since.

THE DIFFERENCE BETWEEN THE SETTING AND THE ACTUAL MILKING TIME IN AN AUTOMATIC MILKING SYSTEM

A. Kageyama[1], S. Morita[1], A. Murakami[1], H. Kawakami[1], M. Komiya[1], S. Hoshiba[1] &
M. Tokida[2]
[1]Faculty of Dairy Science, Rakuno Gakuen University, Ebetu, Japan
[2]Research & Development Center for Dairy Farming, Sapporo, Japan

Farmers are able to set milking times in an automatic milking system. The actual milking times were dependent on the farmers setting and the entering pattern of cows. When there is a difference between the setting and the actual milking time, farmers should set the milking times according to the difference between these times. This study was to examine the difference of the setting and actual milking times in an automatic milking system.

The data was gathered from 10 dairy farms with the automatic milking system. The data collection period was from 8 to 36 days in each farm. The number of milking cows was from 16 to 67 in a farm. The total number of milking cows was 441.Two farms used forced cow traffic and eight farms used free cow traffic with an automatic milking system.

The average number of milking times of the setting was 3.3 times/day. The maximum number of cows was between 2.1 times and 3.0 times of the setting and this was 36.0% of all cows. The percentage of cows from 3.1 to 4.0 times of the setting was 35.9%. The highest number of milking times of the setting was 8.0 times/day, and the least number of milking times of the setting was 1.5 times/day. The average of actual milking times was 2.6 times/day. The highest number of cows ranged 2.0 times/day, the percentage of cows in this range was 44.6%. The highest of actual milking times per day was 7 times. There was a difference between the average of the setting and the actual milking times. The difference between the setting and the actual milking times increased by increasing the number of milking times of the setting ($Y=0.48X-0.91$ $r_s=0.998*$) (Figure 1). The difference of times was increased, which had no relation to both the day after calving and the daily milk yield. It was concluded that suitable milking times of the setting (X) was obtained by following equation:X=expected actual milking times/0.52-1.75.

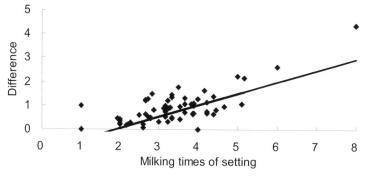

Figure 1. The difference between the setting and the actual milking times with the setting times.Y=0.48X-0.91
rs=0.998*

THE EFFECTS OF DIET COMPOSITION AND FEEDING SYSTEM ON COW ACTIVITY AND VOLUNTARY MILKING IN AN AMS

Arie Klop & Kees Bos
Applied Research, Animal Sciences Group, Wageningen UR, Lelystad, The Netherlands

The number of milkings is an important determinant of the use efficiency of an automatic milk system (AMS). A low rate of voluntary milkings (voluntary milkings/ total milkings) results in longer milking intervals which may impair milk production and udder health. Moreover, forcing cows to visit the AMS to overcome long milking intervals is laborious and time consuming. A low number and rate of voluntary milkings (VM) often coincides with health problems, but is also observed in clinical healthy cows. Clinical healthy, but inactive cows with a low VM rate are defined as "lazy cows". Data from literature indicate clearly that the physical activity of cows and the number and rate of VM is influenced by diet composition and feeding method. However, so far, there is not sufficient understanding to quantify the effects of diet composition and feeding method on the number and rate of VM and use efficiency of an AMS.

Experimental design

Forty-eight high yielding cows were used in a 2×2 factorial design, with two levels of dietary starch and two different feeding methods as treatments. Two different levels of dietary starch were achieved by inclusion of different proportions of grain in the concentrate. The feeding methods were: a) a total mixed ration of concentrate, maize silage and grass silage *ad libitum* (TMR) and b) feeding concentrate with automatic feeders and a mixture of maize silage and grass silage *ad libitum* (ACF) (Table 1). The experimental period consisted of a 3 week pre-treatment period in which the cows were treated equally followed by a 5 week treatment period. After the pre-treatment period the cows were blocked in groups of four according to parity, days in lactation, milk production, number and rate of VM. The cows of each block were randomly assigned to one of the starch level by feeding method combination. Twice daily (7.00 AM and PM), the cows with a milking interval longer than 11 hour were forced to visit the AMS. Milk production, number and rate of VM, feed intake, meal frequency and body weight were recorded daily. Milk composition was recorded during 2 consecutive days per week.

Table 1. Experimental design.

Feeding method	TMR		ACF	
Starch level	Low	High	Low	High
Group	A	B	C	D

OPTIMIZATION OF FEED SPEED IN AUTOMATIC MILKING SYSTEM

Michio Komiya[1], Shigeru Morita[1], Katsumi Kawakami[1] & Mamoru Yokoyama[2]
[1]Rakuno Gakuen University, Ebetsu, Hokkaido, Japan
[2]Cornes eco farm, Tomakomai, Hokkaido, Japan

The feed amount in the milking box is affected by the behavior of the cow and the individual milking ability, because cows are milked by voluntary behavior in automatic milking system (AMS). This research examined the most suitable setting to reduce the rest feed amounts in AMS. The examination was performed on a commercial farm in which 135 cows were milked with three single unit AM-system (Lely Astronaut™) and rest feed amounts, daily milk yield, milk flow, milking frequency were examined by using the logging data of the AMS. Several cows with large rest feed amounts were selected and were incorporated into a group in which the feed speed was changed. After that, their rest feed amounts were examined. Two examination periods of 3 weeks were performed between April and August 2003 at an interval of 3 weeks.

When the feed amount per milking on the milking box and the rest feed rate of individual cow were compared, the rest feed rate in more than 2.5 kg/milking of feed amount had a tendency to increase gradually. In the feed speed change examination, when the feed speed of cows (n=11) which was 12% of the average rest feed rate was changed from 0.55 kg/min to 0.75 kg/min before the first examination period (test 1), their average rest feed rate decreased to 2.4% (p<0.05). But the rest feed rate of 2 cows with 2 milkings per day or less was over 20% after a feed speed change. When the feed speed was changed similarly in the second examination period (test 2), the average rest feed rate of 12 cows decreased from 18% to 7.2%. And the rest feed rate of 3 cows with 2 milkings per day or less showed a high value in 18-36% after a feed speed change. After calculating the optimized feed speed from the milk flow, milk yield, feed amount, and mechanical working time in the AMS, when the rest feed rate and these feed speed from the two examination period were compared (except 2 milkings per day or less), a linear relationship was recognized between these two (Figure 1). As a result of examination, it was possible to reduce the rest feed rate by optimizing the feed speed in the case of 2 milkings per day or more.

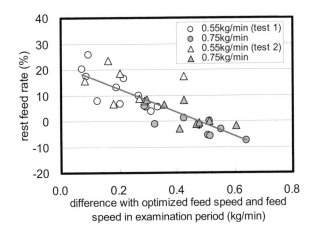

Figure 1. Effect for the rest feed rate by change of the feed speed.

ANALYSIS OF TEAT CUP ATTACHMENT UNDER PRACTICAL CONDITIONS

H. Luther[1], E. Stamer[1], W. Junge[2] & E. Kalm[2]
[1]TiDa Tier und Daten GmbH, Brux, Germany
[2]Institute of Animal Breeding and Husbandry, University Kiel, Kiel, Germany

The main problem to be solved in the early development stage of automatic milking systems (AMS) was that of reliable automatic teat cup attachment. Since the first "prototypes" of AMS were introduced the success rate in terms of teat cup attachment clearly increased. But there are still different problems relating to the attachment process. The objective of the study was to analyse the teat cup attachment processes of a new automatic milking system. Further, we examined the influence of teat cup attachment performance on the milking process.

The study included 127620 milkings performed with the multibox-AMS Leonardo from WestfaliaSurge GmbH at the dairy research farm "Karkendamm" of the University of Kiel. This automatic milking system attaches the four teat cups separately and in a sequential mode. It performs up to four complete new attachment attempts, if the previous attempt was not successful.

5644 observations (≈ 5%) were omitted of the data set, because the teat cups were manually attached to the udder. This affected especially milkings within the first week of lactation. 1068 observations (≈ 1%) were excluded, because attachment of teat cups was not successful at all. All observations of dairy cows with only three functional udder quarters (only 3 teat cups had to be attached) were discarded. 104666 automatic attachment processes with a following milking based on 213 dairy cows remained for the statistical analysis (Table 1).

Table 1. Distribution of attachment attempts.

attachment attempt	n	percent	'repeated attachment attempt' necessary ?	
1st	91749	87.7%	NO	87.7%
2nd	10250	9.8%		
3rd	2077	2.0%	YES	12.3%
4th	590	0.6%		
total	104666			

The threshold trait 'repeated attachment attempt' was analysed using a mixed threshold model. Different fixed effects (number of lactation, days in milk, milkingintervall etc.) affected the occurrence of 'repeated attachment attempts' significantly. But the random animal effect explained only 15.5% of the total variance. Obviously, the occurrence of '*repeated attachment attempts*' was not restricted to special animals. The average '*time required for attachment attempt*' was 73.5 sec and ranged from 51 to 346 sec. The random animal effect explained 39.2% of the total variance within a linear mixed model. Differences between dairy cows in terms of '*time required for attachment attempt*' were more pronounced, but a selection of cows to reduce the required time does not seem to be favourable, because the technical minimum required time (51 sec) is close to the average time (73.5 sec). The milk yield was reduced up to 3.5 kg per milking, if serveral repeated attachment attempts were performed before the start of the milking process.

In conclusion, it is still necessary to reduce the proportion of attachment processes with repeated attachment attempts, because of the negative impact on milk yield. In this context The selection of dairy cows is not a matter of making automatic teat cup attachment more reliable.

DIURNAL CHANGES OF OXYTOCIN RELEASE DURING AUTOMATIC MILKING

Juliana Mačuhová & Rupert M. Bruckmaier
Physiology Weihenstephan, Technical University Munich, Freising, Germany

During long-day photoperiods, i.e. exposure to 16-18 hours of continuous light milk production is increased. This effect can be due to a variety of hormones whose secretion is modified by light. In contrast to conventional twice daily milking at fixed times, robotic milkings occur almost continuously throughout the day. Therefore, we have studied the effect of the photoperiod on the release of oxytocin (OT) in an automatic milking system (AMS) on two farms (farm I and farm II). Experiment on farm I was performed in June, i.e. during long day photoperiod ($\sim$ 16 h) with bright day lighting in the barn. During the night there was dusky lighting in this barn. Experiment on farm II was performed in October, i.e. during short photoperiod of $\sim$ 9 h. Moreover, lighting in this barn was dusky also throughout the day. Twenty Holstein-Friesian cows (ten on each farm) were used for experiments. Cows were milked voluntarily almost during 24 h, i.e. except for the short periods of AMS cleaning. The minimum milking interval was 5 h. During 4 experimental days blood sample collection was performed for analysis of OT concentrations. Altogether blood sample collection was performed during 78 milkings (51 milkings during the light and 27 milkings during the night) on farm I and during 63 milkings (36 milkings during the light and 27 milkings during the night) on farm II. Basal OT levels did not differ between light and night on both farms (4.3 ± 0.2 pg/ml during the light and 3.7 ± 0.4 pg/ml during the night on farm I and 4.6 ± 0.3 pg/ml during the light and 4.8 ± 0.4 pg/ml during night on farm II). Milking related OT release on farm I was higher ($P<0.05$) during light period (AUC/min during entire milking: 59.9 ± 5.9 pg/ml) than during the night (AUC/min during entire milking: 33.4 ± 5.5 pg/ml). However, milking related OT release did not differ between light and night on farm II (AUC/min during entire milking during was 26.6 ± 3.8 pg/ml the light and 20.9 ± 4.0 pg/ml during the night) and was significantly lower ($P<0.05$) during the light and tended to be lower ($P=0.13$) during the night than on farm I. Diurnal changes were only visible if a longer photoperiod and bright day lighting were applied. Possibly, milking related OT release is reduced during the night. Generally low OT release and lacking differences of OT release between day and night are possibly due to the dusky lighting throughout the day. In conclusion, milking related OT release seems to be modified by photoperiod. However, more experiments have to be performed to study the mechanism of the effect of photoperiod on milking related OT release and milk ejection.

Automatic milking – A better understanding

EATING AND RESTING BEHAVIOR OF COWS IN BEDDED PACK TYPE LOOSE HOUSING WITH AN AUTOMATIC MILKING SYSTEM

S. Morita, A. Kageyama, A. Murakami, H. Kawakami, M. Komiya & S. Hoshiba
Faculty of Dairy Science, Rakuno Gakuen University, Ebetsu, Japan

When farmers change their rearing system from tie-stall housing to loose housing, the bedded pack type was preferred for the benefit of manure treatment in Japan. This study was conducted to compare the utilization of automatic milking machine and the eating and resting behavior of cows in two different types of loose housings with an automatic milking system.

The data collections were made in 6 dairy farms with an automatic milking machine (Lely "Astronaut"). The cows (53 cows on average) in two farms were reared in a loose housing of bedded pack type (BP barn). In the BP barn, the sawdust was stacked in the resting area. There were no beddings on the alley. The resting area in the BP barn was separated into two positions by the fence. It was the highest position for both sides of the fence. Four farms with automatic milking used free-stall barn (54 cows on average, FS barn). In each group, the average milking yield per day was about 28 kg. Each of the farms used a free traffic system for cow locomotion in the barn. The position of the cow (TMR trough, stall, and alley in barn) and behavioral type (standing, lying, eating TMR, and ruminating) were observed throughout a 24-hour period. In the BP barn, the position of the cows when lying or standing on the bedded pack and the direction of body axis of the cow were also observed.

There was no difference in the utilization of milking machine (the number of visiting and milking) of cows between loose housing types. The cows visited the automatic milking machine 3.3 times a day in the BP barn and 3.6 times in FS barn and milked 2.4 times and 2.6 times a day, respectively. The maximum number of hourly milkings was 6.0 in the BP barn in afternoon (16:00) and 6.8 in the FS barn (16:00). The minimum number of hourly milkings was 2.7 at 5:00 and 3.8 at 3:00, respectively.

There was no difference in the behavior of cows between the housing types. On average, the cow ate TMR about 5 hours a day, ruminated 7 hours and lay 12 hours a day. The maximum percentage of eating cows was about 50% in BP barn and 62% in FS barn. It was suggested that the TMR trough in the BP barn could be shortened for short trough utilization at same time. The maximum percentage of lying cows was about 95% in the BP barn and 97% in the FS barn. It was suggested that there need to be enough resting space in both housing types for all the cows lying simultaneously. There were no cows lying on the alley in both types of housing. In the BP barn, about 75% cows lay at the highest position and only 2 or 8% at the lowest position of resting area (on bedded pack). The percentage of the cows lying downward was only 3%, almost all the cows lay upward. The percentage of cows that lay from 0 to 45 degrees and from 45 to 90 degrees of the direction to the contour line were equal and about 50% in each. The rough calculation of space allowance for resting in bedded pack type of loose housing was made.

AUTOMATIC MILKING SYSTEMS - SYSTEM PERFORMANCE AND WORKING TIME

Christoph Moriz
Swiss Federal Research Station for Agricultural Economics and Engineering (FAT), Tänikon, Switzerland

Continuing interest in automatic milking systems, the question of maximum possible system performance and a lack of working time data formed the background to an investigation carried out on six Swiss dairy farms with AMS. Process times for AMS with single-stall facilities were recorded by taking measurements on two days on the farms which had a herd size of 37 to 66 cows (average 53). Figures vary considerably from one farm to another, with variations of over 100% for udder preparation and cluster attachment. The process time required for one milking operation averages 2.6 minutes (minimum = 1.8 minutes, maximum = 3.3 minutes). Integrating the process times into a model calculation system enables potential system performance for single-stall facilities milking 50 to 70 cows per day to be calculated as a function of milk yield, type of equipment and average milking rate per minute.

Individual interviews based on a prepared framework formed the basis for recording working time requirements. Once again, marked differences between farms were evident,

Table 1. Process times for automatic milking systems [cmin].

	Average [cmin]	MIN [cmin]	MAX [cmin]	Deviation from average			
				Absolute [cmin]		Relative [%]	
				+	-	+	-
Admission	21.1	17.4	24.7	3.7	3.6	18	17
Preparation	98.3	53.1	130.9	45.1	32.6	46	33
Attachment	97.2	67.6	121.4	29.6	24.1	31	25
Removal	25.2	19.6	31.4	5.7	6.1	22	24
Release	22.5	17.5	25.7	5.1	3.2	22	14
Total	264.3	175.2	334.1	89.2	69.6	33.7	26.3

ranging from 1.6 to 2.1 manpower hours (MPh) per day. The most significant parameter is the length of use of AMS: the more time that has elapsed since AMS was introduced, the smaller the amount of unscheduled or partly scheduled work tends to be. It was not possible, however, to establish any relationship with herd size. All the farms spend about half of their working time on supervision, on average.

References

Moriz, C. (2002): Arbeitszeitbedarf und Arbeitsorganisation auf Betrieben mit automatischen Melksystemen, final report, FAT-Tänikon (CH).

Schick, M. and Moriz C. (2003): Management von automatischen Melksystemen, Agrarforschung 10 (4), pp. 132 - 137.

ELECTRICITY AND WATER CONSUMPTION BY MILKING

Jan Brøgger Rasmussen & Jørgen Pedersen
The Danish Agricultural Advisory Service, National Centre, Building and Technique, Aarhus N, Denmark

Big milking parlours with more technique and the introduction of automatic milking (AMS) leads to increased consumption of electricity and water during the milking. To keep the system at its best, it is important to check up on the consumption of electricity and water.

The consumption of electricity can be registered by installing an electricity meter at all electrical parts concerning milking, such as compressor, vacuum pump. The consumption of water can be registered by installing a water clock at all water supplies for milking, such as main wash; teat wash and floor wash in the milking boxes.

Variation in the results

The results of a survey performed in 2003 vary a lot due to the fact that measurements have been performed under practical conditions in different herds. This means that there are different adjustments for main wash; short wash, the length of pipelines etc.

The total consumption has been registered and then calculated per cow and per milking. The consumption per milking shows much more difference because systems with a lot of milkings have a very low consumption per milking.

A couple of errors

During the measurements different kinds of errors have been noticed at the systems. Naturally it is preferable to avoid this, so that the consumption is kept at a level as low as possible. Please see the examples below:

- A herd with Lely Astronaut - Here the first measurement showed a very big consumption of water compared to the other systems. A water pressure that was too high caused extra water for main wash and short wash. When the error had been corrected the measurements showed the same level as equivalent systems.
- A herd with DeLaval VMS - Here the frequency control for the vacuum pump did not function for a short period. A fall of 20 kWh per twenty-four-hours was registered by utilizing the frequency controlled vacuum pump.

Supervision helps!

By regular supervision of the electricity and water consumption you are always "up-to-date" and can easily follow the consumption. A quick intervention can minimize the expenses.

- Water clocks shall be installed at the water supply for milking systems (parlour/AMS).
- Remember that more water clocks could be needed e.g. one for for warm water; cold water, floor wash.

Automatic milking – A better understanding

- The water clock that is installed at the supply system for hot water must be able to cope with high temperatures.
- Your local power or electrician company can install an electricity meter at the milking system and measure the electricity consumption.
- An alternative solution is to install a permanent electricity meter in connection with the electrical panel. This way you will always be able to follow the consumption.
- When building or rebuilding a parlour or installing AMS, it will only cost you a little amount to have electricity meters installed at the electrical panel in connection with other wiring work. The price of an extra electricity meter will be approximately 270 Euro each + installation.

SURVEY OF AUTOMATIC MILKING SYSTEMS MANAGEMENT PRACTICES IN NORTH AMERICA

Douglas J. Reinemann[1], Yvonne van der Vorst[2], Wilco de Jong[3] & Albrecht Finnema[3]
[1]Univerity of Wisconsin, Wisconsin Madison, USA
[2]Keten Kwaliteit Melk, Leusden, The Netherlands
[3]Van Hall Instituut, Leeuwarden, The Netherlands

A survey was conducted of 10 farms in the USA and 15 farms in Canada using automatic (or robotic) milking systems (AMS). Surveys were conducted in-person with the farm manager during a visit to the farm. The farms surveyed had been using AMS from 6 months to several years. The survey tool was based on a similar survey performed on 120 AMS farms in The Netherlands.

The average number of cows on all farms surveyed was 110, with 16% of farms milking less than fifty cows, 40% of farms between 51 and 100 cows and the remaining 44 percent of farms milking more than 100 cows. The number of cows per milking box ranged from 25 to 70 with an average of 57. Farms reported an average 1.9 full time persons employed. Six-row barns were used by 56% of farms while 32% had a four-row barn and one farm had an eight-row barn. The most common type of bedding used was a combination of mattress and wood shavings used on 52%of farms. The most common form of manure handling was an auto scraper (67%) followed by a slatted floor (20%).

Most farms began by using forced cow traffic, however, at the time of the survey only four farms (16%) used totally force cow traffic while eight farms (30%) used a totally free traffic system with a holding pen in front of the milking box. The AMS farms surveyed averaged 2.6 milkings/cow/day with 36% of the farms below 2.5 milkings/cow/day, 44%, between 2.5 and 3 milkings/cow/day and 20% with more than 3 milkings/cow/day. About 30% reported that they walk through the barn to perform visual inspection of their cows less than three times/day with 42% reporting barn walks four times/day and 29% more than four times/day.

The most common parameter used for putting cows on an attention list was deviation in daily milk yield (84%) followed by time since last successful milking (73%) and milk conductivity (47%). Many farms reported that they were less likely to use milk as they gained more experience with AMS. All but 2 of the 25 farms surveyed used TMR feeding with the amount of concentrate mixed in the TMR adjusted for the amount fed in the robot. More than half (56%) of the users clean the free stalls twice/day. There was an association between more frequent cleaning of free stalls and reduced SCC.

Milk filters were changed once/day on 32% of farms, twice/day on 52% and three times /day on 16% of farms. There was an association between increased frequency of filter change and reduced bacteria count of bulk tank milk. The average self-reported milk production on the 25 AMS farms surveyed was 9500 kg/cow/year (21,000 lb/cow/yr).

Most farmers (84%) indicated that they bought the robot because it allowed for a more flexible work schedule. Farms reported having to cull cows about 4% of cows because they could not adapt to the new milking system, mainly because of unsuitable udder conformation. All of the AMS users surveyed indicated that they were satisfied to very satisfied with their AMS. Most users indicated that AMS has allowed them more time for managerial tasks, and more importantly, more time for themselves and their families and has decreased stress levels for both cows and themselves.

A PROTOCOL FOR TRAINING A DAIRY HERD FOR AUTOMATIC MILKING

Jack Rodenburg
Ontario Ministry of Agriculture and Food, Woodstock, Ontario, Canada

Although automatic milking systems are used world wide, there is little published information describing strategies for training cows to use these systems. A successful training protocol should minimize the amount of labour and time required and should achieve a high level of voluntary milking.

In this field study, a written protocol was developed and used to train 50 cows with no previous automatic milking experience in the voluntary use of a Lely Astronaut. The training facility was a three-row 50-stall freestall barn with an unrestricted crossover at one end, and a holding area accessed by one way gates from both alleys at the other. Cows were selected from a 77-cow commercial parlor trained herd. Cows excluded were 5 just fresh, 4 nearly dry, 3 three-quartered, 3 slow milkers, 2 reproductive culls, 2 treated for mastitis, 2 lame, 2 poor udder shape, and 4 randomly eliminated. Activity was monitored with time lapse video, and paper and computer records. Cows were divided into 2 groups of 25, separated by 2 gates positioned across the alleys. At the first milking, udders were manually cleaned with a paper towel and warm sanitizing solution. A technician assisted in the initial location of teats. As cows were milked the gates were moved to provide continuous access to approximately 25 freestalls. Gates were moved away from the milking box along the manger and toward it in the alley between the stalls. Milking a new group was started at 4-hour intervals so cows were milked three times daily. At the 2nd through 4th milking, the technician made adjustments as needed. Udder preparation was with the robot arm, positioned low so brushes cleaned the teats but had minimal contact with the udder. At the 5th and subsequent milkings the arm was repositioned to wash in the standard position. For the first 7 milkings, all cows were directed into the holding area and into the milking box. Initially two people directed the cows, but by day 3 only one person was needed. At the 8th milking, each 25-cow group was given free access to the holding area and was left undisturbed at the start of each 4-hour period. No cows were directed to the holding area until the time available for milking the remaining cows within the 4-hour window became limiting. On days 4, 5, 6 and 7 respectively, for 68%, 76%, 93% and 93% of milkings, cows entered the holding area and milking box without assistance. For 26, 20, 7 and 7% of milkings, cows had to be directed into the holding area, and for 9, 5, 2 and 2%, cows were directed into the milking box. During this period, relocating the gates did involve chasing some cows out of freestalls. While this was not "fetching cows" in the purest sense, it often led to voluntary entry of the holding area. On day 8 the gates were removed, giving free access to all cows. Cows returning in < 6 hours were refused, and cows not returning in 10 hours were fetched for milking. On day 9, 10 and 11 respectively, 43, 45, and 45 fetched milkings involving 33, 36, and 34 cows occurred. On day 12 the criteria for fetching cows was changed to >12 hours or >15 liters expected production. With less restrictive criteria, fetched milkings and fetched cows gradually decreased from 34 milkings involving 24 cows, on day 12, to 27 milkings involving 22 cows by day 19. Daily activity on day 12 through

19 averaged 2.58 milkings and 2.09 refusals per cow per day. By day 19, 79% of all milkings were voluntary.

The observation that 93% of cows were able to locate and use the milking box after 1 week, when separation gates were used, illustrates that non-attendance is probably not due to inability to learn the required behaviour. Lower attendance when free traffic was established suggests some cows lack adequate motivation for voluntary milking.

Automatic milking – A better understanding

EFFECT OF THE COMPOSITION OF CONCENTRATE FED IN THE MILKING BOX, ON MILKING FREQUENCY AND VOLUNTARY ATTENDANCE IN AUTOMATIC MILKING SYSTEMS

Jack Rodenburg[1], Erik Focker[2] & Karen Hand[3]
[1]Ontario Ministry of Agriculture and Food, Canada
[2]Wageningen University, Wageningen, The Netherlands
[3]Ontario Dairy Herd Improvement Corporation, Canada

In a previous study of Canadian farms with automatic milking systems (AMS), 19.2% ± 12.5 of cows had to be directed to the AMS at least daily, and 12.6% ± 8.6 of all milkings were "involuntary". Fetching these cows was the major labour component of AMS milking (Rodenburg 2002). Since cows may attend more readily when enticed by a palatable concentrate, trials were conducted to compare an experimental pellet against established commercial feeds. A concentrate was formulated for maximum palatability with corn, soyahulls, wheat shorts, barley, bakery meal, soybean meal, corn distillers, extruded soymeal, wet molasses, animal vegetable fat blend, vitamin mineral premix, sodium bicarbonate, salt, pellet binder and fenugreek flavour. Trials with three consecutive 15-day crossover/switchback feeding periods were conducted.

In trial 1, 37 cows in herd A averaged 4.38 visits to the milking box, and 2.75 milkings per cow per day. 2.4% of milkings were involuntary. Visits (3.95 vs 4.80) and milkings (2.69 vs 2.81) were fewer (p <0.05) for the experimental pellet. The number of cows fetched was unaffected. In trial 2, the same experimental pellet was compared to a different commercial pellet in herd B, using 36 cows. Although average number of visits (3.04) and milkings (2.28) were lower than herd A, and a greater number of milkings were involuntary (21.1%), the concentrate fed did not influence attendance. In trial 3, the pellet was reformulated to exclude all mineral ingredients and fed to 36 cows in herd B. Excluding minerals did not affect the number of visits, milkings or cows fetched. In herd C, trial 4, a mixture of 50% commercial pellets and 50% high moisture corn was fed in the AMS. This was compared to our experimental pellet, adjusted to a higher protein content, to make it isonitrogenous with the control. As in trial 1, for the 35 cows in this trial, number of visits (3.06 vs. 3.33) and milkings (2.34 vs 2.49) were lower (p< 0.05) for the experimental pellet. In this trial, pellet strength of the experimental pellet was weaker, 86.9 pdi vs. 97.7 pdi, than the commercial pellet and there was evidence of fines in the feeder when it was fed. In trial 1 weaker pellets (91.2 vs 96.0 pdi) were also associated with lower attendance. In trials 2 and 3 pellet strength was similar (92.5 vs 91.2 pdi) for both of the concentrates fed.

Dried corn distillers is a palatable feed ingredient with poor pelleting characteristics. To test the palatability of a concentrate based on this feed, a mash formulation of 49% distillers, 49% cracked corn, 2% molasses and 0.1% flavouring agent was fed to two herds. In herd D a six-day feeding period resulted in a decrease in number of visits from 3.93 to 3.57, and number of milkings from 2.50 to 2.35. Milk production declined from 25.2 to

23.6 liters and the trial was discontinued. Herd A also fed this formulation in the AMS, after a small amount of the mash was first topdressed at the manger in a five day adjustment period. In a trial with 36 cows number of visits and milkings was not different from the same control ration used in trial 1.

These field trials suggest that small differences in composition of the concentrate in the milking box, or the exclusion of unpalatable mineral ingredients do not alter the frequency of voluntary milking or number of cows that must be fetched. The fines in mash feeds and in low strength pellets were generally associated with less frequent milking and an increase in the number of cows fetched in these field trials. Inclusion of high moisture corn, resulted in more frequent milking and fewer cows fetched.

Rodenburg, J. and B. Wheeler, 2002, Strategies for Incorporating Robotic Milking into North American Herd Management. In: Proceedings of the first North American Conference on Robotic Milking, Toronto, Canada, pp III 18 - III 32

THE EFFECT OF HOUSING ON IMPLEMENTATION OF AUTOMATIC MILKING IN ISRAEL

Ezra Shoshani[1] & Ephraim Ezra[2]
[1]Extension Service, Ministry of Agriculture, Israel
[2]Israel Cattle Breeders Association, Israel

A loosing housing barn is popular in Israel. In the last 20 years it was dramatically changed to adapt it for high producing cows. However, all scientific works on farm design for automatic milking were based on free stall barn and none of them had examined the combination of automatic milking with a loosing housing barn. Therefore, most of Israeli dairy farmers, who implemented automatic milking, also switched to free stall barns. To-date there are six farms with robots (4 farms- 1 to 2 units and 2 farms - 4 to 8 units). In two farms, robots were installed in a loosing housing barn but only one has the experience of more than one year. In this herd, there are 200 cows which are housed in two separate loosing housing barns. Seventy randomly selected cows were transferred to the barn with the robot in March 2002 (AM), while the rest were milked in the conventional system 3 times daily (C). Cows' cooling systems work in both systems.

The effect of automatic milking on milk yield, milk quality and fertility was analyzed. Data of milk records during the years 2002 - 2003 was collected from the Israeli Herd Book of ICBA. Milk yield of all cows in AM group was higher compared to the C group (35.5 Kg vs. 32.2 Kg, respectively, $p<0.07$). The effect of age differed significantly ($p<0.0001$): AM heifers gave 2 Kg less milk while 2nd lactation gave 4 Kg more and =>3rd gave 9 Kg more milk compared to C group. Fat content was significantly higher in C group compared to the AM (3.53 vs. 3.43, $p<0.05$) while protein content was lower (3.06 vs. 3.17, $p<0.05$). The effect of age on milk content was insignificant. The Economic Corrected Milk (ECM) was higher in the AM (35.88 vs. 32.14 Kg, $p<0.05$). The effect of age was also significant (heifers- 30.63 vs. 32.3, 2nd lactation - 36.3 vs. 32.24 and 3rd and higher - 40.7 vs. 31.87, respectively). SCC (transformed to log) and Total Bacteria Count were not significantly different between the two groups. The Pregnancy rate (from first and second insemination) was higher in the AM group but did not significantly differed from C group due to high variation (35% vs. 25%, respectively, $p=0.3$).

The milk yield of cows of C group in this herd was similar to the milk yield of cows milked by other conventional farms in other regions despite the weather conditions of the region this farm is located in (high temperatures during hot seasons (~ 40°C)) due to 3 milkings daily and to efficient cooling system. The production of the AM cows in this particular herd is considerably higher than cows milked by robots of other farms (35.5 Kg vs. 32.8 Kg). The SCC of AM cows did not increase after transfer from conventional system to robotic system, while in other farms it was dramatically increased.

Lactation corrected curve of cows milked by the robot in loosing housing barn was higher compared to the curve of cows from all other farms, housed in free stall barns, during all lactation months. This difference might be attributed to the higher milking frequency.

The introduction of automatic milking did not change the milk quality nor the pregnancy rate but markedly increased the milk yield and ECM. This was not the case in other farms housing the cows in free stall barns. It seems that under hot weather conditions, automatic milking will execute its potential advantages more in a loosing housing barn than in a free stall barn. It correlates well with other Israeli farms which transferred cows from loosing housing barns to free stall barns.

THE EFFECT OF MILKING INTERVAL ON MILK YIELD AND MILK COMPOSITION

K. Svennersten-Sjaunja[1], I. Andersson[2] & H. Wiktorsson[1]
[1]Department of Animal Nutrition and Management, Swedish University of Agricultural Sciences (SLU), Uppsala, Sweden
[2]Swedish Dairy Association, Lund, Sweden

Cows kept in automatic milking systems (AMS) are milked with varying milking intervals and usually more than twice a day. More frequent milking has been reported to increase milk yield, change milk composition and influence raw milk quality, with one indicator being increased content of free fatty acids (FFA). The explanation for increased FFA might be biological and/or technical. The first aim of the present experiment was to study how an extreme increased milking frequency of six times daily, treatment I, influenced the milk composition and raw milk quality compared to twice daily 12-h milking intervals, treatment II. The second aim was to find out whether the increased FFA content was of biological or technical origin.

A four-week change over experiment with two periods and two treatments including 14 cows was carried out at Kungsängens Research Centre, SLU, Uppsala Sweden. The individual milk yield varied between 30 and 43 kg ECM (energy corrected milk) at the onset of the experiment. Each experimental period lasted two weeks and all registrations were made during the second week in each period. Milk samples were analysed for fat, protein, lactose, milk fatty acid composition, free fatty acids (FFA), size of fat globules, content of casein, NPN, Na^+ and K^+. Milk analysed for FFA was collected both from the composite milk (in the TruTest) and from hand-milked milk directly at the teat. FFA was analysed on milk prepared directly after milking and after 24-h storage in the refrigerator.

In treatment I (milking six times a day), the milk yield increased significantly, while the content of fat and protein did not differ between treatments. Nor was there any difference in the fatty acid composition between treatments. However, in treatment II, FFA values were significantly higher in the hand-milked milk compared to treatment I (1.66 and 1.36 mmol/100 g fat, respectively) in the unstored milk. The FFA values increased in both treatments in the milk stored for 24 h. After storage, the FFA content was significantly higher in the stored milk from treatment I. The effect was most pronounced during morning milking and the values observed were 3.21 and 2.36 mmol/100 g fat for treatment I and II, respectively. Fat globules were also significantly larger in treatment I. The casein count and non-protein nitrogen content in the milk did not differ between treatments. An unexpected decrease in Na^+ content in treatment I was observed.

In conclusion, the only negative effects on milk quality in relation to increased milking frequency was that the susceptibility for lipolysis increased in the stored milk. Since these milk samples were taken directly from the teat, the effect of transportation in the milk lines was eliminated. It is likely that there might be a biological explanation, such as size of fat globules, for the increased FFA content.

A STUDY OF PRACTICAL USE OF AUTOMATIC MILKING SYSTEM IN JAPAN

M. Tokida[1], S. Morita[2], M. Komiya[2] & T. Kida[3]
[1]Research and Development Center for Dairy Farming, Sapporo,Japan
[2]Faculty of Dairy Science, Rakuno Gakuen University, Ebetsu, Japan
[3]National Institute of Livestock and Grassland Science, Nishinasuno, Japan

This paper aims at clarifying the practical use of automatic milking system (AM-system) in Japan, and finding the various problems farmers face when transferring from ordinary milking systems. AM-systems in Japan are not popular and are slowly penetrating into dairy farms. We conducted a comprehensive survey of 31 dairy farms equipped with AM-systems and has had experience using them for more than one year. We collected more precise milking data for the period of seven days at 15 farms amongst the 31.

The main findings are as follows: in 31 farms, the average number of cows per AM-system set was 47.8, and 51.7% of the farms also used ordinary milking facilities in combination (combination system) (Figure 1). The number of cows per AM-system set depends on the kind of combination system; because of the number of cows per AM-system set in a combination system with a pipeline (PL) was greater than a combination system with a milking parlor (MP). Data from 15 farms indicate that visiting frequency was 3.8±1.2 per cow a day and milking frequency was 2.7±0.4 per cow a day. In cases of fewer than 30 cows appropriated per AM-system set, visiting and milking frequency were 5.3±2.0 and 3.3±0.6 per cow a day, respectively. We could observe that cow visiting frequency was greater in the cases of farms with fewer than 30 cows. We could not observe that cow visiting frequency was affected by type of combination system, and that cow visiting frequency was affected by type of cow traffic.

This is a preliminary approach to the main objectives and we would like to continue to further research the unresolved problems in AM-system.

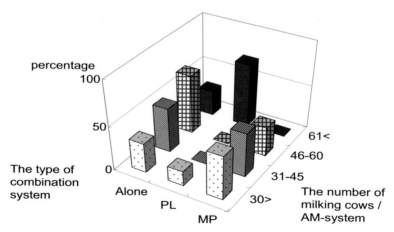

Figure 1. Distribution of farms that classified in the number of cows per AM-system and the type of combination system.

Automatic milking – A better understanding

STRATEGIES TO IMPROVE THE CAPACITY OF AN AUTOMATIC MILKING SYSTEM

H.J.C. van Dooren, C.H. Bos & G. Biewenga
Applied Research, Animal Sciences Group, Wageningen UR, Lelystad, The Netherlands

The High-tech farm is an experimental farm of Applied Research in Lelystad, The Netherlands, with a strong focus on cost price of milk. The farm is equipped with one Lely Astronaut automatic milking system including a pre-selection gate and has a milk production target of 800.000 kg per year. This target can only be met under the condition of optimal use of the AMS. The AMS capacity, in kg per day, is determined by the number of milkings and the milk yield per milking. Too many milkings with a low yield are avoided by introducing a minimum milking interval. The minimum milking interval is adjusted for each cow depending on production level, lactation stage and number of lactations. To avoid production decrease, cows with a milking interval greater than 10 hours are fetched three times a day. This fetching is time consuming, does not reward the cows' own initiative and should therefore be restricted to a minimum. Objective of this study was to optimise the use of the AMS by analysing cow behaviour with a focus on the minimum and realized milking interval and visiting pattern. Data from two three-month periods in 2001 were analysed. Results have been reported by Bos (2003), Dooren (2003) and Biewenga (2003). In the analysed periods the average milking frequency per day was 2,3 with a milk yield of about 13 kg per milking. Average herd size in the first period was 66,9 with a total daily production of 2104 kg and in the second period 66.2 with a total daily production of 1941 kg. Analysis of minimum milk interval showed that animals came shortly before they completed their minimum milk interval. Especially cows with a daily production of 30 kg or more came to visit the AMS only 30 to 60 minutes in advance. Being rejected it lasted 5 to 6 hours before they presented themselves again to the AMS. Introducing a flexible fetching criteria that takes production level, lactation stage and number of lactations into account, reduces the number of fetching with 25% and the number of cows per fetching with 42%. With this new criteria cows were only fetched when the milk interval exceeded the minimum milk interval with at least 30%. An AMS with a high capacity demands a constant visiting pattern throughout the day to be most efficient. Analysis of visiting frequency showed a dip in the number of visits during the late night hours (2 to 6 am). During the day visiting peaks were observed resulting in waiting times before the AMS of more than one hour per milking for heifers and low ranked cows. Generally it can be concluded that if criteria for minimum milking interval and fetching are used with flexibility, the use of AMS capacity can be improved and labour demand can be reduced. A systematically analysis of system data can give input to optimise system performance and can improve AMS output giving the farmer a hand with optimising system capacity.

Biewenga, G., 2003. Wachtruimte effectief benutten, PraktijkKompas Rundvee 17 (5), Praktijkonderzoek, Lelystad, pp. 20-21.

Bos, C.H., 2003. Efficiënter melken met automatisch melksysteem, PraktijkKompas Rundvee 17 (3), Praktijkonderzoek, Lelystad, pp. 10-11.

Dooren, H.J.C. van, 2003. Ophaalcriteria op high-techbedrijf onder de loep, PraktijkKompas Rundvee, 17 (4), Praktijkonderzoek, Lelystad, pp.16-17.

THE CHANGEOVER FROM CONVENTIONAL TO AUTOMATIC MILKING IN DAIRY COWS WITH AND WITHOUT PREVIOUS EXPERIENCE

Daniel Weiss[1], Erich Möstl[2] & Rupert M. Bruckmaier[1]
[1]Physiology Weihenstephan, Technical University Munich, Freising, Germany,
[2]Institute of Biochemistry, University of Veterinary Medicine Vienna, Vienna, Austria

In the present study the effect of the changeover from a conventional parlour to an automatic milking system (AMS) on physiology and behaviour of dairy cows was investigated. The experimental cows had either no previous experience in automatic milking (UC, n = 17) or had been milked during the previous lactation in the experimental automatic milking system (EC, n = 9). EC were housed during the dry period in a separate barn and were milked for 36 ± 5 d after parturition in the parlour before they were changed to automatic milking. AMS and parlour animals were kept in 2 separate herds in one single barn under identical management and feeding conditions. A selectively guided cow traffic was applied in the AMS system.

Milk yield and composition was analysed. Heart rate was measured continuously and faeces samples to determine cortisol metabolites were taken twice daily before the changeover and during the first week of AMS milking. The rate of voluntary visits to the AMS was evaluated. UC obtained a training in the AMS area for 3 d, the first AMS milking was performed on d 4. EC did not need a training period.

Although the visit in the AMS stall was attracted by concentrate UC had to be pushed into the stall for the first one or two visits. In contrast, EC entered the AMS stall instantly without human intervention. Heart rate was higher in UC than in EC during the first AMS visit (35 ± 5 vs. 17 ± 1 beats per min above baseline, P < 0.05). In UC the rate of voluntary visit was 4, 26, 40, 49, 63, 72, 89, 91 and 94% during the first ten days of AMS milking. Concentrations of faecal cortisol metabolites were not effected by the changeover. In UC the milk ejection was disturbed during the first visits. Mean milk yield at the first milking in the AMS was significantly lower than that of the parlour (67 ± 7%, P < 0.05) due to a disturbed milk ejection, whereas milk ejection in EC was not disturbed (96 ± 2%). During the first 10 d of AMS milking frequency was 2.7 ± 0.1 milkings/d in UC and 2.9 ± 0.1 milkings/d in EC. Milk yield of the first 15 milkings was lower in UC than in EC (87 ± 2% and 108 ± 3%, respectively, as a relation of the milk yield obtained 10 d prior to the changeover in the parlour, P < 0.05).

In conclusion, to successfully adapt to automatic milking UC need a period of about 10 d, albeit milk yield was reduced due to the changeover. In practical farming special attention should be taken to a proper training of UC whereas EC do not need an adaptation period even after a transition period of parlour milking. Furthermore EC benefit increased milking frequency by enhanced yields.

AUTHORS INDEX

AUTHORS INDEX

MAIN SPONSORS

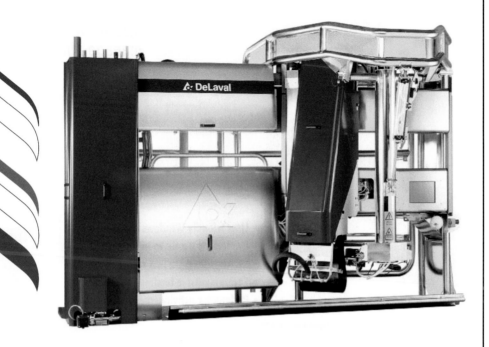

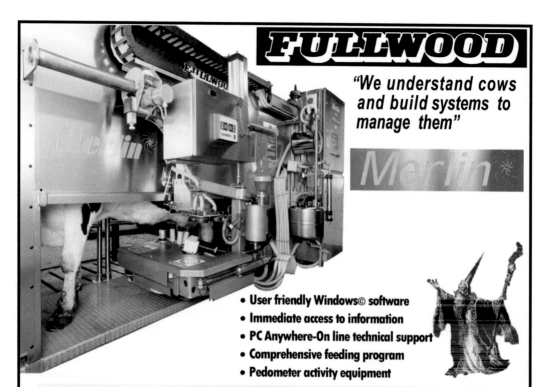
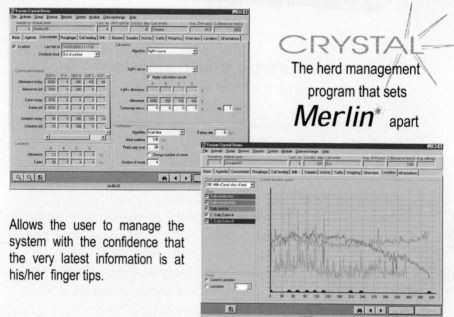

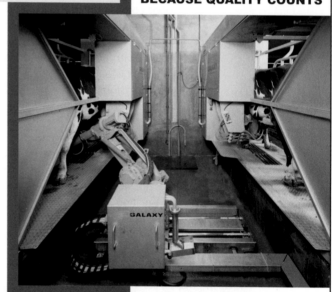

SPONSORS

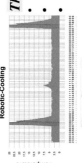

sensortec

Specialists in on-line measurement of biological components in milk

For further details visit
our website at
www.sensortec.co.nz

Sensortec Ltd.,
Waikato Innovation Park
Ruakura Road
Hamilton
New Zealand
Tele: +64 7 859 3364
Fax: +64 7 859 3362
E-mail: sales@sensortec.co.nz

◆ Sensortec is the first company in the world to specialise in the commercial development of intellectual property and product platform technologies for on-line sensing of biological components in milk.

◆ Sensortec has developed a range of technologies capable of detecting an extensive range of milk analytes that are valuable herd management decision support tools in the areas of:

- Milk quality
- Animal health
- Fertility
- Nutrition

◆ Sensortec has a high level of expertise and offers consultancy and contract R&D services in a number of areas of science including:

- Biochemistry
- Optics
- Animal and human physiology
- Electronics engineering
- Mechanical engineering
- Dairy engineering
- Automation

GOODWILL SPONSORS

Avon Rubber p.l.c.
Manvers House, Kingston Road
Bradford on Avon, Wiltshire,
BA15 1AA, England
Tel. +44 (0)1225 86 11 00
Web-site: www.avon-rubber.com

Bou-Matic
PO Box 8050
1919 S. Stougthon Rd.
Madison, WI 53708
United States of America
Tel. +1 608 222 3484
Web-site: www.boumatic.com

City of Lelystad
Stadhuisplein 2
8232 ZX Lelystad
The Netherlands
Tel. +31 (0)320 278911
Web-site: www.Lelystad.nl

Exendis
PO Box 56
6710 BB Ede
The Netherlands
Tel. +31 (0)318 676430
Web-site: www.exendis.com

InterPuls®

Innovative milking components

Interpuls SPA
Via Varisco 18
42020 Albinea (RE)
Italy
Tel. +39 0522 347 511
Web-site: www.interpuls.com

NRS (CR-Delta)
Postbus 454
6800 AL Arnhem
The Netherlands
Tel. +31 (0)26 3898700
Web-site: www.cr-delta.nl